Mathematical Models for the Design of Electrical Machines

Mathematical Models for the Design of Electrical Machines

Editors

Frédéric Dubas

Kamel Boughrara

MDPI • Basel • Beijing • Wuhan • Barcelona • Belgrade • Manchester • Tokyo • Cluj • Tianjin

Editors
Frédéric Dubas
University Bourgogne Franche-Comté
France

Kamel Boughrara
Laboratoire de Rcherche en Electrotechnique (LRE-ENP)
Algeria

Editorial Office
MDPI
St. Alban-Anlage 66 4052 Basel,
Switzerland

This is a reprint of articles from the Special Issue published online in the open access journal *Mathematical and Computational Applications* (ISSN 2297-8747) (available at: https://www.mdpi.com/journal/mca/special_issues/electr_mach).

For citation purposes, cite each article independently as indicated on the article page online and as indicated below:

LastName, A.A.; LastName, B.B.; LastName, C.C. Article Title. *Journal Name* **Year**, *Volume Number*, Page Range.

ISBN 978-3-0365-0398-1 (Hbk)
ISBN 978-3-0365-0399-8 (PDF)

Contents

About the Editors

Frédéric Dubas was born in Vesoul, France, in 1978. He received the M.Sc. degree and the Ph.D. degree from the "Univ. Bourgogne Franche-Comté" (Besançon, France) in 2002 and 2006, respectively, with a focus on the design and optimization of the high-speed surface-mounted permanent-magnet (PM) synchronous motor for the drive of a fuel cell air-compressor. From 2014 to 2016, he was the Head of the "Unconventional Thermal and Electrical Machines" team. Presently, he is the Head of the "Electrical Actuators" group in the "Hybrid & Fuel Cell Systems, Electrical Machines (SHARPAC)" team. He works with ALSTOM Transports (Ornans, France), and RENAULT Technocenter (Guyancourt, France), where he is involved in the modelling, design and optimization of electrical systems and, in particular, induction and PM synchronous (radial and/or axial flux) machines, creative problem-solving, and electrical propulsion/traction. He is currently an Associate Professor with the Dép. ENERGIE, FEMTO-ST Institute, affiliated with the CNRS and jointly with the "Univ. Bourgogne Franche-Comté" (Besançon, France). He has authored over 100 refereed publications and he holds a patent for the manufacturing of axial-flux PM machines with flux-focusing. Dr. Dubas received: i) the Prize Paper Awards in the IEEE Conference Vehicle Power and Propulsion (VPPC) in 2005; ii) the Prize Presentation Awards in the 19th International Conference on Electrical Machines and Systems (ICEMS) in 2017; iii) the RENAULT Internal Award (Direction Engineering Alliance—Innovation) in 2019.

Kamel Boughrara was born in Algiers, Algeria, in 1969. He received the Engineer Diploma degree from Ecole Nationale Polytechnique (ENP), Algiers, in 1994, the magister degree from the University of Sciences and Technology Houari Boumediene, Algiers, in 1997, and the Ph.D. degree from ENP in 2008. He is currently a Professor with ENP and the Director of the Laboratoire de Recherche en Electrotechnique, Algiers. His current research interests include modeling and control of electrical machines.

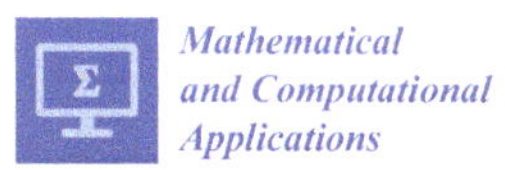

Editorial

Mathematical Models for the Design of Electrical Machines

Frédéric Dubas [1,*] and Kamel Boughrara [2]

1 Département ENERGIE, FEMTO-ST, CNRS, Université Bourgogne Franche-Comté, F90000 Belfort, France
2 Laboratoire de Recherche en Electrotechnique, Ecole Nationale Polytechnique, 16200 Algiers, Algeria; kamel.boughrara@g.enp.edu.dz
* Correspondence: frederic.dubas@univ-fcomte.fr

Received: 8 December 2020; Accepted: 9 December 2020; Published: 9 December 2020

Electrical machines are used in many electrical engineering applications, viz, transports (e.g., electric, hybrid, and fuel cell vehicles, railway traction, and aerospace), energy harvesting (e.g., flywheels), renewable energy (e.g., wind power turbines and hydroelectric power plants) and magnetic refrigeration devices, among others. For decades, numerical methods (e.g., finite element, finite difference or boundary-element analysis) were widely used in research and development (R&D) departments for their accuracy as compared to measurements. Nevertheless, mainly in three-dimensional (3-D) applications, these approaches are time-consuming and not suitable for optimization problems. Nowadays, in order to reduce the computation time, R&D engineers must develop full computer-aided design for electrical machines with accurate and fast models in simulations. Hence, the main objective of this special issue is to bring the latest advances and developments in the mathematical modeling and design of electrical machines to different applications. The main models discussed will be based on the following:

- Equivalent circuits (e.g., electrical, thermal and magnetic);
- The Schwarz–Christoffel mapping method;
- The Maxwell–Fourier method (e.g., multilayer models, eigenvalues models and the subdomain technique).

The interest topics in the mathematical models include, but are not restricted to the following:

- Two-dimensional (2-D), quasi 3-D and 3-D;
- Global and local saturation, slotting and eddy current effects;
- Adaptive generic models;
- Multiphysics modeling with new materials;
- Hybrid models.

The numerical method, as well as the experimental tests, will be used as comparisons or validations.

In this special issue, the authors of selected works contributed to the topics listed above, since contents of their works can be synthesized as follows:

- **Maxwell–Fourier Method** (i.e., the formal resolution of Maxwell's equations by using the separation of variables method and the Fourier's series) in 2-D or 3-D with a quasi-Cartesian or polar coordinate system:
    - Jabbari [1]. In this research, an analytical model was proposed to calculate the magnetic vector potential in surface mounted permanent magnet (PM) machines. It was based on the subdomain technique and applied a hyperbolic function. The saturation effect was neglected. A mathematical expression was also derived for optimizing the PM shape to

reduce the cogging torque and electromagnetic torque components. The analytical results were validated through finite element analysis (FEA);

- ○ Ben Yahia et al. [2]. In this contribution, the authors proposed a 2-D exact subdomain technique in switched reluctance machines, supplied by a sinusoidal current waveform (i.e., variable flux reluctance machines) by applying the Dubas' superposition technique [3]. The global saturation effect was considered with a constant magnetic permeability corresponding to the linear zone of the nonlinear $B(H)$ curve. The comparisons with FEA showed good results for the proposed approach;
- ○ Mendonça et al. [4]. Here, a novel solution for magnetic field calculation in 2-D problems based on the multilayer model, in which one region is defined with a space-varying magnetic parameter was proposed. This contribution was effective at evaluating more realistic magnetic parameters, where measurements of a high-speed PM generator prototype indicated saturation in the retaining sleeve due to pole-to-pole leakage flux. The tsaturation profile is a function of the mechanical angle and can be modeled with the aid of a space-varying relative permeability, expressed in terms of a Fourier's series. The analytical solution was confronted with FEA, which confirmed the validity of the proposed methodology;
- ○ Custers et al. [5]. This work describes the scattering matrix approach to obtain the solution to electromagnetic field quantities in harmonic multilayer models. The method is more memory efficient than classical methods used to solve boundary conditions. The method has been applied to a 3-D electromagnetic configuration for verification and compared to numeric results;
- ○ Vahaj et al. [6]. In this research, a 2-D semi-analytical model based on the subdomain technique was proposed to calculate the magnetic vector potential in outer rotor PM machines with surface-inset PMs. The saturation effect was neglected. The electromagnetic performances were verified by comparing them with those obtained from FEA;

- **Electrical, Thermal or Magnetic Equivalent Circuit (EEC, TEC or MEC)**

- ○ Asfirane et al. [7]. The authors have been developing a 3-D nonlinear MEC for a hybrid excitation synchronous machine. The semi-analytical results have been compared with those obtained from 3-D numeric results, as well as experimental data. The 3-D nonlinear MEC exhibited fairly accurate results when compared to the 3-D FEA, with a significant gain in computation time (viz, 4.7 s and 1560 s for one position, respectively).
- ○ Petitgirard et al. [8]. This original study concerned the winding head thermal design of electrical machines in difficult thermal environments. Based on geometrical assumptions (viz, Delaunay triangulation and Voronoï tessellation), an adaptive generic tool of a 2-D TEC in a steady state was developed, which could be adapted for all basic shapes and solved the thermal behavior of a random wire layout. The network set-up, adaptation, matrix writing and resolution were detailed. The model has been compared with the finite volume method, and several experiments have been planned;
- ○ Mekahlia et al. [9]. In this significant research, a harmonic EEC was presented for multi-phase squirrel-cage induction machines. In order to predict the torque pulsations, the reduced-order model of the rotor was applied. The proposed analysis allowed for avoiding incorrect design with non-sinusoidal magnetomotive forces. The semi-analytical approach has been confirmed by FEA for a three- and five-phase induction machine;

- **Hybrid Models**

- ○ Aleksandrov et al. [10]. In this contribution, the authors developed a 2-D hybrid steady-state magnetic field model, capable of accurately modeling the electromagnetic behavior in a linear induction motor. This model, in a Cartesian coordinate system, integrated a complex harmonic modeling technique (or the Maxwell–Fourier method) with a discretized MEC without the saturation effect. The analytical solution was applied to regions with homogeneous material properties, while the linear MEC approach was used for the regions containing non-homogeneous material properties. The resulting thrust and normal forces showed excellent agreement with respect to FEA and the measurement data;

- Benmessaoud et al. [11,12]. In [11], the authors developed a 2-D hybrid model in Cartesian coordinates, combining an MEC with the Maxwell–Fourier method for eddy current loss calculation. The model coupling was applied to a U-cored static electromagnetic device. Experimental tests and 3-D FEA were compared with the proposed approach on massive conductive parts in aluminum. In [12], the developed hybrid model was extended in polar coordinates to multi-phase synchronous machines for the volumic PM eddy current losses. A global revision on the calculation and analysis of PM eddy current losses can be found in [13].

At this point, as editors of this book, we would like to express our deep gratitude for the opportunity to publish with MDPI. This acknowledgment is deservedly extensive to the MCA Editorial Office and more particularly to Mr. Everett Zhu, who has permanently supported us in this process.

It was a great pleasure to work in such conditions. We look forward to collaborating with MCA in the future.

Conflicts of Interest: The authors declare no conflict of interest.

References

1. Jabbari, A. An Analytical Expression for Magnet Shape Optimization in Surface-Mounted Permanent Machines. *Math. Comput. Appl.* **2018**, *23*, 57. [CrossRef]
2. Ben Yahia, M.; Boughrara, K.; Dubas, F.; Roubache, L.; Ibtiouen, R. Two-Dimensional Exact Subdomain Technique of Switched Reluctance Machines with Sinusoidal Current Excitation. *Math. Comput. Appl.* **2018**, *23*, 59. [CrossRef]
3. Dubas, F.; Boughrara, K. New scientific contribution on the 2-D subdomain technique in polar coordinates: Taking into account of iron parts. *Math. Comput. Appl.* **2017**, *22*, 42. [CrossRef]
4. Mendonça, G.A.; Maia, T.A.C.; Cardoso Filho, B.J. Magnetic Field Analytical Solution for Non-homogeneous Permeability in Retaining Sleeve of a High-Speed Permanent-Magnet Machine. *Math. Comput. Appl.* **2018**, *23*, 72. [CrossRef]
5. Custers, C.H.H.M.; Jansen, J.W.; Lomonova, E.A. Memory Efficient Method for Electromagnetic Multi-Region Model Using Scattering Matrices. *Math. Comput. Appl.* **2018**, *23*, 71. [CrossRef]
6. Vahaj, A.; Rahideh, A.; Moayed-Jahromi, H.; Ghaffari, A. Exact Two-Dimensional Analytical Calculations for Magnetic Field, Electromagnetic Torque, UMF, Back-EMF, and Inductance of Outer Rotor Surface Inset Permanent Magnet Machines. *Math. Comput. Appl.* **2019**, *24*, 24. [CrossRef]
7. Asfirane, S.; Hlioui, S.; Amara, Y.; Gabsi, M. Study of a Hybrid Excitation Synchronous Machine: Modeling and Experimental Validation. *Math. Comput. Appl.* **2019**, *24*, 34. [CrossRef]
8. Petitgirard, J.; Piguet, T.; Baucour, P.; Chamagne, D.; Fouillien, E.; Delmare, J.-C. Steady State and 2D Thermal Equivalence Circuit for Winding Heads—A New Modelling Approach. *Math. Comput. Appl.* **2020**, *25*, 70. [CrossRef]
9. Mekahlia, A.; Semail, E.; Scuiller, F.; Zahr, H. Reduced-Order Model of Rotor Cage in Multiphase Induction Machines: Application on the Prediction of Torque Pulsations. *Math. Comput. Appl.* **2020**, *25*, 11. [CrossRef]
10. Aleksandrov, S.R.; Overboom, T.T.; Lomonova, E.A. 2D Hybrid Steady-State Magnetic Field Model for Linear Induction Motors. *Math. Comput. Appl.* **2019**, *24*, 74. [CrossRef]
11. Benmessaoud, Y.; Dubas, F.; Hilairet, M. Combining the Magnetic Equivalent Circuit and Maxwell–Fourier Method for Eddy-Current Loss Calculation. *Math. Comput. Appl.* **2019**, *24*, 60. [CrossRef]
12. Benmessaoud, Y.; Ouamara, D.; Dubas, F.; Hilairet, M. Investigation of Volumic Permanent-Magnet Eddy-Current Losses in Multi-Phase Synchronous Machines from Hybrid Multi-Layer Model. *Math. Comput. Appl.* **2020**, *25*, 14. [CrossRef]

13. Ouamara, D.; Dubas, F. Permanent-Magnet Eddy-Current Losses: A Global Revision of Calculation and Analysis. *Math. Comput. Appl.* **2019**, *24*, 67. [CrossRef]

Publisher's Note: MDPI stays neutral with regard to jurisdictional claims in published maps and institutional affiliations.

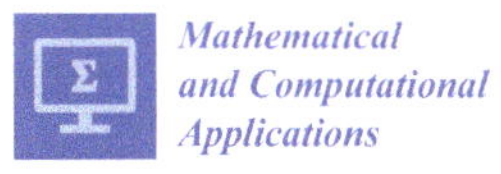

Article

An Analytical Expression for Magnet Shape Optimization in Surface-Mounted Permanent Magnet Machines

Ali Jabbari

Department of Mechanical Engineering, Faculty of Engineering, Arak University, Arak 38156-8-8849, Iran; a-jabbari@araku.ac.ir

Received: 16 July 2018; Accepted: 2 October 2018; Published: 5 October 2018

Abstract: Surface-mounted permanent magnet machines are widely used in low and medium speed applications. Pulsating torque components is the most crucial challenge, especially in low-speed applications. Magnet pole shape optimization can be used to mitigate these components. In this research, an analytical model is proposed to calculate the magnetic vector potential in surface-mounted permanent magnet machines. A mathematical expression is also derived for optimal the magnet shape to reduce the cogging torque and electromagnetic torque components. The presented model is based on the resolution of the Laplace's and Poisson's equations in polar coordinates by using the subdomain method and applying hyperbolic functions. The proposed method is applied to the performance computation of a surface-mounted permanent magnet machine, i.e., a 3-phase 12S-10P motor. The analytical results are validated through the finite element analysis (FEA) method.

Keywords: surface-mounted PM machines; torque pulsation; magnet shape optimization; analytical expression

---

1. Introduction

Surface-mounted permanent magnet machines are interested in high-performance applications because of their high efficiency and power density. However, the noise and vibration caused by pulsating torque components seriously affect the machine performance. Pulsating torque is greatly affected by the distribution of the magnetic field and the configuration of the permanent magnets. Therefore, pulsating torque mitigation can be performed using magnet shape optimization to obtain a better magnetic field waveform and also to reduce the cogging torque and electromagnetic torque, effectively.

An extensive variety of techniques such as magnet skewing [1–4], magnet-arc optimization [5–9], magnet shape optimization [10], and magnet displacing [2–4,7–9] for minimizing cogging torque in permanent magnet motors is documented in the literature.

A variety of techniques including analytical and numerical methods have been conducted to evaluate the pulsating torque components in electrical machines. Numerical methods like the finite element method (FEA) give accurate results and are time-consuming especially in the first step of the design stage. Semi-analytical methods including conformal mapping [11–14] and Magnetic Equivalent Circuit (MEC) [15–17], and analytical methods including the subdomain model [18–33] are reported to model electrical machines and are useful in the design optimization stage. The subdomain model is more accurate than the other analytical models [15].

Indeed, the global or local saturation effect influences the electromagnetic performances, e.g., on the ripple/cogging torque [34]. To overcome that issue, recently, a new technique to account for finite soft-magnetic material permeabilities in the subdomain technique was developed by applying the superposition principle in both directions in polar or Cartesian coordinates [35,36]. According to Reference [37], the Dubas' superposition technique [35,36] is very interesting since it enables the magnetic field calculation in the material of slotted geometries. This technique has been implemented in radial-flux electrical machines considering finite soft-magnetic material permeability [34]. The Dubas's superposition technique could have been used to develop a new model with the consideration of the saturation effect. In References [38–41], an analytical model has been introduced to compute electric machine performance by using the subdomain method.

However, no analytical expression was found at present to calculate the optimal magnet pole shape in surface-mounted permanent magnet machines in order to minimize the pulsating torque components.

The focus of this paper is to derive an analytical expression for the optimal magnet pole shape in surface-mounted permanent magnet machines to reduce pulsating torque components. An analytical model is presented based on the resolution of the Laplace's and Poisson's equations in surface-mounted permanent magnet machines by using the subdomain method whilst considering pole shape optimization. It is shown that the developed model can effectively estimate the magnetic field, cogging torque, electromagnetic torque, back electromotive force and self/mutual inductance. This model is applied to the performance calculation of a surface-mounted permanent magnet motor, i.e., a 3-phase 12S-10P motor. It is shown that the results of the analytical model are in close agreement with the results of the FEA method.

2. Subdomain Definition

The schematic representation of the investigated machines is shown in Figure 1. The machine model is divided into four subdomains. The stator which has two subdomains including the Q_1 slot regions (domain j), the Q_1 slot opening regions (domain i) and the airgap subdomain (region II) are shown in Figure 2. The rotor has one subdomain including the permanent magnet regions (domain I), as shown in Figure 3.

The angular position of the j-th stator slot and i-th stator slot opening are defined as (1) and (2), respectively.

$$\theta_j = -\frac{\beta}{2} + \frac{2j\pi}{Q_1} \quad with \quad 1 \le j \le Q_1 \tag{1}$$

$$\theta_i = -\frac{\alpha}{2} + \frac{2i\pi}{Q_1} \quad with \quad 1 \le i \le Q_1 \tag{2}$$

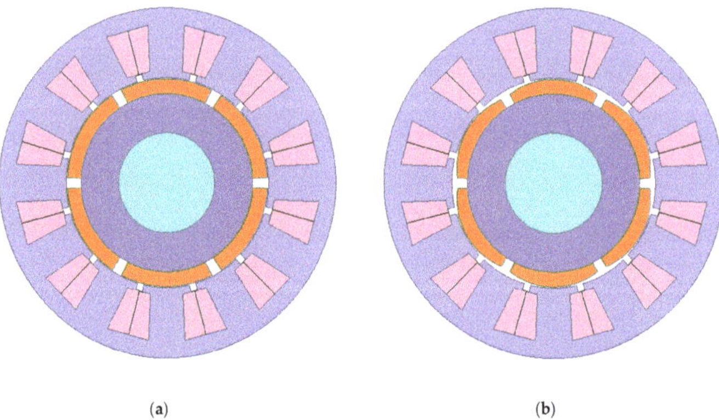

(a) (b)

Figure 1. The geometrical representation of the investigated machines with (**a**) uniform rotor shape, (**b**) non-uniform rotor shape.

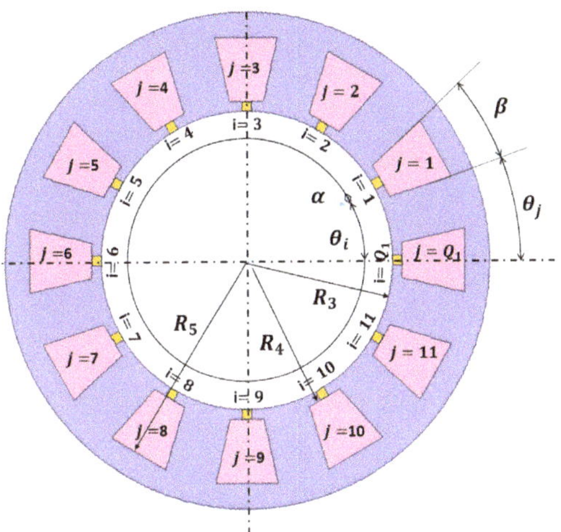

Figure 2. The stator subdomains including the j and i regions.

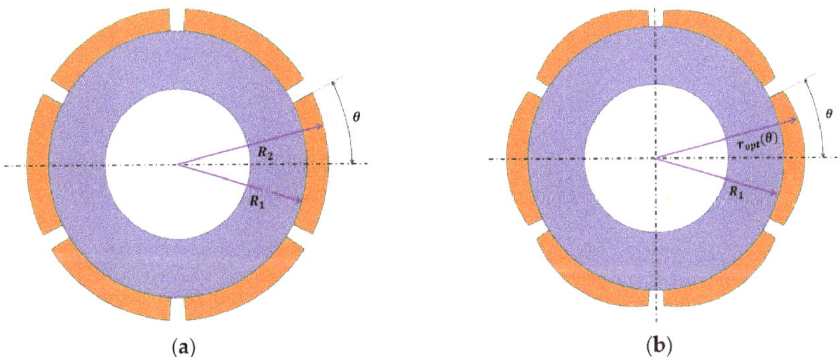

Figure 3. The rotor permanent magnet subdomain. (**a**) Uniform rotor, (**b**) non-uniform rotor.

3. Magnetic Vector Potential Computation

The general solution to Laplace's or Poisson's equations in each subdomain is developed in this section. The Laplace equation can be described in the polar form as

$$\frac{\partial^2 A_z}{\partial r^2} + \frac{1}{r}\frac{\partial A_z}{\partial r} + \frac{1}{r^2}\frac{\partial^2 A_z}{\partial \theta^2} = 0 \quad for \left\{ \begin{array}{l} R_1 \leq r \leq R_2 \\ \theta_1 \leq \theta \leq \theta_2 \end{array} \right. \tag{3}$$

By replacing r by $R_1 e^{-t}$, one obtains

$$\frac{\partial^2 A_z}{\partial t^2} + \frac{\partial^2 A_z}{\partial \theta^2} = 0 \qquad for \qquad \left\{ \begin{array}{l} ln(\frac{R_1}{R_2}) \leq t \leq 0 \\ \theta_1 \leq \theta \leq \theta_2 \end{array} \right. \tag{4}$$

The 2D analytical model in quasi-Cartesian coordinates is formulated with the following assumptions:

- The end effects are neglected (i.e., the machine is infinitely long: the magnetic variables are independent of z).
- The stator is assumed to be infinitely permeable (i.e., the saturation effect is neglected) with zero electrical conductivity.

- The relative magnetic permeability and electrical conductivity of the solid rotor and shaft are assumed to be constant.
- The current density in the slots has only one component along the z-axis.

3.1. Magnetic Vector Potential in the Stator Slot Subdomain (Region j)

The Poisson equation in the stator slot subdomain is given by

$$\frac{\partial^2 A_{zj}}{\partial t^2} + \frac{\partial^2 A_{zj}}{\partial \theta^2} = -\mu_0 J_{zj} R_4^2 e^{-2t} \quad for \quad \begin{cases} t_1 \le t \le t_2 \\ \theta_j \le \theta \le \theta_j + \beta \end{cases} \tag{5}$$

where $t_1 = ln\left(\frac{R_4}{R_5}\right)$, $t_2 = 0$ and J_{zj} is the slot current density.

The Neumann boundary conditions at the bottom and at each side of the slot are obtained as

$$\left.\frac{\partial A_{zj}}{\partial \theta}\right|_{\theta=\theta_j} = 0 \;\; and \;\; \left.\frac{\partial A_{zj}}{\partial \theta}\right|_{\theta=\theta_j+\beta} = 0 \tag{6}$$

$$\left.\frac{\partial A_{zj}}{\partial t}\right|_{t=t_1} = 0 \tag{7}$$

The general solution of Equation (5) using the separation of variables method is given by

$$\begin{aligned} A_{zj}(t,\theta) = a_0{}^j - & \tfrac{1}{2}\mu_0 J_{zj} R_5^2 \left(e^{-2t_1} t + \tfrac{1}{2} e^{-2t+2t_1}\right) \\ & + \sum_{h=1}^{\infty} \left(a_h{}^j \frac{\beta}{h\pi} \frac{Cosh\left(\frac{h\pi}{\beta}(t-t_1)\right)}{Sinh\left(\frac{h\pi}{\beta}(t_2-t_1)\right)} \right) Cos\left(\tfrac{h\pi}{\beta}(\theta-\theta_j)\right) \end{aligned} \tag{8}$$

where h is a positive integer and the coefficients $a_0{}^j$ and $a_h{}^j$ are determined based on the continuity and interface conditions.

The continuity of the magnetic vector potential between the subdomain j and the region i leads to

$$\left.\frac{\partial A_{zj}}{\partial t}\right|_{t=t_2} = f(\theta) = \begin{cases} \left.\frac{\partial A_{zi}}{\partial t}\right|_{t=t_3} for \;\; \theta_i \le \theta \le \theta_i + \alpha \\ 0 \qquad\qquad elsewhere \end{cases} \tag{9}$$

The interface condition (9) gives

$$\mu_0 J_{zj} Sinh(t_1) = \frac{1}{\beta} \int_{\theta_j}^{\theta_j+\beta} f(\theta)\, d\theta \tag{10}$$

$$a_h{}^j = \frac{2}{\beta} \int_{\theta_j}^{\theta_j+\beta} f(\theta)\, Cos\left(\frac{h\pi}{\beta}(\theta-\theta_j)\right) d\theta \tag{11}$$

3.2. Magnetic Vector Potential in the Stator Slot Opening Subdomain (Region i)

The Laplace equation in the stator second inner slot opening subdomain is given by

$$\frac{\partial^2 A_{zi}}{\partial t^2} + \frac{\partial^2 A_{zi}}{\partial \theta^2} = 0 \quad for \quad \begin{cases} t_3 \le t \le t_4 \\ \theta_i \le \theta \le \theta_i + \alpha \end{cases} \tag{12}$$

where $t_3 = ln\left(\frac{R_3}{R_4}\right)$ and $t_4 = 0$.

The Neumann boundary conditions at the bottom and at each side of the slot are obtained as

$$\left.\frac{\partial A_{zi}}{\partial \theta}\right|_{\theta=\theta_i} = 0 \ \ and \ \ \left.\frac{\partial A_{zi}}{\partial \theta}\right|_{\theta=\theta_i+\alpha} = 0 \tag{13}$$

The general solution of Equation (12) using the separation of variables method is given by

$$\begin{aligned} A_{zi}(t,\theta) &= a_0{}^i + b_0{}^i t \\ &+ \sum_{k=1}^{\infty}\left(\frac{Sinh\left(\frac{k\pi}{\alpha}(t-t_4)\right)}{Sinh\left(\frac{k\pi}{\alpha}(t_3-t_4)\right)}a_k{}^i + \frac{Sinh\left(\frac{k\pi}{\alpha}(t-t_3)\right)}{Sinh\left(\frac{k\pi}{\alpha}(t_4-t_3)\right)}b_k{}^i\right) Cos\left(\frac{k\pi}{\alpha}(\theta-\theta_i)\right) \end{aligned} \tag{14}$$

where k is a positive integer and the coefficients, $a_0{}^i$, $b_0{}^i$, $a_k{}^i$ and $b_k{}^i$ are determined based on the continuity and interface conditions.

The continuity of the magnetic vector potential between the subdomain l and the regions i and II leads to

$$Azi(t_4,\theta) = AzII(t_5,\theta) \qquad for \quad \theta_i \le \theta \le \theta_i + \alpha \tag{15}$$

$$Azi(t_3,\theta) = Azj(t_4,\theta) \qquad for \quad \theta_i \le \theta \le \theta_i + \alpha \tag{16}$$

The interface condition (15) gives

$$a_0{}^i = \frac{1}{\alpha}\int_{\theta_l}^{\theta_i+\alpha} AII(t_5,\theta)\, d\theta \tag{17}$$

$$b_k{}^i = \frac{2}{\alpha}\int_{\theta_i}^{\theta_i+\alpha} AII(t_5,\theta)\, Cos\left(\frac{h\pi}{\alpha}(\theta-\theta_i)\right) d\theta \tag{18}$$

The interface condition (16) gives

$$a_0{}^i + ln\left(\frac{R_3}{R_4}\right) b_0{}^i = \frac{1}{\alpha}\int_{\theta_i}^{\theta_i+\alpha} Azj(t_4,\theta)\, d\theta \tag{19}$$

$$a_k{}^i = \frac{2}{\alpha}\int_{\theta_i}^{\theta_i+\alpha} Azj(t_4,\theta)\, Cos\left(\frac{k\pi}{\alpha}(\theta-\theta_i)\right) d\theta \tag{20}$$

3.3. *Magnetic Vector Potential in the Air-Gap Subdomain (Region II)*

The Laplace equation in the air-gap subdomain is given by

$$\frac{\partial^2 A_{zII}}{\partial t^2} + \frac{\partial^2 A_{zII}}{\partial \theta^2} = 0 \qquad for \qquad \begin{cases} t_5 \le t \le t_6 \\ 0 \le \theta \le 2\pi \end{cases} \tag{21}$$

where $t_5 = ln\left(\frac{R_2}{R_3}\right)$ and $t_6 = 0$.

The general solution of Equation (21), considering the periodicity boundary conditions is obtained as

$$\begin{aligned} A_{zII}(t,\theta) &= \sum_{n=1}^{\infty}\left(\frac{1}{n}\frac{Cosh(n(t-t_6))}{Sinh(n(t_5-t_6))}a_n{}^{II} + \frac{1}{n}\frac{Cosh(n(t-t_5))}{Sinh(n(t_6-t_5))}b_n{}^{II}\right)Cos(n\theta) \\ &+ \sum_{n=1}^{\infty}\left(\frac{1}{n}\frac{Cosh(n(t-t_6))}{Sinh(n(t_5-t_6))}c_n{}^{II} + \frac{1}{n}\frac{Cosh(n(t-t_5))}{Sinh(n(t_6-t_5))}d_n{}^{II}\right)Sin(n\theta) \end{aligned} \tag{22}$$

where n is a positive integer.

The coefficients $a_n{}^{II}$, $b_n{}^{II}$, $c_n{}^{II}$ and $d_n{}^{II}$ are determined by considering the continuity of the magnetic vector potential between the internal airgap subdomain II and the region i using a Fourier series expansion of interface condition (23) and (24) over the airgap interval.

The continuity of the magnetic vector potential between the internal airgap subdomain II and the regions i and leads to

$$\left.\frac{\partial A_{zII}}{\partial t}\right|_{t=t_5} = g(\theta) = \begin{cases} \left.\frac{\partial A_{zi}}{\partial t}\right|_{t=t_4} & for \quad \theta_i \le \theta \le \theta_i + \alpha \\ 0 & elsewhere \end{cases} \tag{23}$$

$$\left.\frac{\partial A_{zII}}{\partial t}\right|_{t=t_6} = h(\theta) = \begin{cases} \left.\frac{\partial A_{zI}}{\partial t}\right|_{t=t_7} & for \quad \theta_k \le \theta \le \theta_k + \gamma \\ 0 & elsewhere \end{cases} \tag{24}$$

The interface condition (23) gives

$$a_n{}^{II} = \frac{2}{2\pi}\int_{\theta_i}^{\theta_i+\alpha} g(\theta)\, Cos(n\theta)\, d\theta \tag{25}$$

$$c_n{}^{II} = \frac{2}{2\pi}\int_{\theta_i}^{\theta_i+\alpha} g(\theta)\, Sin(n\theta)\, d\theta \tag{26}$$

The interface condition (24) gives

$$b_n{}^{II} = \frac{2}{2\pi}\int_{\theta_k}^{\theta_k+\gamma} h(\theta)\, Cos(n\theta)\, d\theta \tag{27}$$

$$d_n{}^{II} = \frac{2}{2\pi}\int_{\theta_k}^{\theta_k+\gamma} h(\theta)\, Sin(n\theta)\, d\theta \tag{28}$$

3.4. *Magnetic Vector Potential in the Rotor Permanent Magnet Subdomain (Region I)*

The Poisson equation in the rotor permanent magnet subdomain is given by

$$\frac{\partial^2 A_{zI}}{\partial t^2} + \frac{\partial^2 A_{zI}}{\partial \theta^2} = -\mu_0 R_1 e^{-t}\left(M_\theta - \frac{\partial M_r}{\partial \theta}\right) \quad for \quad \begin{cases} t_7 \le t \le t_8 \\ 0 \le \theta \le 2\pi \end{cases} \tag{29}$$

where $t_7 = ln\left(\frac{R_1}{R_2}\right)$ and $t_8 = 0$, M_θ and M_r are the tangential and radial components of magnetization.

3.4.1. Radial Magnetization

The radial and tangential components of radial magnetization for the surface-mounted design can be expressed as

$$M_{rn} = \frac{4B_r}{\mu_0 n\pi} Sin\left(\frac{n\pi\alpha_p}{2}\right) \tag{30}$$

$$M_{\theta n} = 0 \tag{31}$$

where α_p is the magnet pole width to magnet pitch ratio.

3.4.2. Parallel Magnetization

The radial and tangential components of the parallel magnetization for the surface-mounted design can be expressed as

$$M_{rn} = \frac{B_r}{\mu_0}\alpha_p\left[A_{1n}(\alpha_p) + A_{2n}(\alpha_p)\right] \tag{32}$$

$$M_{\theta n} = \frac{B_r}{\mu_0}\alpha_p\left[A_{1n}(\alpha_p) - A_{2n}(\alpha_p)\right] \tag{33}$$

where

$$A_{1n}(\alpha_p) = \frac{Sin((np+1)\frac{\pi\alpha_p}{2p})}{(np+1)\frac{\pi\alpha_p}{2p}} \tag{34}$$

$$A_{2n}(\alpha_p) = \begin{cases} \frac{Sin((np-1)\frac{\pi\alpha_p}{2p})}{(np-1)\frac{\pi\alpha_p}{2p}} & for \;\; np \neq 1 \\ 1 & for \;\; np = 1 \end{cases} \tag{35}$$

For a surface-mounted design, the Neumann boundary conditions at the bottom of the permanent magnet are obtained as

$$\left.\frac{\partial A_{zI}}{\partial t}\right|_{t=t_8} = 0 \tag{36}$$

The general solution of Equation (29) using the separation of variables method is given by

$$\begin{aligned} A_{zI}(t,\theta) = \sum_{n=1}^{\infty} &\left(\begin{array}{c} a_n^I \frac{\cosh(n(t-t_8))}{\cosh(n(t_7-t_8))} \\ +X_n(t)\, Cos\left(\frac{n\pi\alpha_p}{2\alpha_r}\right) \end{array} \right) Cos(n\theta) \\ &+ \sum_{n=1}^{\infty} \left(\begin{array}{c} c_n^I \frac{\cosh(n(t-t_8))}{\cosh(n(t_7-t_8))} \\ +X_n(t).Sin\left(\frac{n\pi\alpha_p}{2\alpha_r}\right) \end{array} \right) Sin(n\theta) \end{aligned} \tag{37}$$

$$X_n(t) = \left(1 + \frac{1}{n}e^{(n+1)t}\right) f_n(t) - \frac{\mathrm{Cosh}(n(t-t_8))}{\mathrm{Cosh}(n(t_7-t_8))}\left(1 + \frac{1}{n}e^{(n+1)t_7}\right) f_n(t_7) \tag{38}$$

$$f_n(t) = \begin{cases} \mu_0 \frac{npM_{rn}+M_{\theta n}}{1-np^2} R_1 e^{-t} & if \; np \neq 1 \\ -\mu_0 \frac{M_{rn}+M_{\theta n}}{2} R_1 e^{-t} ln(R_1 e^{-t}) & if \; np = 1 \end{cases} \tag{39}$$

where n is a positive integer and the coefficients ${a_n}^I$ and ${c_n}^I$ are determined based on the continuity and interface conditions.

The continuity of the magnetic vector potential between the subdomain I and the regions II leads to

$$A_{zI}(t_7,\theta) = A_{zII}(t_6,\theta) \tag{40}$$

The interface condition (40) gives

$$a_n^I = \frac{2}{2\pi}\int_0^{2\pi} A_{zII}(t_6,\theta).Cos(n\theta)\, d\theta \tag{41}$$

$$c_n^I = \frac{2}{2\pi}\int_0^{2\pi} A_{zII}(t_6,\theta).Sin(n\theta)\, d\theta \tag{42}$$

4. Magnet Pole Shape Optimization

The general solution for the magnetic potential distribution in the air-gap subdomain is

$$A_{zII}(\mathrm{t}.\theta) = \sum_{n=1}^{\infty}\left(\frac{1}{n}\frac{Cosh(n(\mathrm{t}-\mathrm{t}_6))}{Sinh(n(\mathrm{t}_5-\mathrm{t}_6))}{a_n}^{II} + \frac{1}{n}\frac{Cosh(n(\mathrm{t}-\mathrm{t}_5))}{Sinh(n(\mathrm{t}_6-\mathrm{t}_5))}{b_n}^{II}\right)Cos(n\theta) \tag{43}$$

The normal flux density B_r is defined as

$$B_r = -\mu_0\frac{\partial A_{zII}}{\partial r} = -\mu_0\frac{e^{\mathrm{t}_5}}{R_2}\frac{\partial A_{zII}}{\partial t} \tag{44}$$

As the permeability of the stator/rotor iron core is much larger than that of air, the following boundary conditions are employed

- The scalar magnetic potential is expressed as $A_{II} = 0$ in the inner stator surface

$$A_{zII}(t_5, 0) = 0 \tag{45}$$

or

$$\sum_{n=1}^{\infty}\left(\frac{1}{n}\frac{Cosh(n(\mathrm{t}_5))}{Sinh(n(\mathrm{t}_5))}a_n{}^{II} - \frac{1}{n}\frac{1}{Sinh(n(\mathrm{t}_5))}b_n{}^{II}\right) = 0 \tag{46}$$

- The normal flux density waveforms is sinusoidal in the inner stator surface and expressed as $B_r = B_{max}\cos(\theta)$. Therefore,

$$-\mu_0\frac{e^{\mathrm{t}_5}}{R_2}\frac{\partial A_{zII}(t_5, \theta)}{\partial t} = B_{max} \tag{47}$$

or

$$-\mu_0\frac{e^{\mathrm{t}_5}}{R_2}\left(\frac{Sinh(n(\mathrm{t}_5 - \mathrm{t}_6))}{Sinh(n(\mathrm{t}_5 - \mathrm{t}_6))}a_n{}^{II} + \frac{Sinh(n(\mathrm{t}_5 - \mathrm{t}_5))}{Sinh(n(\mathrm{t}_6 - \mathrm{t}_5))}b_n{}^{II}\right) = B_{max} \tag{48}$$

From the boundary conditions (46) and (48), we can get

$$b_1{}^{II} = Cosh((\mathrm{t}_5))a_1{}^{II} \tag{49}$$

$$\begin{cases} a_1{}^{II} = -\frac{B_{max}R_3}{\mu_0} & n = 1 \\ b_1{}^{II} = -\frac{B_{max}R_3}{\mu_0}Cosh((\mathrm{t}_5)) & n = 1 \\ a_n{}^{II} = 0 & n = 3, 5, 7 \\ b_n{}^{II} = 0 & n = 3, 5, 7 \end{cases} \tag{50}$$

At the position of $\theta = 0$, the magnetic potential is expressed as

$$B_{rII}(\mathrm{t}_6, 0) = -\frac{\mu_0}{R_2}\left(\frac{Sinh(n(\mathrm{t}_6 - \mathrm{t}_6))}{Sinh(n(\mathrm{t}_5 - \mathrm{t}_6))}a_1{}^{II} - \frac{Sinh(n(\mathrm{t}_6 - \mathrm{t}_5))}{Sinh(n(\mathrm{t}_5 - \mathrm{t}_6))}b_1{}^{II}\right) \tag{51}$$

or

$$B_{rII}(\mathrm{t}_6, 0) = -\frac{\mu_0}{R_2}b_1{}^{II} = \frac{\mu_0}{R_2}\frac{B_{max}R_3}{\mu_0}Cosh((\mathrm{t}_5)) \tag{52}$$

In the outer surface of the rotor, the magnetic potential can be derived as

$$B_{rII}(\mathrm{t}_6, 0) = B_{rII}(t, 0) \tag{53}$$

or

$$B_{rII}{}^2(\mathrm{t}_6, 0) = B_{rII}{}^2(\mathrm{t}, \theta) \tag{54}$$

or

$$\left(\frac{\mu_0}{R_2}\frac{B_{max}R_3}{\mu_0}Cosh((\mathrm{t}_5))\right)^2 = \left(-\frac{\mu_0}{R_2}\left(\begin{array}{c}\frac{Sinh(n(\mathrm{t}-\mathrm{t}_6))}{Sinh(n(\mathrm{t}_5-\mathrm{t}_6))}a_1{}^{II} \\ -\frac{Sinh(n(\mathrm{t}-\mathrm{t}_5))}{Sinh(n(\mathrm{t}_5-\mathrm{t}_6))}b_1{}^{II}\end{array}\right)Cos(\theta)\right)^2 \tag{55}$$

or

$$t_{opt}(\theta) = Cosh^{-1}\left\{\frac{Cosh(t_5)}{Cos\ (\theta)}\right\} + t_5 \tag{56}$$

Therefore, the optimum magnet radii can be expressed as

$$r_{opt} = \frac{R_1}{exp\left(Cosh^{-1}\left\{\frac{Cosh(t_5)}{Cos\ (\theta)}\right\} + t_5\right)} \tag{57}$$

5. Performance Calculation

The electromagnetic torque is obtained using the Maxwell stress tensor and expressed as

$$T_e = \frac{L_s}{\mu_0} \int_0^{2\pi} BII_r(t_e, \theta)\, BII_\theta(t_e, \theta)\, d\theta \tag{58}$$

where L_s is the axial length of the motor and t_e is calculated by

$$\begin{aligned} t_e &= ln\left(\frac{R_2}{R_e}\right) \\ R_e &= (R_2 + R_3)/2 \end{aligned} \tag{59}$$

The final expression of the electromagnetic torque can be expressed as

$$T_e = \frac{\pi L_s}{\mu_0} \sum_{n=1}^{\infty} (M_n N_n + O_n P_n)$$

where,

$$\begin{aligned} M_n &= -\frac{1}{R_e}\frac{Cosh(n(t_e - t_6))}{Sinh(n(t_5 - t_6))} a_n{}^{II} - \frac{1}{R_e}\frac{Cosh(n(t_e - t_5))}{Sinh(n(t_6 - t_5))} b_n{}^{II} \\ N_n &= -\frac{1}{R_e}\frac{Sinh(n(t_e - t_6))}{Sinh(n(t_5 - t_6))} c_n{}^{II} - \frac{1}{R_e}\frac{Sinh(n(t_e - t_5))}{Sinh(n(t_6 - t_5))} d_n{}^{II} \\ O_n &= \frac{1}{R_e}\frac{Cosh(n(t_e - t_6))}{Sinh(n(t_5 - t_6))} c_n{}^{II} + \frac{1}{R_e}\frac{Cosh(n(t_e - t_5))}{Sinh(n(t_6 - t_5))} d_n{}^{II} \\ P_n &= -\frac{1}{R_e}\frac{Sinh(n(t_e - t_6))}{Sinh(n(t_5 - t_6))} a_n{}^{II} - \frac{1}{R_e}\frac{Sinh(n(t_e - t_5))}{Sinh(n(t_6 - t_5))} b_n{}^{II} \end{aligned}$$

For single layer winding, the phase flux vector is calculated by

$$\begin{bmatrix} \psi_a \\ \psi_b \\ \psi_c \end{bmatrix} = N_c C^T \begin{bmatrix} \varphi_1 & \varphi_2 & \varphi_3 & \cdots & \varphi_{Q_2} \end{bmatrix} \tag{60}$$

where N_c is the number of conductors in the stator slot, C is a matrix connection between the stator slots and phase connections, and φ is the slot flux.

For the stator slots, φ is given by

$$\varphi_i = -\frac{L_s {R_4}^2}{k_f S} \int_0^{\beta} \int_0^{t_8} A_{mi}(t, \theta)\, e^{-2t}\, dt\, d\theta \tag{61}$$

where k_f is the stator fill factor and is the area of the stator slot.

For double-layer winding, the phase flux vector is calculated by

$$\begin{bmatrix} \psi_a \\ \psi_b \\ \psi_c \end{bmatrix} = \begin{bmatrix} \psi 1_a \\ \psi 1_b \\ \psi 1_c \end{bmatrix} + \begin{bmatrix} \psi 2_a \\ \psi 2_b \\ \psi 2_c \end{bmatrix} \tag{62}$$

where

$$\begin{bmatrix} \psi 1_a \\ \psi 1_b \\ \psi 1_c \end{bmatrix} = \frac{N_c}{2} C_1^T \begin{bmatrix} \varphi_{11} & \varphi_{12} & \varphi_{13} & \cdots & \varphi_{1Q_2} \end{bmatrix} \tag{63}$$

and

$$\begin{bmatrix} \psi 2_a \\ \psi 2_b \\ \psi 2_c \end{bmatrix} = \frac{N_c}{2} C_2^T \begin{bmatrix} \varphi_{21} & \varphi_{22} & \varphi_{23} & \cdots & \varphi_{2Q_2} \end{bmatrix} \tag{64}$$

For the stator slots, φ is given by

$$\varphi_{1i} = -\frac{2L_s R_4{}^2}{k_f S} \int_0^{\frac{\beta}{2}} \int_0^{t_8} A_{mi}(t,\theta)\, e^{-2t}\, dt\, d\theta \tag{65}$$

$$\varphi_{2i} = -\frac{2L_s R_4{}^2}{k_f S} \int_{\frac{\beta}{2}}^{\beta} \int_0^{t_8} A_{mi}(t,\theta)\, e^{-2t}\, dt\, d\theta \tag{66}$$

The back-EMF of phase A is given by

$$E_a = \omega \frac{d\psi_a}{d\theta_r} \tag{67}$$

where ω is the rotor angular speed and ψ_a is the flux linkage per phase A.

The stator inductance (self-inductance) of phase A is given by

$$L = \frac{\psi_a}{I_A} \tag{68}$$

where I_A is the peak current in phase A.

The mutual inductance of phase A and phase B is given by

$$M = \frac{N\varphi_{AB}}{I_B} \tag{69}$$

where N is the number of phase turns, φ_{AB} is magnetic flux in phase A, and I_B is the peak current in phase B.

6. Model Evaluation

In this section, the presented analytical model is used to study the magnetic flux density, electromagnetic torque, and back-electromotive force of a 12S-10P motor. The results of the analytical method are then verified by the results of the finite element method. A 2D model of the studied brushless permanent magnet motor is shown in Figure 4 and the motor parameters are given in Table 1. The PM magnetization is radial. The slot contains two coils as shown in Figure 4a. In order to have a good precision in the analytical evaluation, the number of harmonic terms used in the computations is equal to 50 (air-gap and PM subdomains) and 30 (slots and slot-opening subdomain).

We have to solve a system of linear equations with the same number of unknowns (i.e., 12). The matrix connection between the stator slots and phase connections of each layer for the investigated motor are given by

$$C_1 = \begin{bmatrix} 1 & -1 & 0 & 0 & 0 & 0 & -1 & 1 & 0 & 0 & 0 & 0 \\ 0 & 0 & -1 & 1 & 0 & 0 & 0 & 0 & 1 & -1 & 0 & 0 \\ 0 & 0 & 0 & 0 & 1 & -1 & 0 & 0 & 0 & 0 & -1 & 1 \end{bmatrix}$$

$$C_2 = \begin{bmatrix} 0 & -1 & 1 & 1 & 0 & 0 & 0 & 1 & -1 & 0 & 0 & 0 \\ 0 & 0 & 0 & 1 & -1 & 1 & 0 & 0 & 0 & -1 & 1 & 0 \\ -1 & 1 & 0 & 0 & 1 & 0 & 0 & 1 & 0 & 0 & 0 & 1 \end{bmatrix}$$

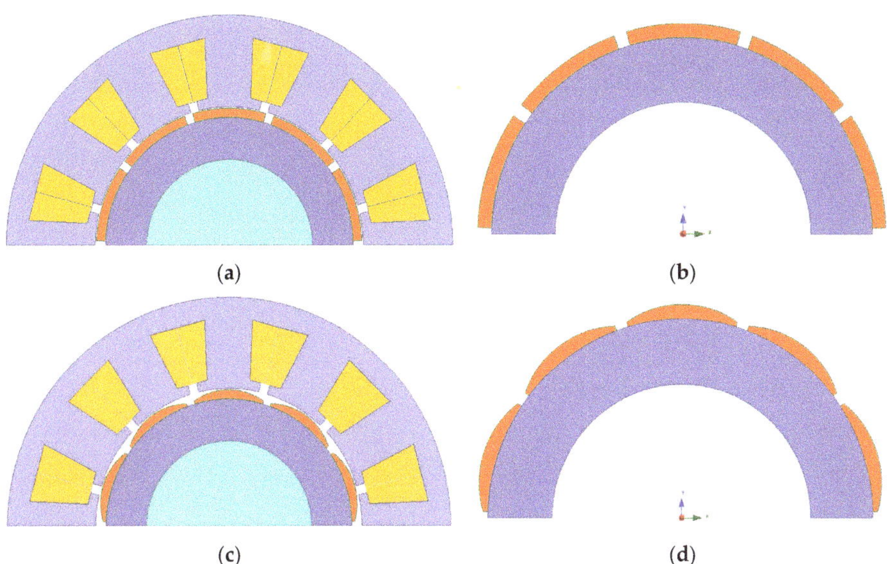

Figure 4. The cross-sections of the studied motor. (**a**) with the uniform rotor; (**b**) uniform rotor, (**c**) with the optimal rotor, (**d**) optimal rotor.

Table 1. The specification of the investigated motors.

Parameter	Value
Rotor Outer Diameter	208 mm
Rotor Inner Diameter	130 mm
Number of poles	10
Pole Arc	35°
Pole Thickness	20 mm
Magnet material	NEO-39SH
Stator Outer Diameter	350 mm
Stator Inner Diameter	210 mm
Number of Slots	12
Stator Tooth Width	30 mm
Stator Yoke Width	26 mm
Slot Open	7 mm
Tip Thickness	2.5 mm
Slot Skew	0°
Stator Length	100 mm
Lamination material	M 19–0.5 mm

The 2D finite element method is applied to the performance calculation of the motor with uniform and non-uniform rotor shapes. The magnetic field distribution in the studied motors is represented in Figure 5. Open circuit analytical and numerical comparisons of the cogging torque for both motors with initial and optimal magnet shapes are shown in Figure 6. The on-load comparison of the back electromotive force of the investigate motors with the initial and optimal magnet shapes is carried out analytically and numerically as shown in Figure 7. An analytical and numerical comparison of the radial flux density for the 12S-10P motor in an open circuit and on-load condition is shown in Figure 8. An analytical and numerical comparison of tangential flux density for the 12S-10P motor in open circuit and on-load condition is shown in Figure 9. An on-load comparison of the electromagnetic torque of the 12S-10P motor with the initial and optimal magnet shapes is shown in Figure 10.

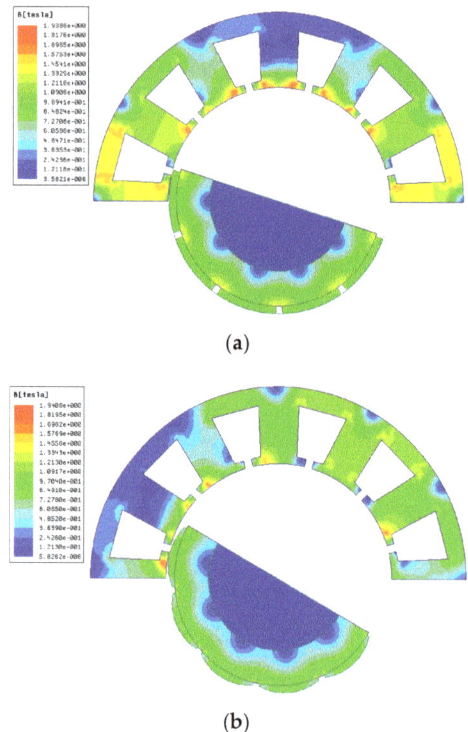

Figure 5. The magnetic field distribution in the 12S-10P motor. (**a**) Initial magnet shape; (**b**) Optimal magnet shape.

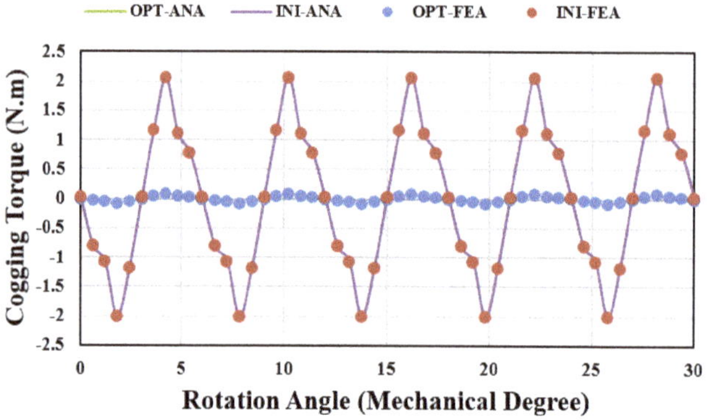

Figure 6. An open circuit analytical and numerical comparison of the cogging torque.

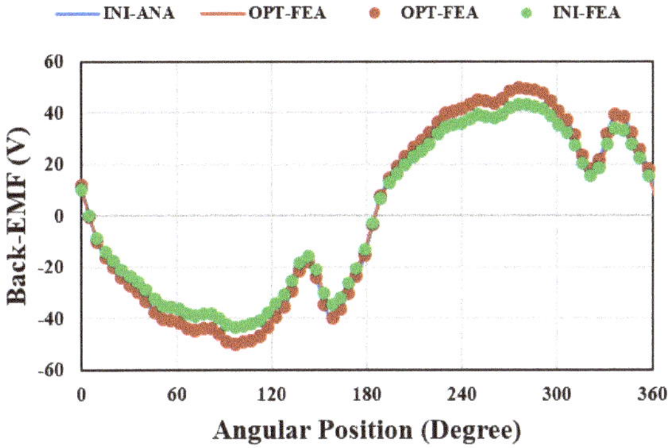

Figure 7. An on-load analytical and numerical comparison of Back-EMF.

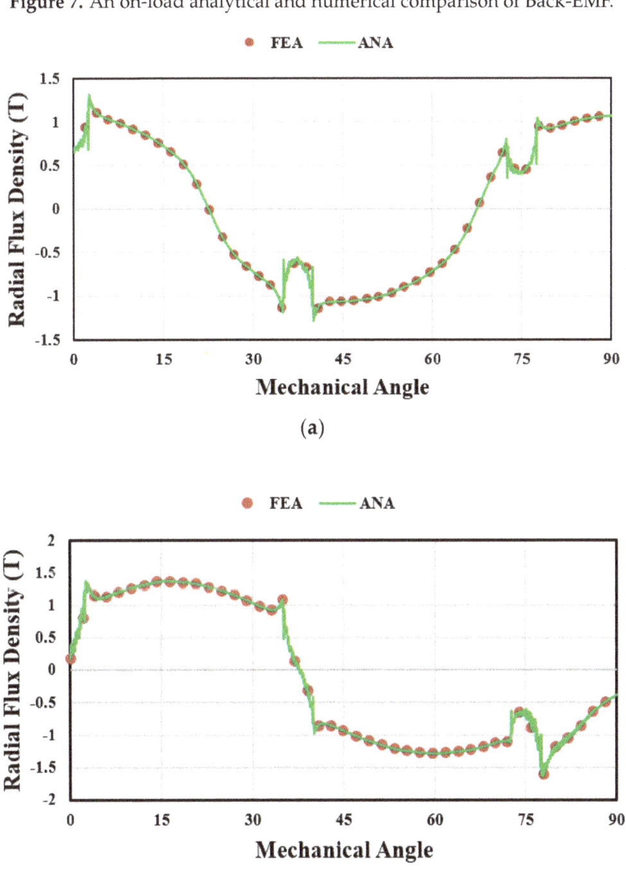

Figure 8. An analytical and numerical comparison of radial flux density for the 12S-10P motor. (**a**) Open circuit condition; (**b**) On-load condition.

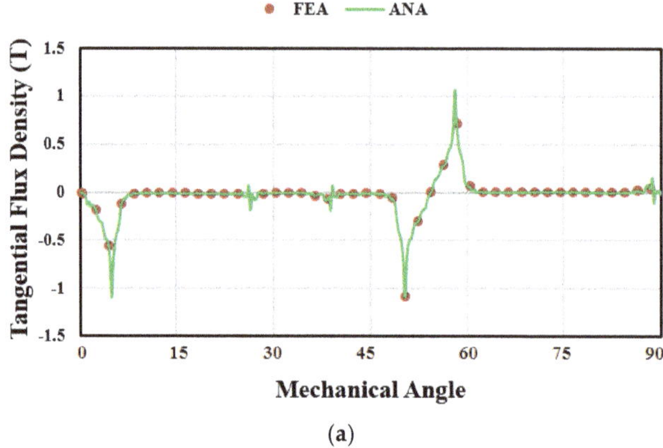

(a)

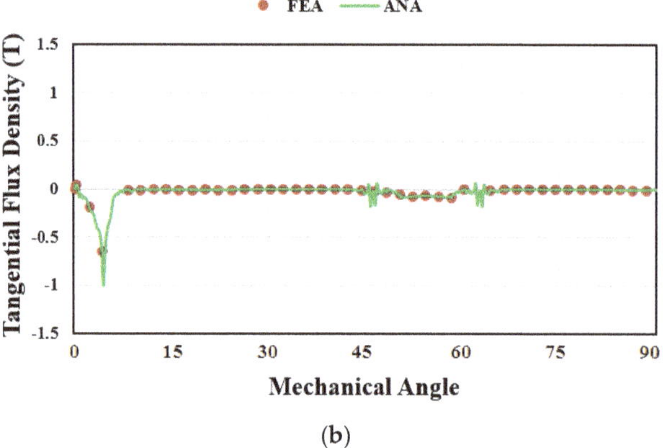

(b)

Figure 9. An analytical and numerical comparison of the tangential flux density for the 12S-10P motor. (**a**) Open circuit condition; (**b**) On-load condition.

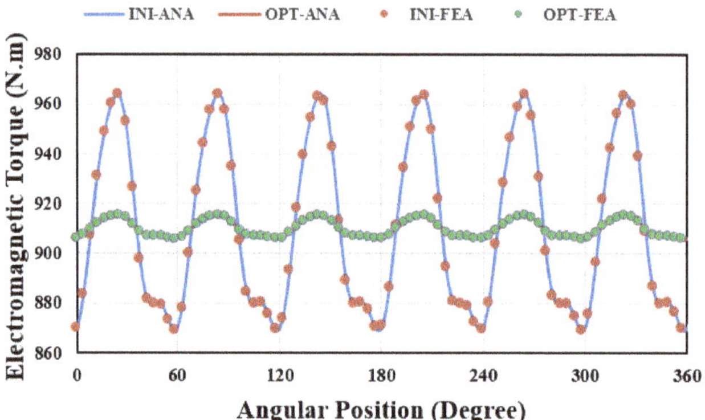

Figure 10. An on-load analytical and numerical comparison of the electromagnetic torque for the 12S-10P motor.

7. Conclusions

A mathematical expression for the optimal magnet shape in surface mounted permanent magnet machines was considered in this paper. The Fourier analysis method based on the subdomain method using hyperbolic functions is applied to derive the analytical expressions for the calculation of magnetic vector potential, magnetic flux density, cogging torque, electromagnetic torque and back-electromotive force in surface-mounted permanent magnet machines. This model is applied for the performance computation of a 12S-10P surface-mounted permanent magnet motor. The results of the proposed model have been verified thanks to the FEA results. In future work, the iron permeability for global saturation can be considered in the analytical model by Dubas' superposition technique [35,36].

Conflicts of Interest: The author declares no conflict of interest.

References

1. Jahns, T.M.; Soong, W.L. Pulsating torque minimization techniques for permanent magnet AC motor drives—A review. *IEEE Trans. Ind. Electron.* **1996**, *43*, 321–330. [CrossRef]
2. Bianchi, N.; Bolognani, S. Design techniques for reducing the cogging torque in surface-mounted PM motors. *IEEE Trans. Ind. Appl.* **2002**, *38*, 1259–1265. [CrossRef]
3. Lukaniszyn, M.; Jagiela, M.; Wrobel, R. Optimization of permanent magnet shape for minimum cogging torque using a genetic algorithm. *IEEE Trans. Magn.* **2004**, *40*, 1228–1231. [CrossRef]
4. Lateb, R.; Takorabet, N.; Meibody-Tabar, F. Effect of magnet segmentation on the cogging torque in surface-mounted permanent-magnet motors. *IEEE Trans. Magn.* **2006**, *42*, 442–445. [CrossRef]
5. Ackermann, B.; Sottek, R.; Janssen, J.H.H.; van Steen, R.I. New technique for reducing cogging torque in a class of brushless DC motors. *IEE Proc. B Electric Power Appl.* **1992**, *139*, 315–320. [CrossRef]
6. Ishikawa, T.; Slemon, G.R. A method of reducing ripple torque in permanent magnet motors without skewing. *IEEE Trans. Magn.* **1993**, *29*, 2028–2031. [CrossRef]
7. Keyhani, C.B.; Studer, T.; Sebastian, S.K.; Murthy, A. Study of cogging torque in permanent magnet machines. *Electric Mach. Power Syst.* **1999**, *27*, 665–678. [CrossRef]
8. Li, T.; Slemon, G. Reduction of cogging torque in permanent magnet motors. *IEEE Trans. Magn.* **1988**, *24*, 2901–2903.
9. Hwang, C.C.; John, S.B.; Wu, S.S. Reduction of cogging torque in spindle motors for CD-ROM drive. *IEEE Trans. Magn.* **1998**, *34*, 468–470. [CrossRef]
10. Jabbari, A.; Shakeri, M.; Nabavi Niaki, S.A. Pole shape optimization of permanent magnet synchronous motors using the reduced basis technique. *Iran. J. Electron. Electr. Eng.* **2010**, *6*, 48–55.
11. Markovic, M.; Jufer, M.; Perriard, Y. Reducing the cogging torque in brushless DC motors by using conformal mappings. *IEEE Trans. Magn.* **2004**, *40*, 451–455. [CrossRef]
12. Zarko, D.; Ban, D.; Lipo, T.A. Analytical calculation of magnetic field distribution in the slotted air gap of a surface permanent-magnet motor using complex relative air-gap permeance. *IEEE Trans. Magn.* **2006**, *42*, 1828–1837. [CrossRef]
13. Boughrara, K.; Zarko, D.; Ibtiouen, R.; Touhami, O.; Rezzoug, A. Magnetic field analysis of inset and surface-mounted permanent-magnet synchronous motors using Schwarz–Christoffel transformation. *IEEE Trans. Magn.* **2009**, *45*, 3166–3178. [CrossRef]
14. Ilhan, E.; Gysen, B.L.J.; Paulides, J.J.H.; Lomonova, E.A. Analytical hybrid model for flux switching permanent magnet machines. *IEEE Trans. Magn.* **2010**, *46*, 1762–1765. [CrossRef]
15. Tang, Y.; Motoasca, T.E.; Paulides, J.J.H.; Lomonova, E.A. Analytical modeling of flux-switching machines using variable global reluctance networks. In Proceedings of the 2012 XXth International Conference on Electrical Machines, Marseille, France, 2–5 September 2012; pp. 2792–2798.
16. Hua, W.; Zhang, G.; Cheng, M.; Dong, J. Electromagnetic performance analysis of hybrid-excited flux-switching machines by a nonlinear magnetic network model. *IEEE Trans. Magn.* **2011**, *47*, 3216–3219. [CrossRef]
17. Gu, Q.; Gao, H. Effect of slotting in PM electric machines. *Electric Mach. Power Syst.* **1985**, *10*, 273–284.

18. Boules, N. Prediction of no-load flux density distribution in permanent magnet machines. *IEEE Trans. Ind. Appl.* **1985**, *3*, 633–643. [CrossRef]
19. Ackermann, B.; Sottek, R. Analytical modeling of the cogging torque in permanent magnet motors. *Electr. Eng. Archiv fur Elektrotechnik* **1995**, *78*, 117–125. [CrossRef]
20. Radun, A. Analytical calculation of the switched reluctance motor's unaligned inductance. *IEEE Trans. Magn.* **1999**, *35*, 4473–4481. [CrossRef]
21. Rasmussen, K.F.; Davies, J.H.; Miller, T.J.E.; McGelp, M.I.; Olaru, M. Analytical and numerical computation of air-gap magnetic fields in brushless motors with surface permanent magnets. *IEEE Trans. Ind. Appl.* **2000**, *36*, 1547–1554.
22. Wang, X.; Li, Q.; Wang, S.; Li, Q. Analytical calculation of air-gap magnetic field distribution and instantaneous characteristics of brushless DC motors. *IEEE Trans. Energy Convers.* **2003**, *18*, 424–432. [CrossRef]
23. Liu, Z.J.; Li, J.T. Analytical solution of air-gap field in permanent-magnet motors taking into account the effect of pole transition over slots. *IEEE Trans. Magn.* **2007**, *43*, 3872–3883. [CrossRef]
24. Liu, Z.J.; Li, J.T.; Jiang, Q. An improved analytical solution for predicting magnetic forces in permanent magnet motors. *J. Appl. Phys.* **2008**, *103*, 07F135. [CrossRef]
25. Liu, Z.J.; Li, J.T. Accurate prediction of magnetic field and magnetic forces in permanent magnet motors using an analytical solution. *IEEE Trans. Energy Convers.* **2008**, *23*, 717–726. [CrossRef]
26. Kumar, P.; Bauer, P. Improved analytical model of a permanent-magnet brushless DC motor. *IEEE Trans. Magn.* **2008**, *44*, 2299–2309. [CrossRef]
27. Lubin, T.; Mezani, S.; Rezzoug, A. Exact analytical method for magnetic field computation in the air gap of cylindrical electrical machines considering slotting effects. *IEEE Trans. Magn.* **2010**, *46*, 1092–1099. [CrossRef]
28. Gysen, B.L.; Ilhan, E.; Meessen, K.J.; Paulides, J.J.; Lomonova, E.A. Modeling of flux switching permanent magnet machines with fourier analysis. *IEEE Trans. Magn.* **2010**, *46*, 1499–1502. [CrossRef]
29. Boughrara, K.; Ibtiouen, R.; Lubin, T. Analytical prediction of magnetic field in parallel double excitation and spoke-type permanent-magnet machines accounting for tooth-tips and shape of polar pieces. *IEEE Trans. Magn.* **2012**, *48*, 2121–2137. [CrossRef]
30. Boughrara, K.; Lubin, T.; Ibtiouen, R. General subdomain model for predicting magnetic field in internal and external rotor multiphase flux-switching machines topologies. *IEEE Trans. Magn.* **2013**, *49*, 5310–5325. [CrossRef]
31. Tiang, T.L.; Ishak, D.; Jamil, M.K.M. Complete subdomain model for surface-mounted permanent magnet machines. In Proceedings of the 2014 IEEE Conference on Energy Conversion, Johor Bahru, Malaysia, 11–13 October 2014; pp. 140–145.
32. Liu, X.; Hu, H.; Zhao, J.; Belahcen, A.; Tang, L.; Yang, L. Analytical solution of the magnetic field and EMF calculation in ironless BLDC motor. *IEEE Trans. Magn.* **2016**, *52*, 1–10. [CrossRef]
33. Dubas, F.; Espanet, C. Analytical solution of the magnetic field in permanent-magnet motors taking into account slotting effect: No-load vector potential and flux density calculation. *IEEE Trans. Magn.* **2009**, *45*, 2097. [CrossRef]
34. Roubache, L.; Boughrara, K.; Dubas, F.; Ibtiouen, R. New subdomain technique for electromagnetic performances calculation in radial-flux electrical machines considering finite soft-magnetic material permeability. *IEEE Trans. Magn.* **2018**. [CrossRef]
35. Dubas, F.; Boughrara, K. New scientific contribution on the 2-D subdomain technique in Cartesian coordinates: Taking into account of iron parts. *Math. Comput. Appl.* **2017**, *22*, 17.
36. Dubas, F.; Boughrara, K. New scientific contribution on the 2-D subdomain technique in polar coordinates: Taking into account of iron parts. *Math. Comput. Appl.* **2017**, *22*, 42.
37. Hannon, B.; Sergeant, P.; Dupré, L. Two-dimensional Fourier-based modeling of electric machines. In Proceedings of the 2017 IEEE International Electric Machines and Drives Conference, Miami, FL, USA, 21–24 May 2017; pp. 1–8.
38. Jabbari, A. 2D Analytical Modeling of Magnetic Vector Potential in Surface Mounted and Surface Inset Permanent Magnet Machines. *Iran. J. Electr. Electron. Eng.* **2017**, *13*, 362–373.
39. Jabbari, A. Exact analytical modeling of magnetic vector potential in surface inset permanent magnet DC machines considering magnet segmentation. *J. Electr. Eng.* **2018**, *69*, 39–45. [CrossRef]

40. Jabbari, A. Analytical Modeling of Magnetic Field Distribution in Inner Rotor Brushless Magnet Segmented Surface Inset Permanent Magnet Machines. *Iran. J. Electr. Electron. Eng.* **2018**, *14*, 259–269.
41. Jabbari, A. Analytical Modeling of Magnetic Field Distribution in Multiphase H-Type Stator Core Permanent Magnet Flux Switching Machines. *Iran. J. Sci. Tech. Trans. Electr. Eng.* **2018**, 3, 1–13. [CrossRef]

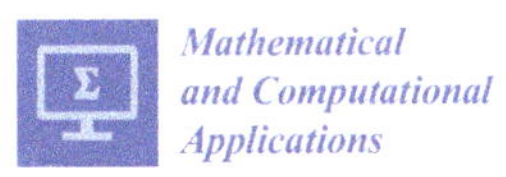

Article

Two-Dimensional Exact Subdomain Technique of Switched Reluctance Machines with Sinusoidal Current Excitation

Mohammed Ben Yahia [1,*], Kamel Boughrara [1], Frédéric Dubas [2], Lazhar Roubache [1] and Rachid Ibtiouen [1]

1 Laboratoire de Recherche en Electrotechnique, Ecole Nationale Polytechnique, 16200 Algiers, Algeria; kamel.boughrara@g.enp.edu.dz (K.B.); lazhar.roubache@g.enp.edu.dz (L.R.); rachid.ibtiouen@gmail.com (R.I.)

2 Département ENERGIE, FEMTO-ST, CNRS, Univ. Bourgogne Franche-Comté, F90000 Belfort, France; frederic.dubas@univ-fcomte.fr

* Correspondence: mohammed.ben_yahia@g.enp.edu.dz; Tel.: +213-66-693-7080

Received: 18 September 2018; Accepted: 10 October 2018; Published: 11 October 2018

Abstract: This paper presents a two-dimensional (2D) exact subdomain technique in polar coordinates considering the iron relative permeability in 6/4 switched reluctance machines (SRM) supplied by sinusoidal waveform of current (aka, variable flux reluctance machines). In non-periodic regions (e.g., rotor and/or stator slots/teeth), magnetostatic Maxwell's equations are solved considering non-homogeneous Neumann boundary conditions (BCs). The general solutions of magnetic vector potential in all subdomains are obtained by applying the interface conditions (ICs) in both directions (i.e., r- and θ-edges ICs). The global saturation effect is taken into account, with a constant magnetic permeability corresponding to the linear zone of the nonlinear *B(H)* curve. In this investigation, the magnetic flux density distribution inside the electrical machine, the static/dynamic electromagnetic torques, the magnetic flux linkage, the self-/mutual inductances, the magnetic pressures, and the unbalanced magnetic forces (UMFs) have been calculated for 6/4 SRM with two various non-overlapping (or concentrated) windings. One of the case studies is a M1 with a non-overlapping all teeth wound winding (double-layer winding with left and right layer) and the other is a M2 with a non-overlapping alternate teeth wound winding (single-layer winding). It is important to note that the developed semi-analytical model based on the 2D exact subdomain technique is also valid for any number of slot/pole combinations and for non-overlapping teeth wound windings with a single/double layer. Finally, the semi-analytical results have been performed for different values of iron core relative permeability (viz., 100 and 800), and compared with those obtained by the 2D finite-element method (FEM). The comparisons with FEM show good results for the proposed approach.

Keywords: 2D; electromagnetic performances; finite iron relative permeability; numerical; sinusoidal current excitation; subdomain technique; switched reluctance machine

1. Introduction

Benefiting from the advantages of a simple mechanical structure—the rotor does not carry any windings, commutators, or permanent magnets (PMs)—and a robust, fault-tolerant nature, low-cost maintenance, high-thermal capability, and high-speed potential [1–3], SRM is receiving renewed attention as a viable candidate for various adjustable-speed and high-torque applications such as in the automotive and traction fields [4–8].

However, a major disadvantage of this machine is the undesirable electromagnetic vibration and acoustic noise, which are mainly excited by the radial UMF acting on the salient stator and rotor poles [9,10]. Moreover, it poses a drawback for SRM in noise-sensitive applications and still creates bottlenecks in vehicle propulsion. It is important to consider noise and vibration problems during the process of electrical machine design. Electrical machine noise and vibration are mainly of electromagnetic, aerodynamic, and mechanical origin, the most important of which are generated by electromagnetic sources [11]. Also, in SRM, the attraction magnetic force can be divided into tangential and radial components relative to the rotor. The tangential magnetic force is converted into rotational torque, and the radial magnetic force converts into magnetic pressure equal to the radial magnetic force per unit area of the stator tooth and UMFs, which contributes to the radial vibration behaviour and therefore the motor noise [9,12]. A perfect machine with balanced stator windings should have net zero UMFs on the stator structure. However, UMFs can be present in machines having diametrically asymmetric disposition of slots and phase windings [13,14]. This magnetic force acts on the stator of these machine configurations due to an asymmetric magnetic field distribution in the air gap.

In the interest for design and optimization of electrical machines, there are various modelling methods; the first step in these is the magnetic field calculation. Some comprehensive reviews of the models of electrical machines for magnetic field prediction along with their (dis)advantages can be found in [15–24] and their references. Currently, the Maxwell-Fourier method is one of the most used semi-analytical methods, and combines the very accurate electromagnetic performances calculation with a reduced computation time compared to numerical methods. In models from this method (viz., multi-layer models, eigenvalues model, and subdomain technique), the magnetic field solutions are based on the formal resolution of Maxwell's equations by using the separation of variables method and the Fourier's series. In electromagnetic devices, the major assumption is that an infinite permeability of iron parts has to be assumed [25]. Therefore, the global and/or local saturation effect is neglected. It is interesting to note that an overview of the existing (semi-)analytical models in the Maxwell-Fourier method with a global and/or local saturation effect has been realized in [24], where some details and the (dis)advantages of these techniques can be found. To overcome that issue, Spranger et al. (2016) [21] and Dubas et al. (2017) [24,26] have recently developed new techniques to account for finite soft-magnetic material permeabilities:

- multi-layer models using the convolution theorem (i.e., Cauchy's product theorem). The adjacent regions (e.g., rotor and/or stator slots/teeth) are assumed to be one homogeneous region with a relative permeability developed as a Fourier's series expansion;
- the subdomain technique using a superposition that allows for any non-periodic subdomain. The subdomain connection is performed directly in both directions. The general solutions of Maxwell's equations are deduced by applying the principle of superposition by respecting the BCs on the various edges of subdomains.

For the same reason, another technique based on subdomain technique and Taylor polynomial has been developed and only applied in spoke-type PM synchronous machines (PMSM) [27,28]. Spranger's approach has been extended and used in different machines with only the global saturation effect. It has been applied with the finite soft-magnetic material permeability in synchronous reluctance machine [29], surface-mounted PMSM [30], and many structures of PMSMs (i.e., for inset-/surface-/spoke-type PMSMs with different PM magnetization patterns and internal/external rotor) [31], with the nonlinear $B(H)$ curve in switched reluctance machine [32,33]. The Dubas superposition technique has been implemented in radial-flux electrical machines with(out) PMs supplied by a direct or alternate current (with any waveforms) [34]. This technique has been extended to: (i) the thermal modelling for the steady-state temperature distribution in rotating electrical machines [35], and (ii) elementary subdomains in the rotor and stator regions for full prediction of magnetic field in rotating electrical machines with the local saturation effect solving by the Newton-Raphson iterative algorithm [36]. The Dubas superposition technique is very interesting since, like Spranger's approach, it enables the magnetic field calculation in iron parts of slotted structures. Apart from its complexity, the main downfall of Spranger's approach is that it suffers from the Gibb's phenomenon at boundaries between slots and teeth. This introduces inaccuracies in the computation of the field and results in higher computational times [37].

In this paper, the authors propose applying the Dubas superposition technique in polar coordinates [26] to SRM with sinusoidal current excitation, which has not yet been realized in the literature. The soft magnetic material permeability is constant corresponding to the linear zone of the *B(H)* curve. Nevertheless, as in [33,36], it should be mentioned that the material properties could be updated iteratively to take the nonlinear *B(H)* curve of the material into account. However, this is beyond the scope of the paper. In this investigation, the magnetic flux density distribution inside the machine, electromagnetic performances and non-intrinsic UMFs have been calculated for 6/4 SRM supplied by sinusoidal waveform of current with two various non-overlapping (or concentrated) windings. One of the case studies is a M1 with a non-overlapping all teeth wound winding (double-layer winding with left and right layer) and the other is a M2 with a non-overlapping alternate teeth wound winding (single-layer winding). All results obtained with the proposed semi-analytical model are verified by 2D FEM [38] for different values of iron core relative permeability (viz., 100 and 800). The comparisons with FEM show good results.

2. Studied SRMs and Magnetic Field Solutions

2.1. Machine Geometry and Assumptions

Figure 1 represents the studied SRMs having two various non-overlapping windings: (i) M1 with a non-overlapping all teeth wound winding (double-layer winding with left and right layer) (see Figure 1a), and (ii) M2 with a non-overlapping alternate teeth wound winding (single-layer winding) (see Figure 1b). The three-phase SRMs have six stator slots and four rotor slots, and do not contain any stator tooth tips. The main geometrical parameters of two studied SMRs are shown in Figure 1 and are given in Table 1 for the semi-analytical and numerical comparisons. These machines have been partitioned into nine regions as shown on Figure 2, viz.,

- Region I is the air gap;
- Regions II and III are the rotor yoke (i.e., between rotor shaft and rotor slots/teeth) and the stator yoke, respectively;
- Region IV is the rotor slots;
- Region V is the rotor teeth;
- Regions VI and VII are the stator slots of the first layer (i.e., right in the slot) and second layer (i.e., left in the slot), respectively;
- Region VIII is the stator teeth;
- Region XI is the non-periodic air gap (i.e., between the two-layer winding of the stator slots).

The semi-analytical model, based on the exact subdomain technique, is formulated in 2D, in polar coordinates, and in magnetic vector potential with the following assumptions:

- The end-effects are neglected, i.e., $\boldsymbol{A} = \{0;\ 0;\ A_z\}$;
- The eddy-current effects in the materials are neglected;
- The current density in the stator slots has only one component along the z-axis, i.e., $\boldsymbol{J} = \{0;\ 0;\ J_z\}$;
- The magnetic materials are considered as isotropic with constant magnetic permeability corresponding to linear zone of the *B(H)* curve;
- The stator and rotor slots/teeth have radial sides (see Figure 2).

However, it accounts for:

- The internal/external rotor topology;
- The saturation, slotting and curvature effect;
- The (non-)overlapping winding distribution;

- Any current waveform (i.e., sinusoidal or rectangular).

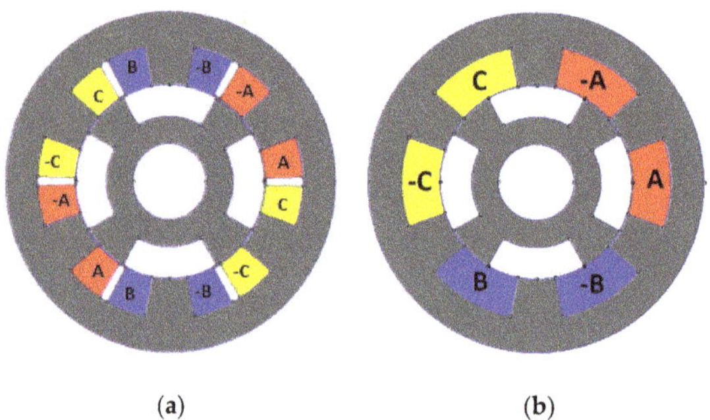

Figure 1. Studied 6/4 SRM with sinusoidal current excitation having a non-overlapping (or concentrated) winding, viz., (**a**) M1: all teeth wound (double-layer winding with left and right layer), and (**b**) M2: alternate teeth wound (single-layer winding).

Table 1. Parameters of the studied SRMs.

Parameters, Symbols [Units]	Values	
	M1 Double Layer	**M2 Single Layer**
Winding distribution	**Concentrated**	
Number of stator slots, Q_s [-]	6	
Number of rotor poles, Q_r [-]	4	
Radius of the external stator surface, R_{ext} [mm]	45	
External radius of stator slot, R_5 [mm]	36	
Radius of the internal stator surface, R_4 [mm]	25.7	
Radius of the rotor surface, R_3 [mm]	25.5	
Internal radius of rotor slot, R_2 [mm]	17.3	
Radius of the shaft, R_1 [mm]	10	
Air gap thickness, g [mm]	0.2	
Axial length of the machine, L_u [mm]	60	
Rotor slot-opening, a [deg.]	60	
Rotor tooth-opening, b [deg.]	30	
Stator slot opening, c [deg.]	38	
Stator tooth opening, d [deg.]	22	
Non-periodic air gap (i.e., between the two-layer winding of stator slots) opening, e [deg.]	4	0
Opening of a slot coil, f [deg.]	17	38
Number of conductor of slot coil, N_c [-]	20	40
Phase current, I [A]	15	
Current density of the coil, J [A/mm^2]	3.18	

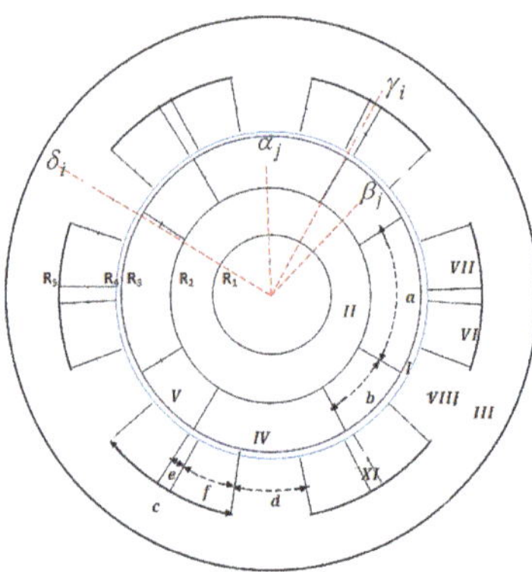

Figure 2. Simplified model of SRM.

2.2. General Solution with Non-Homogeneous Neumann BCs

Magnetic vector potential A is calculated analytically with solving the magnetostatic Maxwell's equations with the separation of variables method, viz.,

$$\Delta A = 0 \text{ in Region I, II, III, IV, V, VIII and XI,} \tag{1}$$

$$\Delta A = -\mu_0 \cdot J \text{ in Region VI and VII,} \tag{2}$$

where μ_0 is the vacuum permeability.

According to [24,26], the solutions to A in all regions of conventional SRM are:

- **Air gap subdomain (Region I)**: The solution of (1) in Region I, $r \in [R_3;\ R_4]$ & $\forall\theta$, is defined by:

$$\begin{aligned} A_{zI} &= A_{10} + A_{20} \cdot \ln(r) \\ &\cdots + \sum_{n=1}^{\infty}\left[A_{1n}\cdot\left(\tfrac{r}{R_3}\right)^{n} + A_{2n}\cdot\left(\tfrac{r}{R_2}\right)^{-n}\right]\cdot\sin(n\theta) + \sum_{n=1}^{\infty}\left[A_{3n}\cdot\left(\tfrac{r}{R_3}\right)^{n} + A_{4n}\cdot\left(\tfrac{r}{R_2}\right)^{-n}\right]\cdot\cos(n\theta) \end{aligned} \tag{3}$$

  where n is a positive integer, and $\{A_{10};\ A_{20};\ A_{1n};A_{4n}\}$ are the integration constants of Region I.
- **Stator and rotor yoke subdomain (Region II and III)**: In adding Dirichlet BC of A at $r = R_1$ and $r = R_{ext}$, viz., $A_{zII}(R_1,\theta) = 0$ & $A_{zIII}(R_{ext},\theta) = 0$, the solution of (1) in Region II, $r \in [R_1;\ R_2]$ & $\forall\theta$, can be written as:

$$\begin{aligned} A_{zII} &= A_{50} \cdot \ln\left(\tfrac{r}{R_1}\right) \\ &\cdots + \sum_{n=1}^{\infty} A_{5n}\cdot\left[\left(\tfrac{r}{R_1}\right)^{n} - \left(\tfrac{r}{R_1}\right)^{-n}\right]\cdot\sin(n\theta) + \sum_{n=1}^{\infty} A_{6n}\cdot\left[\left(\tfrac{r}{R_1}\right)^{n} - \left(\tfrac{r}{R_1}\right)^{-n}\right]\cdot\cos(n\theta) \end{aligned} \tag{4}$$

  where $\{A_{50};\ A_{5n};\ A_{6n}\}$ are the integration constants of Region II.

  The solution of Region III, $r \in [R_5;\ R_{ext}]$ & $\forall\theta$, is similar to (4) by replacing $\{A_{50};\ A_{5n};\ A_{6n}\}$ with $\{A_{70};\ A_{7n};\ A_{8n}\}$ and R_1 with R_{ext}.

- ***i*-th Stator slot subdomain (Region VI and VII)**: The solution of (2) in Region VI, $r \in [R_4;R_5]$ & $\theta \in [\gamma_{1i} - f/2;\gamma_{1i} + f/2]$, is defined by:

$$\begin{aligned} A_{zVIi}(r,\theta) = {} & C_{1i0} + C_{2i0}\cdot\ln(r) - \tfrac{1}{4}\cdot\mu_0\cdot J1(i)_z\cdot r^2 \\ & \cdots + \sum_{m=1}^{\infty}\left[C_{1im}\left(\tfrac{r}{R_5}\right)^{v_{mf}} + C_{2im}\left(\tfrac{r}{R_4}\right)^{-v_{mf}}\right]\cdot\cos\left[v_{mf}\cdot\left(\theta - \gamma_{1i} + \tfrac{f}{2}\right)\right] \\ & \cdots + \sum_{k=1}^{\infty}\left\{C_{3ik}\cdot\frac{\mathrm{sh}\left[\lambda_{ks}\cdot\left(\theta-\gamma_{1i}+\frac{f}{2}\right)\right]}{\mathrm{sh}(\lambda_{ks}\cdot f)} + C_{4ik}\cdot\frac{\mathrm{sh}\left[\lambda_{ks}\cdot\left(\theta-\gamma_{1i}-\frac{f}{2}\right)\right]}{\mathrm{sh}(\lambda_{ks}\cdot f)}\right\}\cdot\sin\left[\lambda_{ks}\cdot\ln\left(\tfrac{r}{R_4}\right)\right] \end{aligned} \tag{5}$$

  where m and k are positive integers, $\gamma_{1i} = \gamma_i - (e+f)/2$ and f are respectively the position and opening width of first layer winding in the i-th stator slot, $\{C_{1i0};\ C_{2i0};\ C_{1im};\ C_{2im};\ C_{3ik};\ C_{4ik}\}$ are the integration constants of Region VI, $v_{mf} = m\pi/f$ and $\lambda_{ks} = k\pi/\ln(R_5/R_4)$ are respectively the periodicity of A_{zVIi} in θ-and r-edges.

  The solution of Region VII, $r \in [R_4;R_5]$ & $\theta \in [\gamma_{2i} - f/2;\ \gamma_{2i} + f/2]$, is similar to (5) by replacing $\{C_{1i0};\ C_{2i0};\ C_{1im};\ C_{2im};\ C_{3ik};\ C_{4ik}\}$ with $\{C_{5i0};\ C_{6i0};\ C_{5im};\ C_{6im};\ C_{7ik};\ C_{8ik}\}$, $J1(i)_z$ with $J2(i)_z$, and γ_{1i} with $\gamma_{2i} = \gamma_i + (e+f)/2$.

- ***i*-th Non-periodic air gap and *i*-th stator tooth subdomain (Region XI and VIII)**: The solution of (1) in Region VIII, $r \in [R_4;R_5]$ & $\theta \in [\gamma_i - e/2;\gamma_i + e/2]$, and in Region XI, $r \in [R_4;R_5]$ & $\theta \in [\delta_i - d/2;\delta_i + d/2]$, can be obtained directly from (5) with $J1(i)_z = 0$.

  For Region VIII, $\{C_{1i0};\ C_{2i0};\ C_{1im};\ C_{2im};\ C_{3ik};\ C_{4ik}\}$ is replaced by $\{D_{1i0};\ D_{2i0};\ D_{1im};\ D_{2im};\ D_{3ik};\ D_{4ik}\}$, γ_{1i} by γ_i, f by e, and v_{mf} by $v_{me} = m\pi/e$.

For Region IX, $\{C_{1i0};\ C_{2i0};\ C_{1im};\ C_{2im};\ C_{3ik};\ C_{4ik}\}$ is replaced by $\{D_{1i0};\ D_{2i0};\ D_{1im};\ D_{2im};\ D_{3ik};\ D_{4ik}\}$, γ_{1i} by δ_i, f by d, and v_{mf} by $v_{md} = m\pi/d$.

- ***j*-th Rotor slot and *j*-th rotor tooth subdomain (Region IV and V)**: The solution of (1) in Region IV, $r \in [R_2; R_3]$ & $\theta \in [\alpha_j - a/2;\ \alpha_j + a/2]$, is defined by:

$$\begin{aligned} A_{zIVj}(r,\theta) &= B_{1j0} + B_{2j0} \cdot \ln(r) \\ &\cdots + \sum_{m=1}^{\infty} \left[B_{1jm} \left(\frac{r}{R_3}\right)^{v_{ma}} + B_{2jm} \left(\frac{r}{R_2}\right)^{-v_{ma}} \right] \cdot \cos\left[v_{ma} \cdot \left(\theta - \alpha_j + \tfrac{a}{2}\right)\right] \\ &\cdots + \sum_{k=1}^{\infty} \left\{ B_{3jk} \cdot \frac{\mathrm{sh}\left[\lambda_{kr} \cdot \left(\theta - \alpha_j + \frac{a}{2}\right)\right]}{\mathrm{sh}(\lambda_{kr} \cdot a)} + B_{4jk} \cdot \frac{\mathrm{sh}\left[\lambda_{kr} \cdot \left(\theta - \alpha_j - \frac{a}{2}\right)\right]}{\mathrm{sh}(\lambda_{kr} \cdot a)} \right\} \cdot \sin\left[\lambda_{kr} \cdot \ln\left(\frac{r}{R_2}\right)\right] \end{aligned} \quad (6)$$

where α_j and a are respectively the position and opening width of j-th rotor slot, $\left\{B_{1j0};\ B_{2j0};\ B_{1jm};\ B_{2jm};\ B_{3jk};\ B_{4jk}\right\}$ are the integration constants of Region IV, $v_{ma} = m\pi/a$ and $\lambda_{kr} = k\pi/\ln(R_3/R_2)$ are respectively the periodicity of A_{zIVj} in θ-and r-edges.

The solution of Region V, $r \in [R_2; R_3]$ & $\theta \in [\beta_j - b/2;\ \beta_j + b/2]$, is similar to (6) by replacing $\left\{B_{1j0};\ B_{2j0};\ B_{1jm};\ B_{2jm};\ B_{3jk};\ B_{4jk}\right\}$ with $\left\{B_{5j0};\ B_{6j0};\ B_{5jm};\ B_{6jm};\ B_{7jk};\ B_{8jk}\right\}$, a with b, and α_j with β_j.

2.3. Magnetic Flux Density

The field vectors $\boldsymbol{B} = \{B_r;\ B_\theta;\ 0\}$ and $\boldsymbol{H} = \{H_r;\ H_\theta;\ 0\}$ are coupled by:

$$\boldsymbol{B} = \mu_0 \cdot \boldsymbol{H} \text{ in Region I, IV, VI, VII and XI,} \quad (7)$$

$$\boldsymbol{B} = \mu_0 \cdot \mu_{rc} \cdot \boldsymbol{H} \text{ in Region II, III, V and VIII,} \quad (8)$$

where μ_{rc} is the relative recoil permeability of iron parts.

Using $\boldsymbol{B} = \nabla \times \boldsymbol{A}$, the components of $\boldsymbol{B} = \nabla \times \boldsymbol{A}$ can be deduced by

$$B_r = \frac{1}{r} \cdot \frac{\partial A_z}{\partial \theta} \text{ and } B_\theta = -\frac{\partial A_z}{\partial r}. \quad (9)$$

2.4. Stator Current Density Source

The stator current densities in the stator slots for double-layer concentrated winding are defined as [33]:

$$J1(i) = \frac{N_c}{S} \cdot C_{(1)}^T \cdot i_g \text{ and } J2(i) = \frac{N_c}{S} \cdot C_{(2)}^T \cdot i_g. \quad (10)$$

where $i_g = \begin{bmatrix} i_a & i_b & i_c \end{bmatrix}$ is the vector of phase currents whose currents' waveform is sinusoidal with a phase shift of $2\pi/3$ electric, $S = f \cdot ({R_5}^2 - {R_4}^2)/2$ is the surface of the stator slot coil, and $C_{(1)}^T$ & $C_{(2)}^T$ are the transpose of the connection matrix between the three phases and the stator slots that represent the distribution of stator windings in the slots of the M1 with all teeth wound (double-layer winding with left and right layer) (see Figure 1a) is given by [33]:

$$C_{(1)} = \begin{bmatrix} -1 & 0 & 0 & 1 & 0 & 0 \\ 0 & 1 & 0 & 0 & -1 & 0 \\ 0 & 0 & -1 & 0 & 0 & 1 \end{bmatrix} \text{ and } C_{(2)} = \begin{bmatrix} 0 & 0 & -1 & 0 & 0 & 1 \\ -1 & 0 & 0 & 1 & 0 & 0 \\ 0 & 1 & 0 & 0 & -1 & 0 \end{bmatrix}. \quad (11)$$

For the M2 with alternate teeth wound (single-layer winding) (see Figure 1b), the same model is used with few modifications:

- The opening of the non-periodic air gap will be equal to zero (i.e., e = 0);

- The stator current density will be equal to $J1(i) = J2(i) = \frac{N_c}{S} \cdot C_{(1)}^T \cdot i_g$ with $S = c \cdot (R_5{}^2 - R_4{}^2)/2$ and

$$C_{(1)} = \begin{bmatrix} 1 & -1 & 0 & 0 & 0 & 0 \\ 0 & 0 & 0 & 0 & 1 & -1 \\ 0 & 0 & 1 & -1 & 0 & 0 \end{bmatrix}. \tag{12}$$

These connection matrices can be generated automatically by using the ANFRACTUS TOOL developed in [39].

2.5. Boundary Conditions

The ICs in this semi-analytical model can be divided into two types, viz.,

- **θ-edges ICs:** over angle interval for given radius value $\{R_2;\ R_3;\ R_4; R_5\}$;
- **r-edges ICs:** over radius interval for given angle $\{\alpha_j \pm a/2;\ \beta_j \pm b/2;\ \gamma_i \pm c/2;\ \delta_i \pm d/2;\ \gamma_i \pm e/2\}$.

Therefore, we obtain on the:

- **θ-edges ICs:**

- The ICs between Region II, IV and V at $r = R_2$ as:

$$A_{zII}(R_2, \theta) = A_{zIVj}(R_2, \theta) \text{ for } \theta \in [\alpha_j - a/2, \alpha_j + a/2], \tag{13}$$

$$H_{\theta II}(R_2, \theta) = H_{\theta IVj}(R_2, \theta) \text{ for } \theta \in [\alpha_j - a/2, \alpha_j + a/2], \tag{14}$$

$$H_{\theta II}(R_2, \theta) = H_{\theta IVj}(R_2, \theta) \text{ for } \theta \in [\alpha_j - a/2, \alpha_j + a/2], \tag{15}$$

$$H_{\theta II}(R_2, \theta) = H_{\theta Vj}(R_2, \theta) \text{ for } \theta \in [\beta_j - b/2, \beta_j + b/2], \tag{16}$$

- The ICs between Region I, IV and V at $r = R_3$ are similar to (13)–(16) by replacing II with I and R_2 with R_3.
- The ICs between Region I, VI, VII, VIII and XI at $r = R_4$ as:

$$A_{zI}(R_4, \theta) = A_{zVIIi}(R_4, \theta) \text{ for } \theta \in [\gamma_i + c/2 - f, \gamma_i + c/2], \tag{17}$$

$$A_{zI}(R_4, \theta) = A_{zVIIIi}(R_4, \theta) \text{ for } \theta \in [\delta_i - d/2, \delta_i + d/2], \tag{18}$$

$$A_{zI}(R_4, \theta) = A_{zVIIIi}(R_4, \theta) \text{ for } \theta \in [\delta_i - d/2, \delta_i + d/2], \tag{19}$$

$$A_{zI}(R_4, \theta) = A_{zXIi}(R_4, \theta) \text{ for } \theta \in [\gamma_i - e/2,\ \gamma_i + e/2], \tag{20}$$

$$H_{\theta I}(R_4, \theta) = A_{\theta VIi}(R_4, \theta) \text{ for } \theta \in [\gamma_i - c/2, \gamma_i - c/2 + f], \tag{21}$$

$$H_{\theta I}(R_4, \theta) = A_{\theta VIIi}(R_4, \theta) \text{ for } \theta \in [\gamma_i + c/2 - f, \gamma_i + c/2], \tag{22}$$

$$H_{\theta I}(R_4, \theta) = A_{\theta VIIi}(R_4, \theta) \text{ for } \theta \in [\delta_i - d/2, \delta_i + d/2], \tag{23}$$

$$H_{\theta I}(R_4, \theta) = A_{\theta XIi}(R_4, \theta) \text{ for } \theta \in [\gamma_i - e/2, \gamma_i + e/2], \tag{24}$$

- The ICs between Region III, VI, VII, VIII and XI at $r = R_5$ are similar to (17)–(24) by replacing I with III and R_4 with R_5.

- **r-edges ICs:**

The ICs between Region IV and V at $\alpha_j + a/2 = \beta_j - b/2$ and $\alpha_{j+1} - a/2 = \beta_j + b/2$ for $r \in [R_2;\ R_3]$:

$$A_{zIVj}(r, \alpha_j + a/2) = A_{zVj}(r,\ \beta_j - b/2)\ , \tag{25}$$

$$H_{rIVj}(r, \alpha_j + a/2) = H_{rVj}(r,\ \beta_j - b/2)\ , \quad (26)$$

$$A_{zIV(j+1)}(r, \alpha_{j+1} - a/2) = A_{zVj}(r,\ \beta_j + b/2)\ , \quad (27)$$

$$H_{rIV(j+1)}(r, \alpha_{j+1} - a/2) = H_{rVj}(r,\ \beta_j + b/2), \quad (28)$$

- The ICs between Region VII and VIII at $\gamma_i + c/2 = \delta_i - d/2$ and between Region VI and VIII at $\gamma_{i+1} - c/2 = \delta_i + d/2$ for $r \in [R_4;\ R_5]$:

$$A_{zVIIi}(r, \gamma_i + c/2) = A_{zVIIIi}(r,\ \delta_i - d/2)\ , \quad (29)$$

$$H_{rVIIi}(r, \gamma_i + c/2) = H_{r\ VIIIi}(r,\ \delta_i - d/2), \quad (30)$$

$$A_{zVI(i+1)}(r, \gamma_{i+1} - c/2) = A_{zVIIIi}(r,\ \delta_i + d/2)\ , \quad (31)$$

$$H_{rVI(i+1)}(r, \gamma_{i+1} - c/2) = H_{rVIIIi}(r,\ \delta_i + d/2)\ , \quad (32)$$

- The ICs between Region VI and XI at $\gamma_i - e/2 = \gamma_i - c/2 + f$ and between Region VII and XI at $\gamma_i + e/2 = \gamma_i + c/2 - f$ for $r \in [R_4;\ R_5]$:

$$A_{zVIi}(r, \gamma_i - c/2 + f) = A_{zXIi}(r,\ \gamma_i - e/2)\ , \quad (33)$$

$$H_{rVIi}(r, \gamma_i - c/2 + f) = H_{r\ XIi}(r,\ \gamma_i - e/2), \quad (34)$$

$$A_{zVIIi}(r, \gamma_i + c/2 - f) = A_{zXIi}(r,\ \gamma_i + e/2)\ , \quad (35)$$

$$H_{rVIIi}(r, \gamma_i + c/2 - f) = H_{rXIi}(r,\ \gamma_i + e/2)\ , \quad (36)$$

The system of the 36 BCs matrix (Equations (13)–(36)) is used to determine the coefficients of A in nine regions.

Figure 3 briefly represents a flowchart of the subdomain technique.

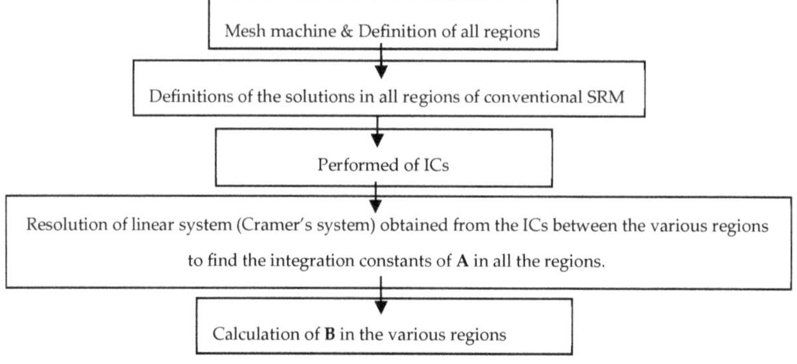

Figure 3. Flowchart of subdomain technique.

To solve the Cramer's system, the number of integration constants is equal to $2 \cdot (4N + 2) + 2Q_r \cdot (2 + 2M + 2K) + 4Q_s \cdot (2 + 2M + 2K)$ where N, M and K are the finite numbers of spatial harmonic terms in the various regions.

3. Electromagnetic Performance Calculations

3.1. Torque, Flux Linkage and Inductance Calculations

The semi-analytical model, based on the 2D exact subdomain technique and taking into account the iron core relative permeability, is used to determine the static/dynamic electromagnetic torque, the magnetic flux linkage, and the self-/mutual inductances whose various formulas have been clarified in [32,33].

3.2. Magnetic Pressure and UMF Calculations

The magnetic pressure P is the magnetic force per unit area of the stator tooth. It can be calculated both by Maxwell's stress tensor and by finite element analysis. The r-and θ-components of P are calculated from the spatial magnetic field in the air gap middle at $R_g = (R_4 + R_3)/2$ [14]:

$$P_r(\theta_r, \theta) = \frac{1}{2\mu_0} \cdot \left[B_{rI}(R_g, \theta)^2 - B_{\theta I}(R_g, \theta)^2\right], \quad (37a)$$

$$P_\theta(\theta_r, \theta) = \frac{1}{\mu_0} \cdot B_{rI}(R_g, \theta) \cdot B_{\theta I}(R_g, \theta), \quad (37b)$$

where $\theta_r = \Omega \cdot t + \theta_{rs0}$ is the temporal rotor angle with Ω the mechanical pulse and θ_{rs0} the initial mechanical angular position between the rotor and the stator at the instant $t = 0$ s.

The x- and y-components of non-intrinsic UMF $\boldsymbol{F}$ are calculated at R_g over $\theta = [0;\ 2\pi]$ as [14]:

$$F_x(\theta_r) = -R_g \cdot L_u \cdot \int_0^{2\pi} [P_r(\theta) \cdot \cos(\theta) - P_\theta(\theta) \cdot \sin(\theta)] \cdot d\theta, \quad (38a)$$

$$F_y(\theta_r) = -R_g \cdot L_u \cdot \int_0^{2\pi} [P_r(\theta) \cdot \sin(\theta) + P_\theta(\theta) \cdot \cos(\theta)] \cdot d\theta, \quad (38b)$$

where L_u is the axial length of the machine.

The acoustic noise and vibration is primarily due to the rotor eccentric position with respect to the stator bore, the UMF, if present in a motor even with perfectly aligned shaft, can create the rotor eccentricity. Moreover, $\boldsymbol{P}$ and $\boldsymbol{F}$ are transmitted through the teeth from the air gap to the yoke, which may cause deformation on the stator rings resulting from the rotor displacement and result in excessive acoustic noise and vibration. Different vibration modes are commonly called "mode shapes" having their own natural mode frequency. Any particular mode shape is excited when its natural mode frequency matches with any of the harmonics of $\boldsymbol{P}$ and $\boldsymbol{F}$ [40].

4. Results and Validations

The developed model (see Section 2) considering finite soft-magnetic material permeability is used to determine the magnetic flux density distribution inside the electrical machines as well as the electromagnetic performances for 6/4 SRM with two various non-overlapping (or concentrated) windings. The main dimensions and parameters of studied machines are given in Table 1. The results of the semi-analytical model are verified by 2D FEM.

4.1. Magnetic Flux Density Distribution

The waveforms of r- and θ-components of the magnetic flux density in the various regions are computed with a finite number of harmonic terms, viz., $N = 200$ and $M = K = 30$. The analytic calculation of magnetic flux density distribution in all regions is done considering the same relative permeability in all iron parts (i.e., stator/rotor yoke and teeth). The soft magnetic material permeability is constant corresponding to the linear zone of the *B(H)* curve. However, it is possible to use a different relative permeability value for each region [28,33,36].

In Figures 4 and 5, a comparison between the numerical results and semi-analytical predictions is shown the r- and θ-components of $\boldsymbol{B}$ in the air gap middle (i.e., Region I at $R_g = (R_4 + R_3)/2$) for two studied SRMs (i.e., M1 and M2). The simulations are done for two different values of iron core relative permeability (viz., 100 and 800). It can be seen an asymmetric distribution in the r- and θ-component of the air gap magnetic flux density in M2 in contrary to M1, this is due to the diametrically asymmetric disposition of slots and phase windings as shown in Figure 1b.

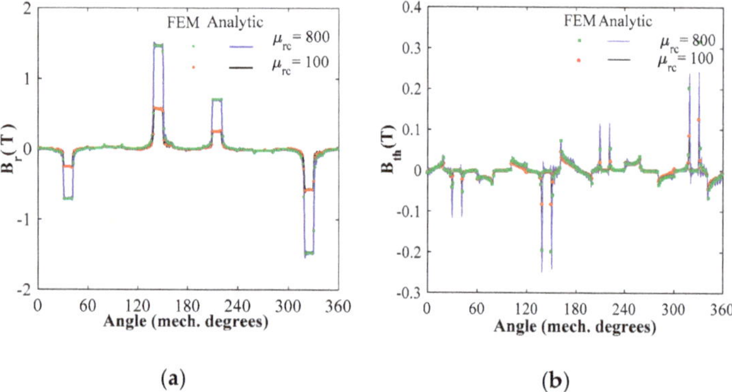

Figure 4. Waveform of the magnetic flux density in the air gap middle (i.e., Region I) for M1: (**a**) r- and (**b**) θ-component.

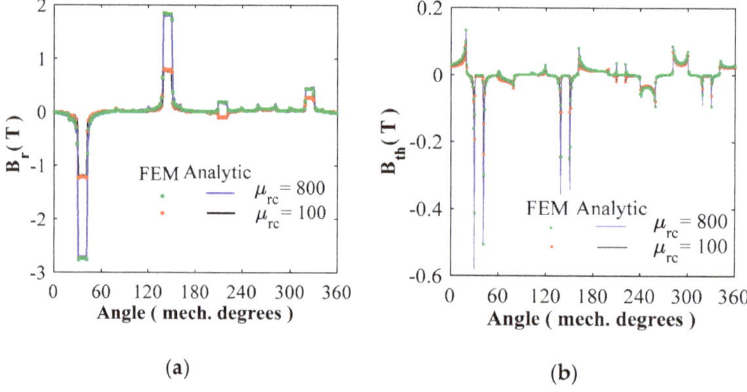

Figure 5. Waveform of the magnetic flux density in the air gap middle (i.e., Region I) for M2: (**a**) r- and (**b**) θ-component.

Figures 6 and 7 show the magnitude of B in all machine's regions for two studied SRMs (i.e., M1 and M2), we can notice that M2 is more saturated than M1.

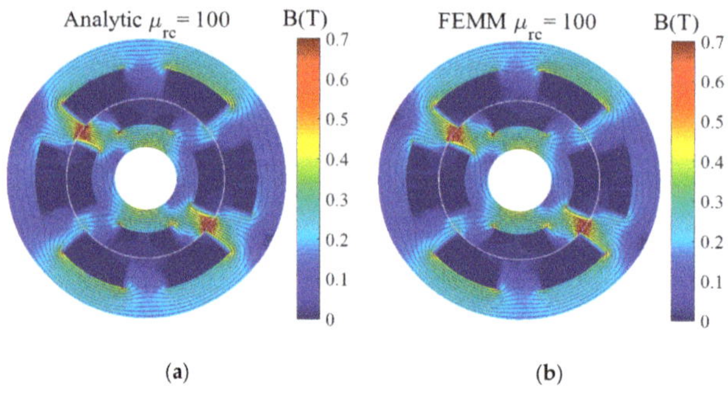

Figure 6. Flux density inside the machine in M1: (**a**) analytic and (**b**) FEM.

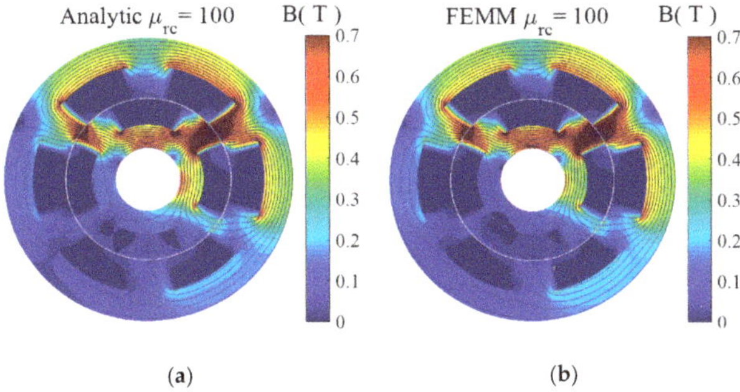

Figure 7. Flux density inside the machine in M2: (a) analytic and (b) FEM.

In Figures 8–15, for two studied SRMs (i.e., M1 and M2), a numerical and semi-analytical comparison is shown the r- and θ-components of B in the

- stator yoke middle (see Figure 8 for M1 and Figure 9 for M2);
- rotor yoke middle (see Figure 10 for M1 and Figure 11 for M2);
- stator slots/non-periodic air gap/teeth middle (see Figure 12 for M1 and Figure 13 for M2);
- rotor slots/teeth middle (see Figure 14 for M1 and Figure 15 for M2).

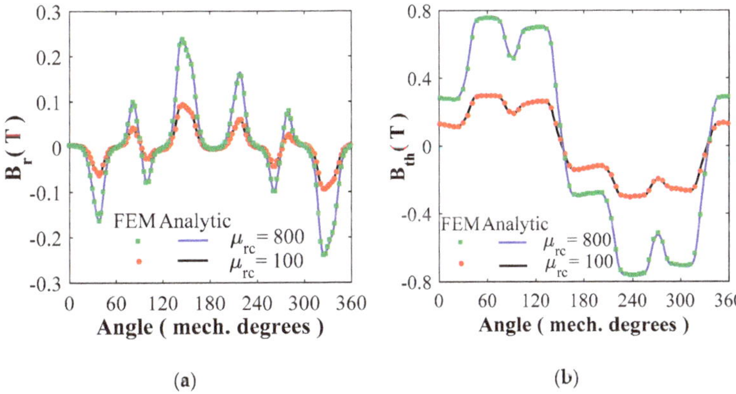

Figure 8. Waveform of the magnetic flux density in the stator yoke middle for M1: (**a**) r- and (**b**) θ-component.

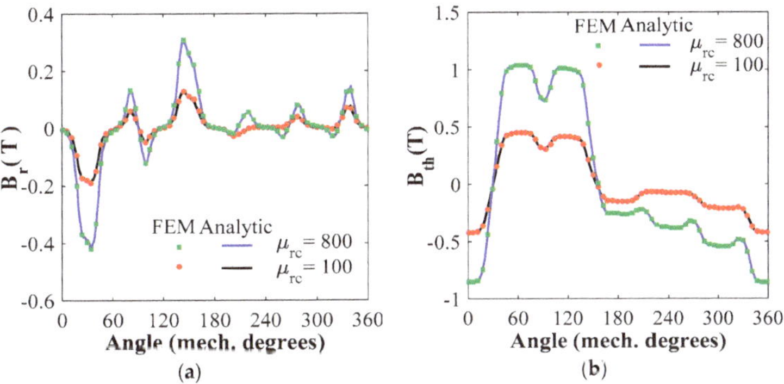

Figure 9. Waveform of the magnetic flux density in the stator yoke middle for M2: (**a**) r- and (**b**) θ-component.

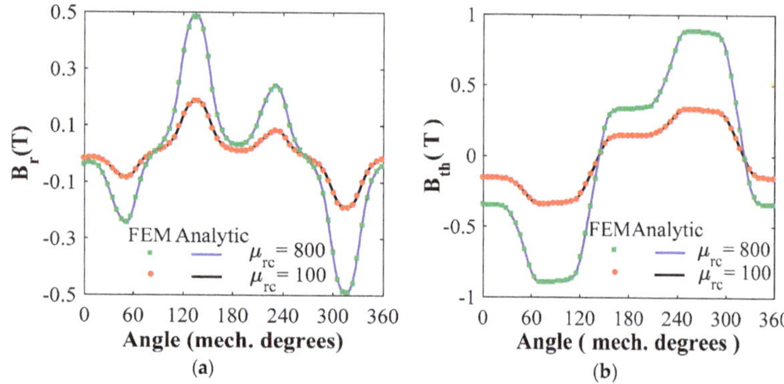

Figure 10. Waveform of the magnetic flux density in the rotor yoke middle for M1: (**a**) r- and (**b**) θ-component.

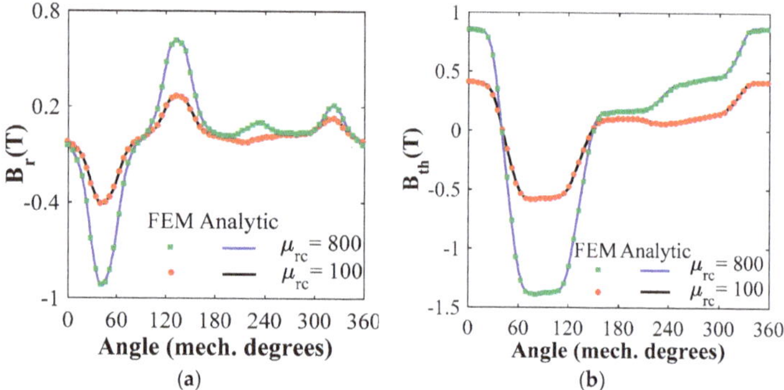

Figure 11. Waveform of the magnetic flux density in the rotor yoke middle for M2: (**a**) r- and (**b**) θ-component.

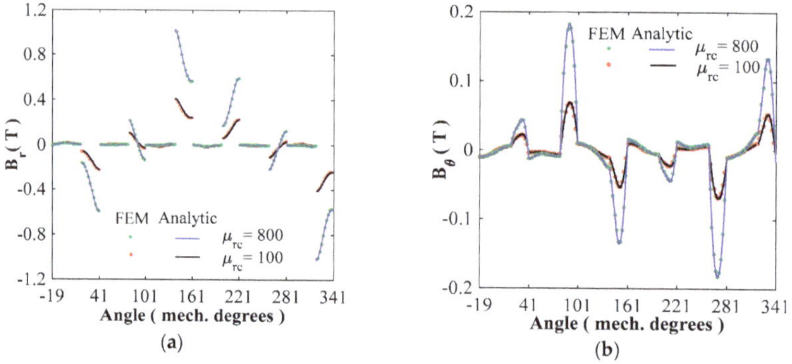

Figure 12. Waveform of the magnetic flux density in the stator slots/non-periodic air gap/teeth for M1: (**a**) r- and (**b**) θ-component.

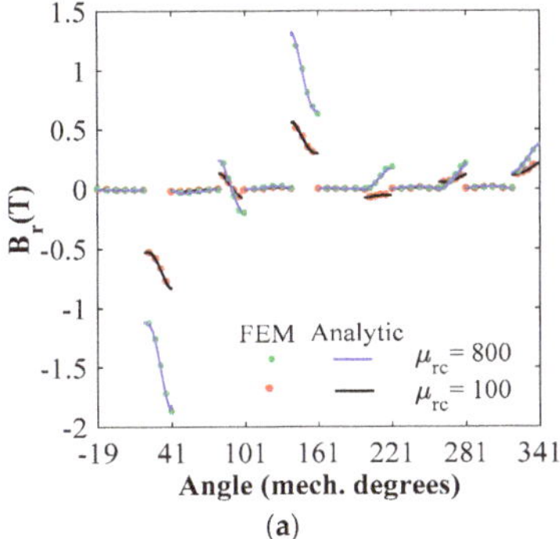

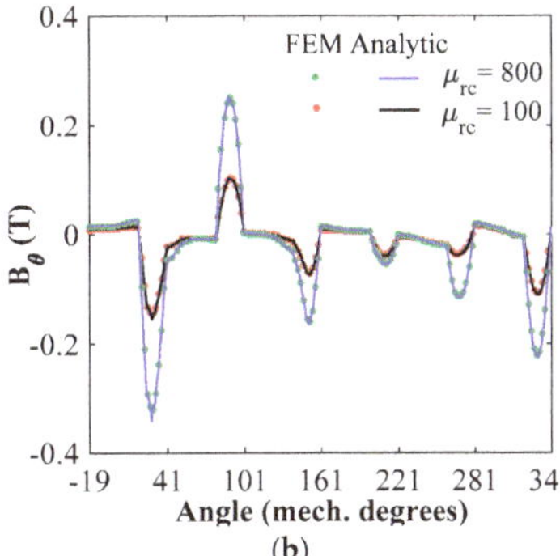

(a) (b)

Figure 13. Waveform of the magnetic flux density in the stator slots/non-periodic air gap/teeth for M2: (**a**) r- and (**b**) θ-component.

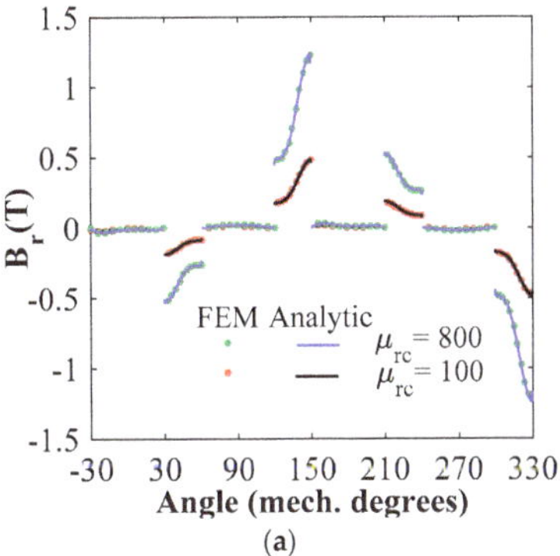

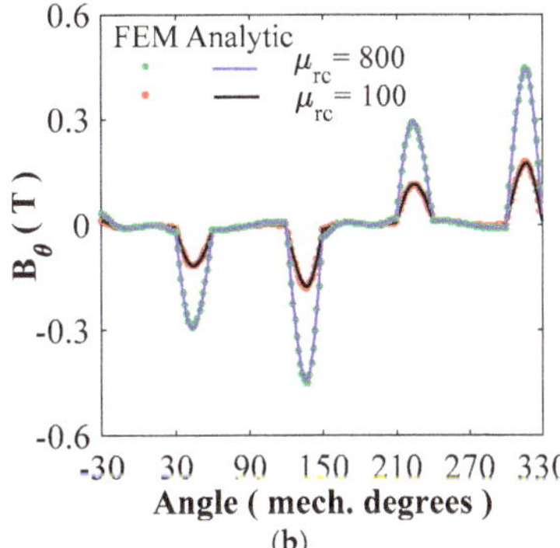

(a) (b)

Figure 14. Waveform of the magnetic flux density in the rotor slots/teeth for M1: (**a**) r- and (**b**) θ-component.

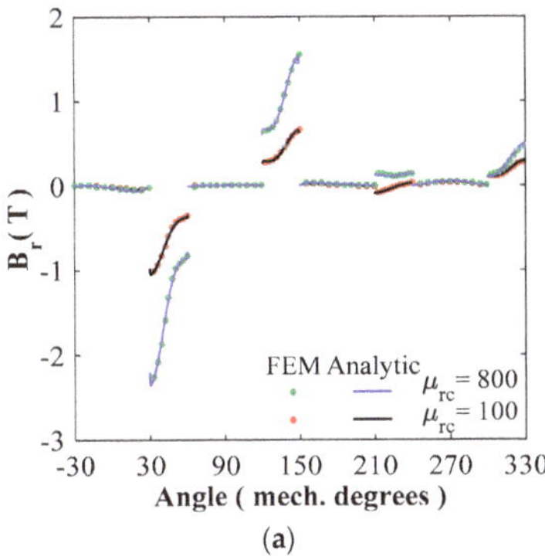

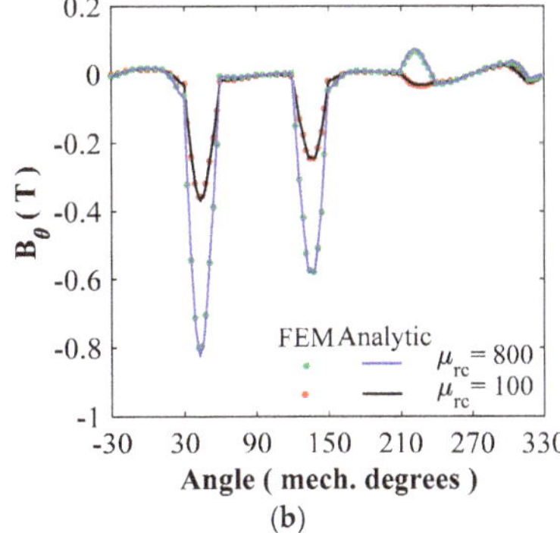

(a) (b)

Figure 15. Waveform of the magnetic flux density in the rotor slots/teeth for M2: (**a**) r- and (**b**) θ-component.

The simulations were done for both values of iron core relative permeability. One can see that a very good agreement is obtained for the various components of B in all regions.

4.2. Static/Dynamic Electromagnetic Torques

For two studied SRMs (i.e., M1 and M2) and for both values of iron core relative permeability, Figures 16–19 show the waveform as well as the harmonic spectrum of the static/dynamic electromagnetic torques for full-load condition (viz., 15 A @ 1500 rpm). The static electromagnetic torque represents the torque due to a single phase of the electrical machine (e.g., due to phase A). The dynamic electromagnetic torque represents the torque when the three phases are powered or due to the combination of three static electromagnetic torques. The good agreement between the results from 2D FEM and the proposed semi-analytical model can be seen. It is interesting to note that the ripple torques are more important for M2 (see Figure 19) with the same operating point.

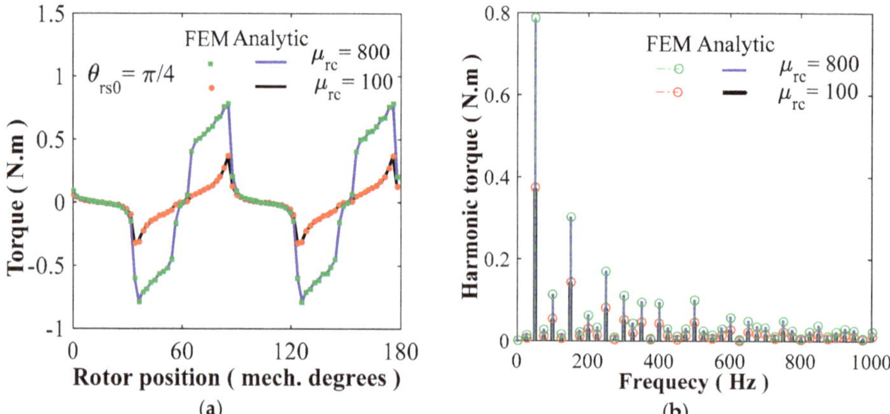

Figure 16. The static electromagnetic torque due to phase-A for M1: (**a**) waveform; (**b**) harmonic spectrum.

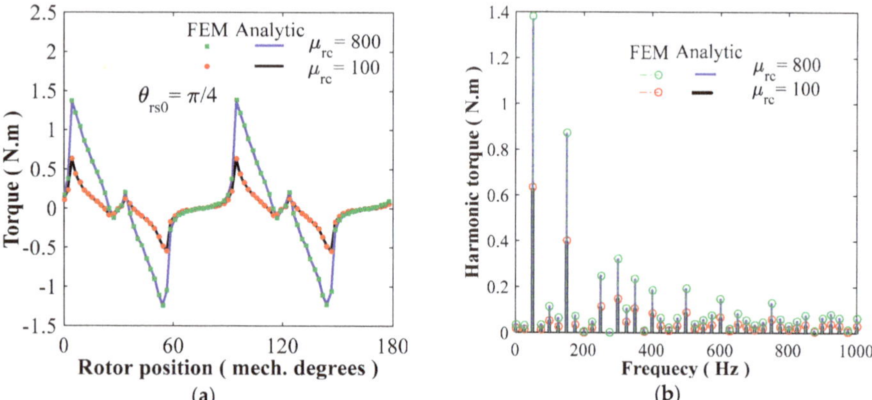

Figure 17. The static electromagnetic torque due to phase-A for M2: (**a**) waveform; (**b**) harmonic spectrum.

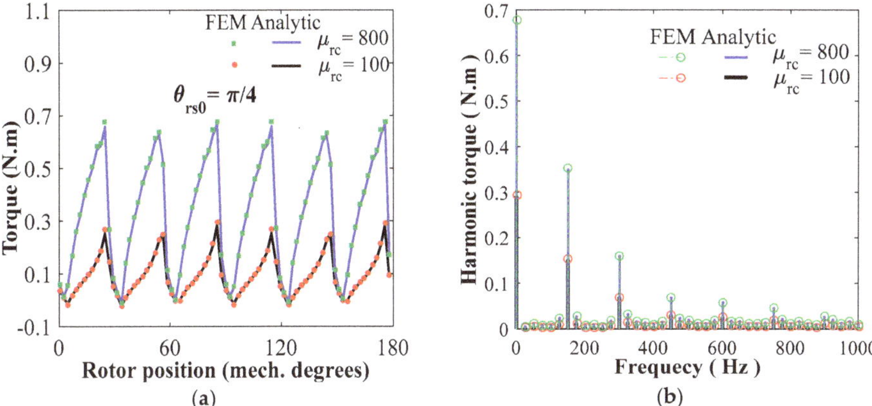

Figure 18. The dynamic electromagnetic torque (for full-load condition) for M1: (**a**) waveform; (**b**) harmonic spectrum.

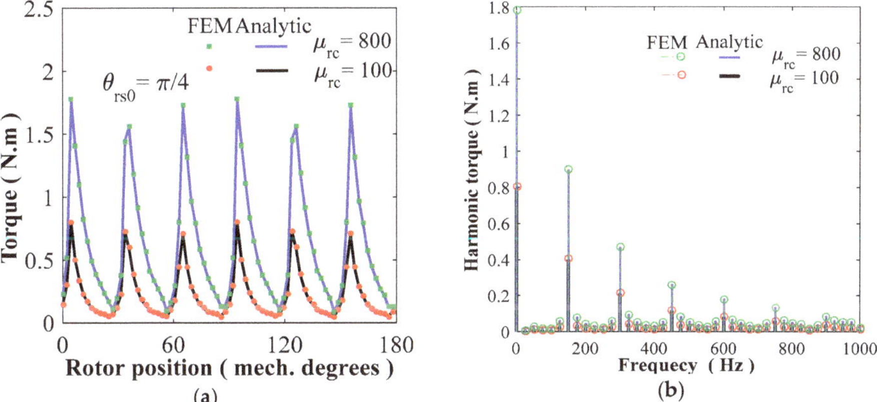

Figure 19. The dynamic electromagnetic torque (for full-load condition) for M2: (**a**) waveform; (**b**) harmonic spectrum.

4.3. Magnetic Flux Linkage and Self-/Mutual Inductances

For full-load condition (viz., 15 A @ 1500 rpm), the induced magnetic flux linkage per phase of two studied SRMs (i.e., M1 and M2) are given in Figure 20. The simulations were done for both values of iron core relative permeability.

Figures 21 and 22 show the self- and mutual inductance for M1 and M2, respectively; the simulation is done for nominal current. One can see that the self-inductance is slightly more important, while the mutual inductance is a much more important and negative value for M2. The obtained results confirm the accuracy of the proposed semi-analytical model, considering both amplitude and waveform.

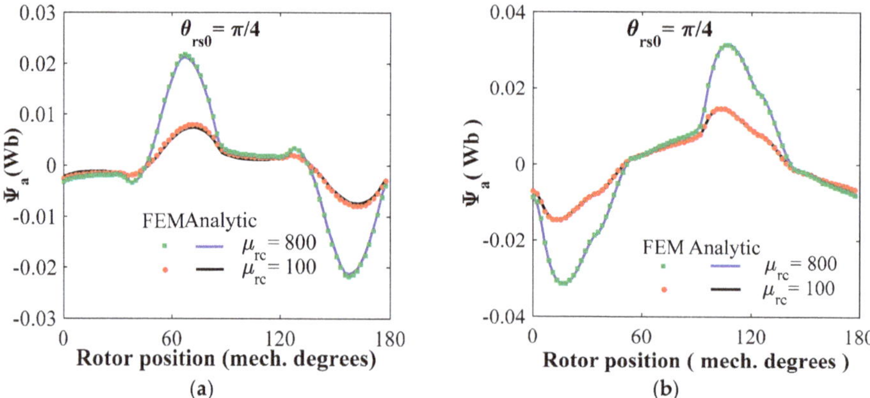

Figure 20. Waveform of the magnetic flux linkage at full-load condition (15 A @ 1500 rpm) for $\theta_{rs0} = \pi/Q_r$ in (**a**) M1 and (**b**) M2.

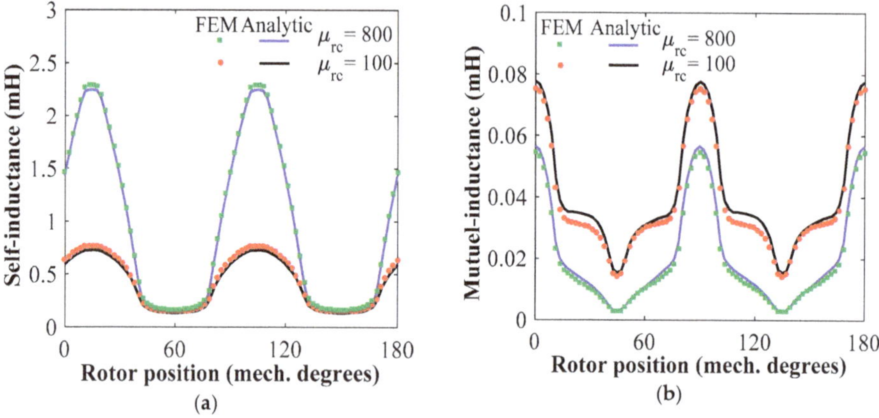

Figure 21. Waveform of the (**a**) self- and (**b**) mutual inductance for M1.

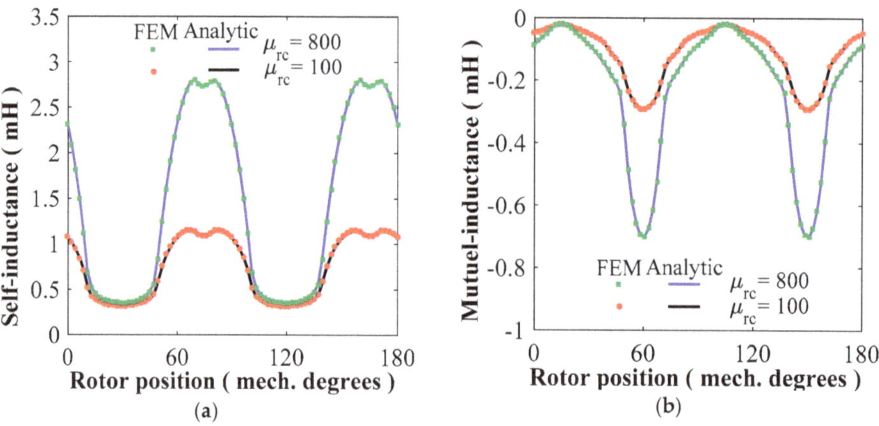

Figure 22. Waveform of the (**a**) self- and (**b**) mutual inductance for M2.

4.4. Magnetic Pressure and Non-Intrinsic UMFs

Figures 23 and 24 show the r- and θ-components of $\boldsymbol{P}$ in function of space angle for the two studied SRMs (i.e., M1 and M2) and both values of iron core relative permeability. The analytical radial magnetic pressure in function of temporal rotor angle and the spatial angle is represented in Figure 25 for a no-load condition. Figure 26 shows the fast Fourier transform in 2D (FFT2D) of analytical radial magnetic pressure P_r for M1 and M2. The x- and y-components of $\boldsymbol{F}$ in M1 and M2, for a no-load condition, are shown in Figure 27. It is clear that the non-intrinsic UMFs can be significant in SRMs, having diametrically asymmetric disposition of non-overlapping winding and due to the asymmetric magnetic field distribution in the air gap (see Figure 5).

The UMFs in M1 are null due to the proper choice of the armature winding type in the stator with same phase windings in diametrically opposite slots (see Figure 1a). Figure 28 shows the locus of the non-intrinsic UMF in M2. It is interesting to note that the UMFs and the magnetic pressures increases with the increase of the iron core relative permeability. One can see that the proposed semi-analytical model taking into account the iron core relative permeability gives good results compared to FEM.

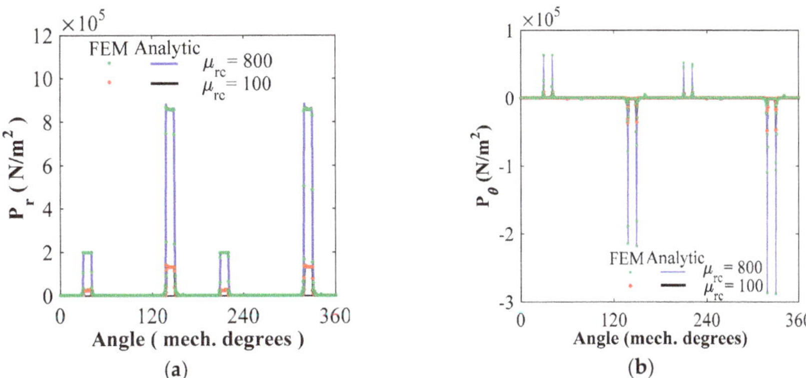

Figure 23. Waveform of magnetic pressures (for no-load condition) in M1: (**a**) r- and (**b**) θ-component.

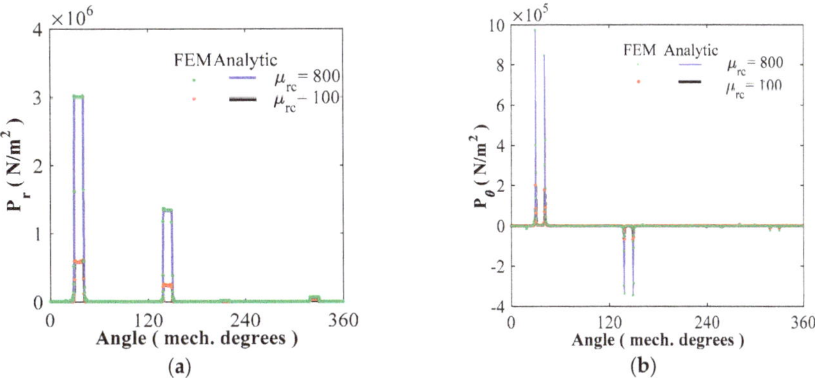

Figure 24. Waveform of the magnetic pressures (for no-load condition) in M2: (**a**) r- and (**b**) θ-component.

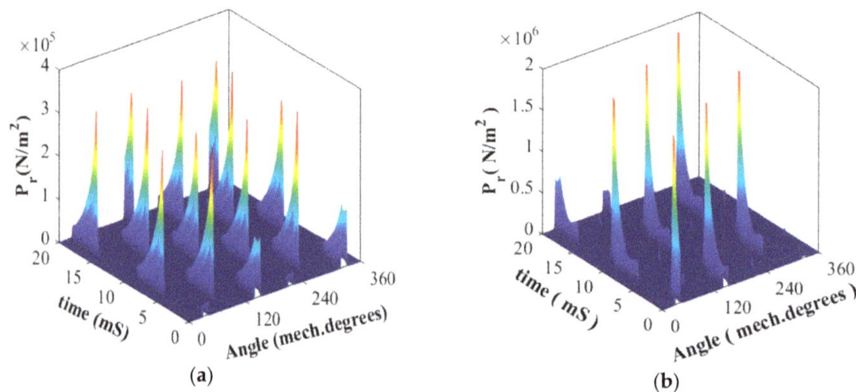

Figure 25. Analytical radial magnetic pressure versus time and space angle (for no-load condition) in (**a**) M1 and (**b**) M2.

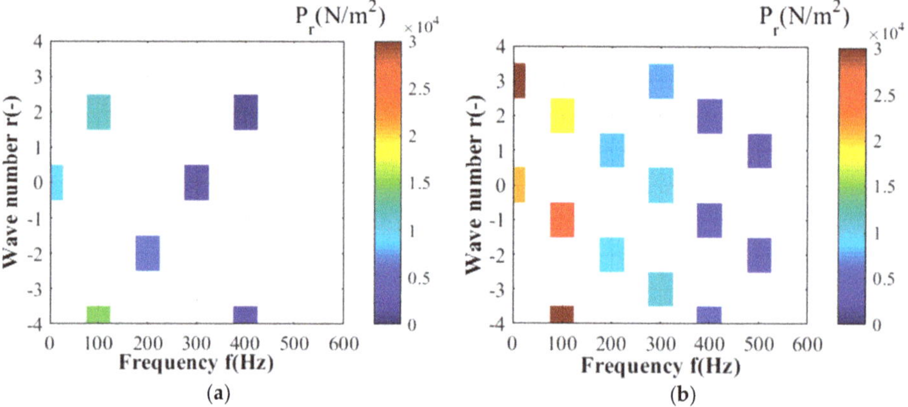

Figure 26. Analytical radial magnetic pressure with FFT2D for no-load condition in (**a**) M1 and (**b**) M2.

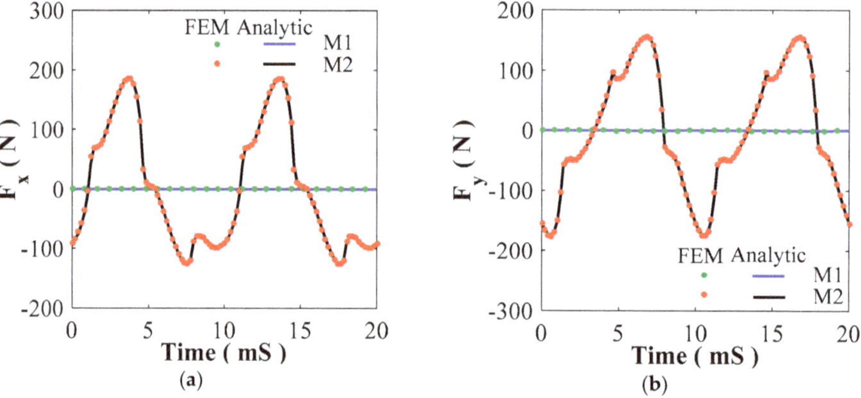

Figure 27. Waveform of UMFs (for no-load condition) in the two studied SRMs (i.e., M1 and M2): (**a**) x- and (**b**) y-component.

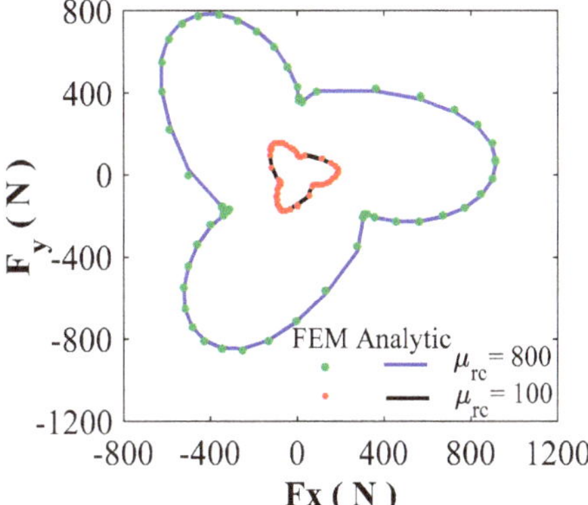

Figure 28. Locus of the non-intrinsic UMF in M2 for no-load condition.

The obtained results confirm the previous interpretation of the UMF. The component of the mode shapes one correspond to the wave number (i.e., $r = 1$) of the UMF does not appear in M1 due to the symmetric distribution of the air gap magnetic field with respect to the space angle p, contrary to M2. The component corresponds to the wave number (i.e., $r = 2$) is appeared in both SRMs, because of the asymmetric distribution magnetic field in the air gap with respect to the space angle $p/2$, as seen in Figures 4 and 5. Moreover, it can be seen that the other modes that appeared in the two SRMs are multiples of the least non-null wave number $r_{\text{min}} = 1$ in M2 and $r_{\text{min}} = 2$ in M1.

5. Conclusions

In this paper, we have developed a 2D exact subdomain technique in polar coordinates considering the iron relative permeability for SRM supplied by a sinusoidal waveform of the current (aka, variable flux reluctance machines). This semi-analytical model, based on the scientific works of [24,33], predicts the magnetic flux density distribution inside the electrical machine as well as the electromagnetic performances. It has been applied to 6/4 SRM with two various non-overlapping (or concentrated) windings. These two configurations of non-overlapping winding have been considered to show their effect on the UMFs. Moreover, the spectrum of UMF permit us to study the effect of each harmonic on the vibrations of these machines. However, this research proved that the SRM with an asymmetric disposition of winding is more prone to higher levels of vibration than the SRM with a symmetric disposition due to the UMFs' presence, which is the main source of vibration and acoustic noise. All results confirmed the accuracy of the proposed model. It can be considered a reliable alternative to FEM for analysis of SRMs.

Author Contributions: The work presented here was a cooperative effort by all the authors.

Conflicts of Interest: The authors declare no conflict of interest.

References

1. Radimov, N.; Ben-Hail, N.; Rabinovici, R. Switched reluctance machines as three-phase AC autonomous generator. *IEEE Trans. Magn.* **2006**, *42*, 3760–3764. [CrossRef]
2. Cheng, H.; Chen, H.; Yang, Z. Design indicators and structure optimisation of switched reluctance machine for electric vehicles. *IET Electric Power Appl.* **2015**, *9*, 319–331. [CrossRef]
3. Takeno, M.; Chiba, A.; Hoshi, N.; Ogasawara, S.; Takemoto, M.; Rahman, A. Test results and torque improvement of the 50-kW switched reluctance motor designed for hybrid electric vehicles. *IEEE Trans. Ind. Appl.* **2012**, *48*, 1327–1334. [CrossRef]

4. Li, G.J.; Ojeda, J.; Hlioui, S.; Hoang, E.; Lecrivain, M.; Gabsi, M. Modification in rotor pole geometry of mutually coupled switched reluctance machine for torque ripple mitigating. *IEEE Trans. Magn.* **2012**, *48*, 2025–2034. [CrossRef]
5. Lawrenson, P.J.; Stephenson, J.M.; Blenkinsop, P.T.; Korda, J.; Fulton, N.N. Variable-speed switched reluctance motors. *IEE Proc. B Electric Power Appl.* **1980**, *127*, 253–265. [CrossRef]
6. Sahin, C.; Amac, A.E.; Karacor, M.; Emadi, A. Reducing torque ripple of switched reluctance machines by relocation of rotor moulding clinches. *IET Electric Power Appl.* **2012**, *6*, 753–760. [CrossRef]
7. Lin, C.; Fahimi, B. Prediction of acoustic noise in switched reluctance motor drives. *IEEE Trans. Energy Convers.* **2014**, *29*, 250–258. [CrossRef]
8. Lee, D.H.; Pham, T.H.; Ahn, J.W. Design and operation characteristics of four-two pole high-speed SRM for torque ripple reduction. *IEEE Trans. Ind. Electron.* **2013**, *60*, 3637–3643. [CrossRef]
9. Husain, I.; Radun, A.; Nairus, J. Unbalanced force calculation in switched reluctance machines. *IEEE Trans. Magn.* **2000**, *36*, 330–338. [CrossRef]
10. Shen, L.; Wu, J. Switched reluctance motor vibration prediction: From low frequency to high frequency. In Proceedings of the IEMDC, Miami, FL, USA, 3–6 May 2009.
11. Gieras, J.F.; Lai, J.C.; Wang, C. *Noise of Polyphase Electric Motors*; CRC/Taylor & Francis: Boca Raton, FL, USA, 2006.
12. Lin, C.; Fahimi, B. Prediction of radial vibration in switched reluctance machines. *IEEE Trans. Energy Convers.* **2013**, *28*, 1072–1081. [CrossRef]
13. Zhu, Z.Q.; Ishak, D.; Howe, D.; Chen, J. Unbalanced magnetic forces in permanent-magnet brushless machines with diametrically asymmetric phase windings. *IEEE Trans. Ind. Appl.* **2007**, *43*, 1544–1553. [CrossRef]
14. Boughrara, K.; Ibtiouen, R.; Dubas, F. Analytical prediction of electromagnetic performances and unbalanced magnetic forces in fractional-slot spoke-type permanent-magnet machines. In Proceedings of the ICEM, Lausanne, Switzerland, 4–7 September 2016.
15. Yilmaz, M.; Krein, P.T. Capabilities of finite element analysis and magnetic equivalent circuits for electrical machine analysis and design. In Proceedings of the PESC, Rhodes, Greece, 15–19 June 2008.
16. Dubas, F.; Espanet, C. Analytical solution of the magnetic field in permanent-magnet motors taking into account slotting effect: No-load vector potential and flux density calculation. *IEEE Trans. Magn.* **2009**, *45*, 2097–2109. [CrossRef]
17. Zhu, Z.Q.; Wu, L.J.; Xia, Z.P. An accurate subdomain model for magnetic field computation in slotted surface-mounted permanent-magnet machines. *IEEE Trans. Magn.* **2010**, *46*, 1100–1115. [CrossRef]
18. Tiegna, H.; Amara, Y.; Barakat, G. Overview of analytical models of permanent magnet electrical machines for analysis and design purposes. *Math. Comput. Simul.* **2013**, *90*, 162–177. [CrossRef]
19. Dubas, F.; Rahideh, A. Two-dimensional analytical permanent-magnet eddy-current loss calculations in slotless PMSM equipped with surface-inset magnets. *IEEE Trans. Magn.* **2014**, *50*, 54–73. [CrossRef]
20. Curti, M.; Paulides, J.J.H.; Lomonova, E.A. An overview of analytical methods for magnetic field computation. In Proceedings of the EVER, Grimaldi Forum, Monte Carlo, Monaco, 31 March–2 April 2015.
21. Sprangers, R.L.J.; Paulides, J.J.H.; Gysen, B.L.J.; Lomonova, E.A. Magnetic saturation in semi-analytical harmonic modeling for electric machine analysis. *IEEE Trans. Magn.* **2016**, *52*, 1–10. [CrossRef]
22. Pfister, P.-D.; Yin, X.; Fang, Y. Slotted permanent-magnet machines: General analytical model of magnetic fields, torque, eddy currents, and permanent-magnet power losses including the diffusion effect. *IEEE Trans. Magn.* **2016**, *52*, 1–13. [CrossRef]
23. Devillers, E.; Besnerais, J.L.; Lubin, T.; Hecquet, M.; Lecointe, J.-P. A review of subdomain modeling techniques in electrical machines: Performances and applications. In Proceedings of the ICEM, Lausanne, Switzerland, 4–7 September 2016.
24. Dubas, F.; Boughrara, K. New scientific contribution on the 2-D subdomain technique in Cartesian coordinates: Taking into account of iron parts. *Math. Comput. Appl.* **2017**, *22*, 17. [CrossRef]
25. Boughrara, K.; Lubin, T.; Ibtiouen, R. General subdomain model for predicting magnetic field in internal and external rotor multiphase flux-switching machines topologies. *IEEE Trans. Magn.* **2013**, *49*, 5310–5325. [CrossRef]
26. Dubas, F.; Boughrara, K. New scientific contribution on the 2-D subdomain technique in polar coordinates: Taking into account of iron parts. *Math. Comput. Appl.* **2017**, *22*, 42. [CrossRef]

27. Roubache, L.; Boughrara, K.; Dubas, F.; Ibtiouen, R. Semi-analytical modeling of spoke-type permanent-magnet machines considering the iron core relative permeability: Subdomain technique and Taylor polynomial. *Prog. Electromagn. Res. B* **2017**, *77*, 85–101. [CrossRef]
28. Roubache, L.; Boughrara, K.; Dubas, F.; Ibtiouen, R. Semi-analytical modeling of spoke-type permanent-magnet machines considering nonlinear magnetic saturation: Subdomain technique and Taylor polynomial. *Math. Comput. Simul.* **2018**. to be published.
29. Sprangers, R.L.J.; Paulides, J.J.H.; Gysen, B.L.J.; Waarma, J.; Lomonova, E.A. Semi-analytical framework for synchronous reluctance motor analysis including finite soft-magnetic material permeability. *IEEE Trans. Magn.* **2015**, *51*, 1–4. [CrossRef]
30. Ramakrishnan, K.; Curti, M.; Zarko, D.; Mastinu, G.; Paulides, J.J.H.; Lomonova, E.A. Comparative analysis of various methods for modelling surface permanent magnet machines. *IET Electric Power Appl.* **2017**, *11*, 540–547. [CrossRef]
31. Djelloul, K.Z.; Boughrara, K.; Dubas, F.; Kechroud, A.; Souleyman, B. Semi-analytical magnetic field predicting in many structures of permanent-magnet synchronous machines considering the iron permeability. *IEEE Trans. Magn.* **2018**, *54*, 8103921.
32. Djelloul, K.Z.; Boughrara, K.; Ibtiouen, R.; Dubas, F. Nonlinear analytical calculation of magnetic field and torque of switched reluctance machines. In Proceedings of the CISTEM, Marrakech, Morocco, 26–28 October 2016.
33. Djelloul, K.Z.; Boughrara, K.; Dubas, F.; Ibtiouen, R. Nonlinear analytical prediction of magnetic field and electromagnetic performances in switched reluctance machines. *IEEE Trans. Magn.* **2017**, *53*, 1–11. [CrossRef]
34. Roubache, L.; Boughrara, K.; Dubas, F.; Ibtiouen, R. New subdomain technique for electromagnetic performances calculation in radial-flux electrical machines considering finite soft-magnetic material permeability. *IEEE Trans. Magn.* **2018**, *54*, 8103315. [CrossRef]
35. Boughrara, K.; Dubas, F.; Ibtiouen, R. 2-D exact analytical method for steady-state heat transfer prediction in rotating electrical machines. *IEEE Trans. Magn.* **2018**, *54*, 1–19. [CrossRef]
36. Roubache, L.; Boughrara, K.; Dubas, F.; Ibtiouen, R. Elementary subdomain technique for magnetic field calculation in rotating electrical machines with local saturation effect. *Int. J. Comput. Math. Electr. Electron. Eng.* **2018**. [CrossRef]
37. Hannon, B.; Sergeant, P.; Dupré, L. Two-dimensional Fourier-based modeling of electric machines. In Proceedings of the IEMDC, Miami, FL, USA, 21–24 May 2017.
38. Meeker, D.C. Finite Element Method Magnetics ver. 4.2. Available online: www.femm.info/wiki/download (accessed on 10 October 2018).
39. Ouamara, D.; Dubas, F.; Benallal, M.N.; Randi, S.A.; Espanet, C. Automatic winding generation using matrix representation—ANFRACTUS TOOL 1.0. *Acta Polytech.* **2018**, *58*, 37–46. [CrossRef]
40. Islam, R.; Husain, I. Analytical model for predicting noise and vibration in permanent-magnet synchronous motors. *IEEE Trans. Ind. Appl.* **2010**, *46*, 2346–2351. [CrossRef]

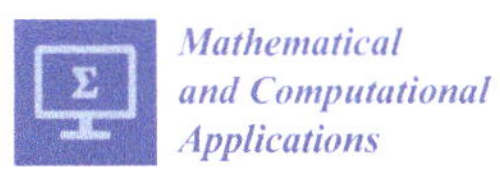

Article

Magnetic Field Analytical Solution for Non-homogeneous Permeability in Retaining Sleeve of a High-Speed Permanent-Magnet Machine

Gabriel A. Mendonça *, Thales A. C. Maia and Braz J. Cardoso Filho

Departamento de Engenharia Elétrica (DEE), Universidade Federal de Minas Gerais (UFMG), Belo Horizonte, MG 31270-901, Brazil; thalesmaia@gmail.com (T.A.C.M.); cardosob@ufmg.br (B.J.C.F.)

* Correspondence: gforti@gmail.com; Tel.: +55-31-98823-6263

Received: 9 October 2018; Accepted: 7 November 2018; Published: 10 November 2018

Abstract: This work presents a novel solution for magnetic field calculation in two-dimensional problems in which one region is defined with space-varying magnetic parameter. The proposed solution extends the well-established Maxwell–Fourier method for calculating magnetic fields in surface-mounted cylindrical high-speed permanent-magnet machines. This contribution is effective to evaluate more realistic magnetic parameters, where measurements of a high-speed permanent-magnet generator prototype indicate saturation in the retaining sleeve due to pole-to-pole leakage flux. The saturation profile is a function of mechanical angle and can be modeled with the aid of a space-varying relative permeability, expressed in terms of a Fourier series. As an example, the presented solution has been applied to a surface-mounted PM machine at no-load condition. Magnetic field calculations show that a simple saturation profile, with low order space-varying permeability in the retaining sleeve significantly affects the magnetic flux density distribution in the air-gap. The analytical solution is confronted with finite-element method, which confirms validity of the proposed methodology.

Keywords: analytical model; high-speed; sleeve; non-homogeneous permeability; permanent-magnet

---

1. Introduction

An increasing interest in high-speed (HS) machines are noticeable throughout several applications, ranging from distributed energy resources to medical equipment [1,2]. These machines commonly use permanent-magnet (PM) due to high torque per unit volume and low rotor losses [3].

For HS applications, surface-mounted PM machines usually require banding with a sleeve for retaining the PMs against centrifugal forces [4]. One important disadvantage of introducing such a component is an increased effective air-gap, where the added reluctance decreases PM excitation performance [5]. Furthermore, the addition of a conducting retaining sleeve (RS), which works as damper winding or shielding cylinder, reduces PM eddy-current losses in HS operating range [6]. Therefore, an accurate modeling of the RS is crucial for the evaluation of the PM machine overall electromagnetic behavior [7].

The modeling method plays an important role in HS machines electromagnet performance analysis. Numerical methods, such as finite-elements, provide accurate results for magnetic field, but with great computational cost. Alternatively, (semi-)analytical modeling based on the formal solution of Maxwell's equations provides precise calculation of magnetic field distribution in different machine regions. Compared with finite-element method (FEM), this approach is based on strong and limiting assumptions for both geometrical and physical parameters, however, with lower computation time and greater insight into the problem. (Semi-)analytical techniques are considered powerful tools, being vastly used in HS machine early design stage [1–3,6,8,9].

The available literature on (semi-)analytical technique for PM machine evaluation is extensive, with published papers ranging over twenty-five years, to the best of the authors' knowledge [10–12]. Moreover, there is still great effort towards methods with better accuracy and less restricting assumptions [4,13–22] to list a few recently published works. Dubas et al. [23] realized an overview on the existing (semi-)analytical models in Maxwell–Fourier methods (vis., multi-layer models, eigenvalues model, subdomain technique and hybrid models) with the effect of local/global saturation. Further details, advantages and disadvantages of these techniques can be found in [23–25]. Ramakrishnan et al. [13] presented a comprehensive comparison of analytical methods, where a subdomain technique exceeded the others in terms of accuracy with reasonable calculation time.

Furthermore, extending the limitations discussed in [25], recent developments include more realistic geometric structures, such as tooth-tips [14–16,26–28] or rotor eccentricity [27,29]. Chebak et al. [30] and Rahdeh et al. [31] defined more realistic magnetic parameters, but in slotless topologies. Qian et al. [29] and Ortega et al. [32] evaluate several geometries and physical imperfections expected from manufacturing processes. Spranger et al. [33] and Dubas et al. [23,24] have recently developed new techniques to account for finite soft-magnetic permeabilities, respectively: (i) in the multi-layer model using the Cauchy's product theorem [33], and (ii) in the subdomain technique by applying the superposition principle in both directions [23,24]. As discussed by Hannon et al. [34], both methodologies are very effective since they enable the magnetic field calculation in the ferromagnetic material of slotted geometries. The subdomain technique has been improved to consider soft-magnetic material permeability in radial-flux electrical machines, such as switched reluctance machines [21] and spoke-type PM machines [4]. In addition, this technique has been extended to the thermal modeling for the steady-state temperature distribution in rotating electrical machines [22]. According to Pfister et al. [35], the first work on the semi-analytical subdomain technique was published by [36] with the state-of-the-art in the introduction. Moreover, the method based on the Cauchy's product theorem has been improved to account for saturation in stator tooth [19] and a configurable model, adaptable to many machine geometries [20].

However, among the presented references, a small number of authors give a proper realistic evaluation for RS [3,6,9,14,18,28]. Furthermore, only a few of these works evaluate all electromagnetic properties of the retaining sleeve, where relative permeability is different from the air [14,18] and eddy-currents are considered [3,6,9,18,28].

In this paper, a novel formulation is presented to extend a two-dimensional analytical solution from Laplace's and Poisson's equation in polar coordinates. The proposed technique allows the definition of periodic space-varying magnetic parameter, which is defined through Fourier series. A similar approach was used by Sprangers et al. [33], where permeability variation in slotted regions was evaluated directly into the field solution. Such method, based on the convolution theorem, allows the calculation of magnetic field in the tooth region, where interactive solutions can be applied to evaluation of magnetic saturation [19]. The technique presented in this paper, based on well-established Maxwell–Fourier method solutions, evaluates the space-varying permeability by reexamining the boundary conditions. Therefore, the proposed analysis provides an intuitive and fast solution to problems involving magnetic saturation of an annular region and can be extended to other analytical techniques, e.g., the subdomain method for evaluating slotting effects.

Finally, validation of the proposed model is carried out based on a two-pole surface-mounted permanent-magnet generator prototype [37], where experimental data indicated saturation in the retaining sleeve. The geometry and magnetic parameters of the PM generator are approximated, thus providing simplified formulation. The derived model is, then, evaluated with different saturation characteristics and the results are compared with those obtained from FEM.

2. Geometry and Mathematical Formulation

Throughout the literature, great efforts are invested in developing and improving analytical models focused on the early design stages, where optimization routines play an important role.

This work is based on the performance of a HS machine prototype and justifies the reevaluation of common approximations. Thus, the model presented is implemented to include a space-varying permeability to evaluate magnetic saturation effects of the retaining sleeve. Moreover, several other machine aspects not related to the present method are simplified, such as machine geometry and some magnetic parameters, as discussed in this section.

The proposed method is derived according to well-established analytical solution of Maxwell's equations. The saturation phenomenon in the RS is mainly due to pole-to-pole leakage flux, thus changing the RS relative permeability in the tangential direction. In a first approximation, this behavior is considered as independent on loading conditions and the machine is studied evaluating only PM magnetization. Furthermore, stator iron core is assumed to present linear and very large relative permeability, assumed to be infinite. The PM do not demagnetize. All electrical conductivities are assumed to be null. Finally, the magnetic field solutions are derived considering a slotless geometry, as illustrated in Figure 1.

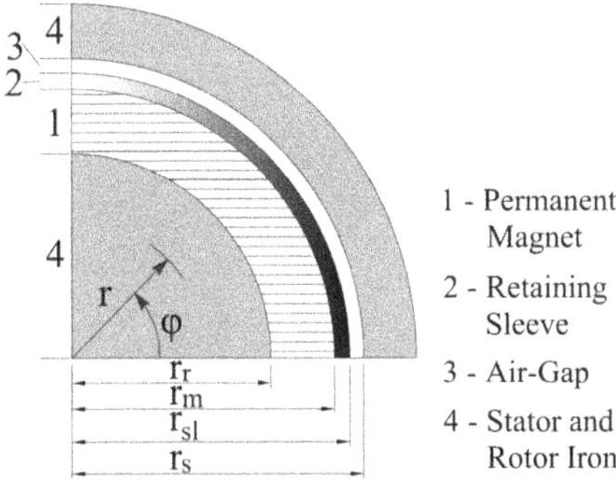

Figure 1. A quarter cross section of PM synchronous machine for analytical calculation.

These approximations are necessary to simplify the analytical model and must be used with care. The objective is to derive simpler mathematical formulation, allowing better insight of the proposed method. However, to fully investigate PM machine performance, further research must be done to evaluate more realistic geometry and electromagnetic parameters, such as slotting effects, conducting regions and loading influence.

The remainder of assumptions are necessary to evaluate the two-dimensional analytical problem using Fourier series representation. Magnetic field end effects are neglected and all parameters are periodic in respect to the tangential coordinate. In the RS region, saturation occurs from pole-to-pole leakage flux, where the resultant relative permeability is a function of rotor mechanical angle and varies periodically from pole-to-pole. Therefore, it can be characterized as periodic in the circumferential direction with fundamental space period equals to $2p$, where p is the number of pole pairs. The relative permeability distribution is defined in terms of a Fourier series as

$$\mu_r(\varphi) = \sum_{k=0,1,2,3,\ldots}^{\infty} \hat{\mu}_{r,k} \cos(2pk\varphi_r), \tag{1}$$

where μ_r is the space-varying relative permeability.

The Fourier coefficients $\hat{\mu}_{r,k}$ can be calculated accordingly for the evaluated phenomenon. For example, to estimate saturation in an annular region, an iterative method could be used to account for the nonlinear magnetic material.

The Poisson's Equation Solution

Considering previous approximations, the basis for the analytical solution for the air-gap magnetic field is set. Using the magnetic vector potential, and using the Coulomb gauge, the general governing equation for the defined regions is

$$\nabla^2 \mathbf{A} = -\nabla \times \mathbf{B}_{rem}. \tag{2}$$

The general solution to Poisson's Equation (2) is determined in each annular region depending on their electromagnetic characteristic and can be expressed as

$$A_z^{(v)}(r,\varphi) = \sum_{k=1,3,5,\ldots}^{\infty} \hat{A}_{z,k}^{(v)}(r)\sin(kp\varphi_r), \tag{3a}$$

where:

$$\hat{A}_{z,k}^{(v)}(r) = \left(C_k^{(v)} r^{kp} + D_k^{(v)} r^{-kp} + \hat{A}_{part,z,k}^{(v)} r\right). \tag{3b}$$

In the present investigation, where focus is given to characterize saturation in the RS region, PM magnetization pattern and topology does not need to be determined. These aspects are needed, however, to obtain the particular solution, $\hat{A}_{part,z,k}^{(v)}$, in (3b). Further discussion on this particular solution from different magnetization profiles (e.g., parallel, radial, Halbach, etc.) can be found in [3,6,10,31,36]. Regions where no magnetic field sources are defined, namely air-gap and RS, the Poisson's equation (1) simplifies to the corresponding Laplace's equation. The solution is also obtained from Equations (3a) and (3b), where the particular solution is null, i.e., $\hat{A}_{part,z,k}^{(v)} = 0$.

The magnetic vector potential in region v, $A_z^{(v)}$, is described through an infinite series of space harmonics k. The regions are defined according to Figure 1, where $v = 1, 2$ and 3 for PM, RS and air-gap, respectively. Rotor reference frame, φ_r, is defined from magnetization axis. Finally, the constants $C_k^{(v)}$ and $D_k^{(v)}$ are determined from the boundary conditions, which are defined by the continuity of the tangential field intensity, H_φ, and the normal flux density, B_r and expressed as

$$H_\varphi^{(v+1)}(r_v,\varphi) - H_\varphi^{(v)}(r_v,\varphi) = K_z^{(v)}(\varphi), \tag{4}$$

$$B_r^{(v+1)}(r_v,\varphi) - B_r^{(v)}(r_v,\varphi) = 0. \tag{5}$$

The surface density current, $K_z^{(v)}$, is usually defined along the z-axis, being positive in the out-of-the-page direction. Equations (3a) and (3b)–(5) are the basis to the field solution procedure. The analysis is usually simplified, since space harmonics of different orders are orthogonal. Hence, the constants $C_k^{(v)}$ and $D_k^{(v)}$ are related only to the field source space harmonic component of the k-th order, being either from PM or current sources.

The present formulation is based on the reevaluation of such criteria. Reexamining the magnetic flux density tangential component boundary condition for the interfaces of the RS, Equation (4), with the proposed relative permeability in (1), that is

$$\frac{B_\varphi^{(2)}(r_{sl},\varphi)}{\mu_r(\varphi)\mu_0} = \frac{B_\varphi^{(3)}(r_{sl},\varphi)}{\mu_0}, \tag{6}$$

$$\frac{B_\varphi^{(1)}(r_m,\varphi) - B_{rem,\varphi}(\varphi)}{\mu_0} = \frac{B_\varphi^{(2)}(r_m,\varphi)}{\mu_r(\varphi)\mu_0}. \tag{7}$$

With the tangential component for magnetic flux density calculated from the magnetic vector potential Equations (3a) and (3b), Equation (6) can be rewritten as

$$\frac{1}{\mu_r(\varphi)} \sum_{k=1,3,5,...}^{\infty} \hat{B}_{\varphi,k}^{(2)}(r_{sl}) \sin(kp\varphi_r) = \sum_{k=1,3,5,...}^{\infty} \hat{B}_{\varphi,k}^{(3)}(r_{sl}) \sin(kp\varphi_r), \tag{8}$$

where $\hat{B}_{\varphi,k}^{(v)}(r) = kp\left(C_k^{(v)} r^{kp-1} - D_k^{(v)} r^{-kp-1} + \frac{\partial\left(\hat{A}_{part,z,k}^{(v)} r\right)}{\partial r}\right)$. The solution, with relative permeability defined with (1), includes multiple summation terms. For generalization purposes, these summation indexes are defined with different variables, k, m and n. Thus, from (8), we have

$$\sum_{k=1,3,5,...}^{\infty} \hat{B}_{\varphi,k}^{(2)}(r_{sl}) \sin(kp\varphi_r) = \sum_{m=0,1,2,3,...}^{\infty} \hat{\mu}_{r,m} \cos(2mp\varphi_r) \sum_{n=1,3,5,...}^{\infty} \hat{B}_{\varphi,n}^{(3)}(r_{sl}) \sin(np\varphi_r). \tag{9}$$

The calculation of the constants and proper interaction of the different space harmonic indexes are found from the definition of Fourier series. Therefore, if we establish

$$f(\varphi_r) = \sum_{k=1,3,5,...}^{\infty} \hat{B}_{\varphi,k}^{(2)} \sin(kp\varphi_r), \tag{10}$$

the constants in (10) are determined from

$$\hat{B}_{\varphi,k}^{(2)} = \frac{1}{\pi} \int_0^{2\pi} f(\varphi_r) \sin(kp\varphi_r)\, d\varphi. \tag{11}$$

Rearranging the right-hand side of (11) with the aid of (9) yields

$$\hat{B}_{\varphi,k}^{(2)}(r_{sl}) = \sum_{m=0,1,2,3,...}^{\infty} \sum_{n=1,3,5,...}^{\infty} \frac{\hat{\mu}_{r,m} \hat{B}_{\varphi,n}^{(3)}(r_{sl})}{\pi} \int_0^{2\pi} \sin(np\varphi_r) \cos(2mp\varphi_r) \sin(kp\varphi_r)\, d\varphi, \tag{12}$$

where the space harmonic index k can be thought as a reference index. The integration on the right-hand side of (12) will not be null for specific values of k, m and n.The same reasoning is used for (7), that is,

$$\begin{aligned} \hat{B}_{\varphi,k}^{(2)}(r_m) = &\sum_{m=0,1,2,3,...}^{\infty} \sum_{i=1,3,5,...}^{\infty} \frac{\hat{\mu}_{r,m} \hat{B}_{\varphi,i}^{(1)}(r_m)}{\pi} \int_0^{2\pi} \sin(ip\varphi_r) \cos(2mp\varphi_r) \sin(kp\varphi_r)\, d\varphi \\ &- \sum_{m=0,1,2,3,...}^{\infty} \sum_{l=1,3,5,...}^{\infty} \frac{\hat{\mu}_{r,m} \hat{B}_{rem,\varphi,l}}{\pi} \int_0^{2\pi} \sin(lp\varphi_r) \cos(2mp\varphi_r) \sin(kp\varphi_r)\, d\varphi. \end{aligned} \tag{13}$$

The definition of a relative permeability from Fourier series expansion results in an interaction of magnetic fields of different orders. The integrations in (12) and (13) will be different from zero only for certain values of i, k, l, m and n. The trigonometric transformations needed to evaluate integration terms in Equations (12) and (13) are well known and will not be reproduced here. Summarizing the results, Table 1 presents the relations between the indexes which results non-zero integration values in Equation (12). Though there are different indexes relations in Equation (13), the results can be easily extended by replacing n with i or l.

Table 1. Space harmonic index relations for non-zero integral solutions.

Condition	Index Relation	Integral Result
1	$k - n - 2m = 0$	$\frac{\pi}{2}$
2	$k - n + 2m = 0$	$\frac{\pi}{2}$
3	$k + n - 2m = 0$	$-\frac{\pi}{2}$

One first conclusion derived from Table 1 is that, assuming homogeneous and constant RS magnetic parameter, the solution is determined from interaction of magnetic fields with the same space harmonic order. In such case, for $m = 0$, the linear equation systems are obtained with $k = i = l = n$. In addition, it is important to note that the results from the index relations must be added up accordingly with the summation notations. Therefore, from the index relations in Table 1, the result for Equations (12) and (13) can be derived as

$$\hat{B}^{(2)}_{\varphi,k}(r_{sl}) = \sum_{m=0,1,2,3,\ldots}^{\infty} \frac{\hat{\mu}_{r,m}}{2}\left[\hat{B}^{(3)}_{\varphi,k-2m}(r_{sl}) + \hat{B}^{(3)}_{\varphi,k+2m}(r_{sl}) - \hat{B}^{(3)}_{\varphi,2m-k}(r_{sl})\right], \tag{14}$$

$$\begin{aligned}\hat{B}^{(2)}_{\varphi,k}(r_m) = &\sum_{m=0,1,2,3,\ldots}^{\infty} \frac{\hat{\mu}_{r,m}}{2}\left[\hat{B}^{(1)}_{\varphi,k-2m}(r_m) + \hat{B}^{(1)}_{\varphi,k+2m}(r_m) - \hat{B}^{(1)}_{\varphi,2m-k}(r_m)\right] \\ &- \sum_{m=0,1,2,3,\ldots}^{\infty} \frac{\hat{\mu}_{r,m}}{2}\left[\hat{B}_{rem,\varphi,k-2m} + \hat{B}_{rem,\varphi,k+2m} - \hat{B}_{rem,\varphi,2m-k}\right].\end{aligned} \tag{15}$$

From (14) and (15), it can be verified that the Fourier coefficients of the magnetic flux density in the non-homogeneous region, $v = 2$, are related to those of different space harmonic orders of the surrounding regions, i.e., $v = 1$ and 3. For example, $\hat{B}^{(2)}_{\varphi,k}$ is expressed in terms of $\hat{B}^{(3)}_{\varphi,k-2m}$, $\hat{B}^{(3)}_{\varphi,k+2m}$ and $\hat{B}^{(3)}_{\varphi,2m-k}$, which are the Fourier coefficients of space harmonic orders $(k-2m)$, $(k+2m)$ and $(2m-k)$ for flux density in region $v = 3$ and where m is the space harmonic order of the relative permeability Fourier series. These relations will be further detailed in the application example (Section 4).

The mathematical formulation presented in this section is general and can be applied to several PM machine configurations. However, to help illustrate the proposed methodology and to show the importance of evaluating the RS saturation, the following sections will discuss its application to a HS machine prototype.

3. Evaluated Prototype

The machine used for verifying the proposed methodology is part of a micro-compressed air energy storage (micro-CAES) system [37]. It was built from a conventional turbocharger system adapted to operate only with air [37]. Due to its high speeds, the power density is considerably higher. It is suitable for small scale applications, such as operating with distributed generation to improve power supply reliability. The prototype is illustrated in Figure 2, where Figure 2a shows the dissembled system, and Figure 2b illustrates the mounted system.

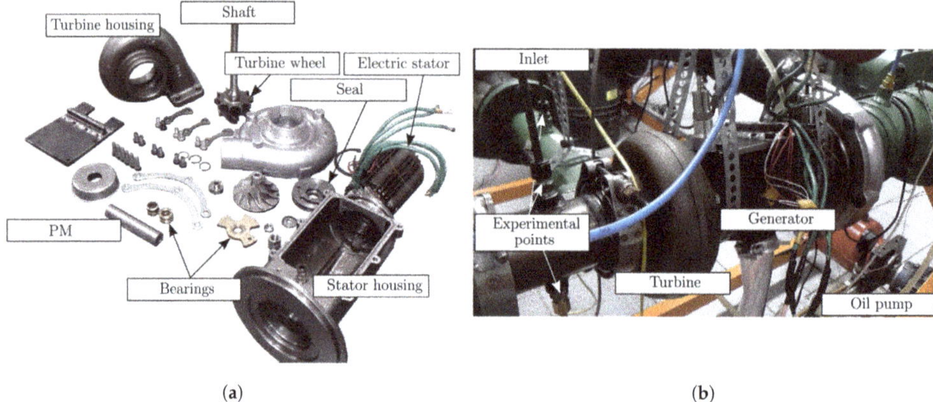

Figure 2. Low cost high-speed micro-CAES system: (**a**) disassembled components; and (**b**) system in operation [37].

In such application, generator design is carried out based on some overall restrictions, such as reduced size and cost. A single high-speed shaft eliminates the need of gear-box, thus improving system reliability [2]. However, a high-speed drive system design is critical, with difficulties associated with generator topology, material selection, manufacturing procedure, bearings types and many others. All of these concerns for the evaluated machine are addressed by Maia et al. [37].

The constructed machine consists of a two-pole surface-mounted PM generator with parallel magnetization. In order to fulfill the banding requirement, a stainless steel AISI 310 sleeve was adopted for its additional properties. The high electrical conductivity is expected to help shielding the magnets and rotor iron from high-frequency magnetic fields [3], thus reducing rotor losses. Moreover, behaving like a damper winding, it could help improving transient stability. The stator, manufactured with electric steel with 0.35 mm lamination thickness, was designed with full-pitch concentrated winding, as pictured in Figure 3a. The assembled prototype is presented in Figure 3b and final machine characteristics are shown in Table 2.

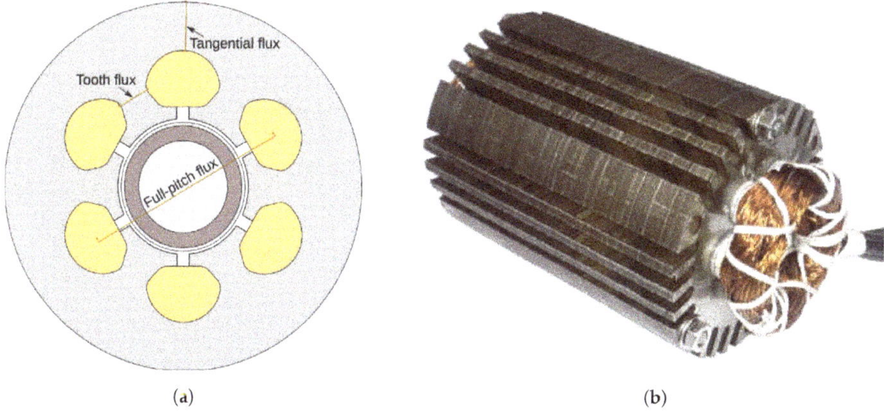

(**a**) (**b**)

Figure 3. PM generator: (**a**) schematic view; and (**b**) assembled prototype stator.

Table 2. Prototype PM machine characteristics.

Parameters	Value
Induced voltage	220 V_{RMS}
Rated speed	70,000 RPM
Rated current	9.2 A
Rated power	3.5 kW
Stator length	100 mm
PM diameter	21 mm

Maia et al. [37,38] successfully evaluated the electromechanical design of a HS generator directly driven by a micro-CAES system. Additionally, experimental measurements carried out helped reevaluate the assumptions usually made to simplify the mathematical formulation. Using flux sensing coils, as indicated in Figure 3a, measurements pointed that the RS had higher magnetic permeability than expected from the material specification. The RS machining, needed to achieve proper mechanical tolerances, changed the relative permeability to approximately 300. With the adjusted values, numerical simulations using FEMM [39] were performed. The results, compared with measurements, are presented in Figure 4.

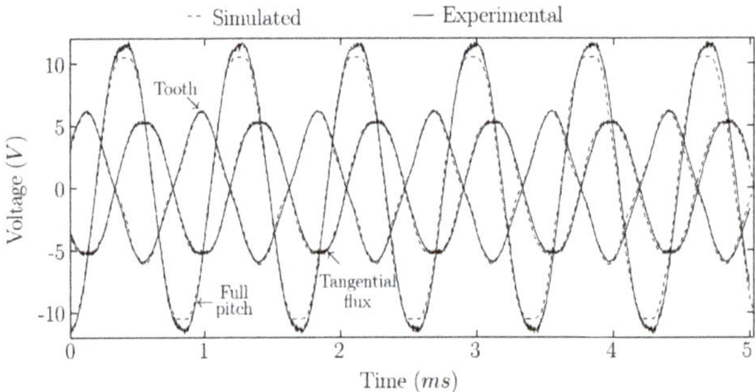

Figure 4. Measured and simulated results for induced voltage in exploring coils.

Discarding any saturation effects in the RS region, the higher relative permeability increases the pole-to-pole leakage flux, reducing significantly the no-load stator flux linkage and electromotive force [5]. Experimental no-load measurements did not match the results from FEM simulations which assumed linear RS relative permeability. Reevaluating the numerical simulation with nonlinear relative permeability for the RS region yields accurate results compared with measurements, as illustrated in Figure 5. The saturated RS provides a preferential path for the magnetic flux in the magnetization axis, diminishing the pole-to-pole leakage flux. The saturation curve for the stainless-steel RS was based on [40] and adjusted to provide the unsaturated relative permeability of approximately 300. The relative permeability profile in the RS for the PM generator at no-load condition is illustrated in Figure 6.

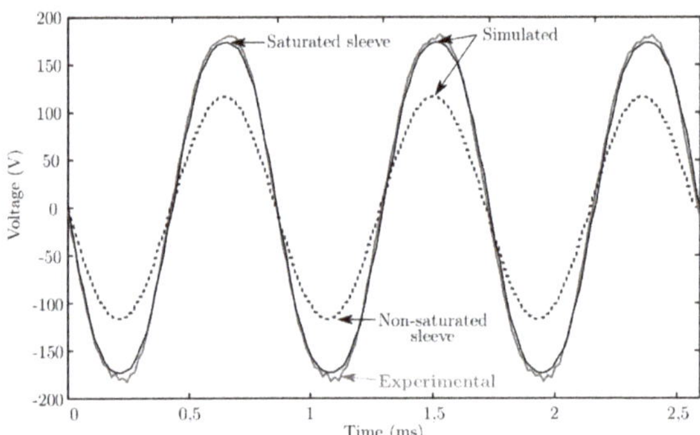

Figure 5. Measured and simulated results for no-load voltage.

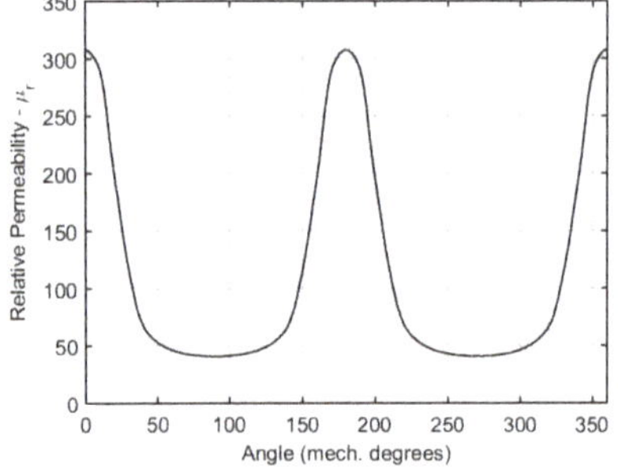

Figure 6. RS space-varying relative permeability profile.

Another aspect observed from the manufactured system is that, though not measured, cogging torque was perceived by manually rotating the machine shaft. This is an interesting result because it also leads to the assumption that saturation was taking place in the RS. A linear magnetic parameter, with this type of PM excitation and slot geometry, would result in a constant torque, without any harmonic content.

The interaction between magnetic flux produced by the PM and the stator non-uniform structure governs the cogging torque phenomenon [29]. This relationship can be clearly visualized through the complex permeance modeling method [8,13,41,42]. From closed form solutions for calculating the cogging torque using conformal mapping [41], no cogging torque would be expected for the studied motor. The rotor was manufactured with a two-pole NdFeB ring shaped with parallel magnetization, which produces a sinusoidal air-gap magnetic flux density with no space harmonic content, and the slotting configuration results in a constant torque profile, with no oscillation. Using the revised relative permeability for the RS, simulations using FEMM [39] were performed to illustrate the effect of saturation in the cogging torque.

To allow the evaluation of the stator core saturation effect on the cogging torque, the simulation was made using different remanence values for the PM. Figure 7a,b illustrate the results for different remanence values, $\hat{B}_{rem} = 1.2$ T and $\hat{B}_{rem} = 1.3$ T, respectively. For each PM remanent magnetization value, simulation evaluated effects of using linear or nonlinear permeability of RS, stator and rotor core.

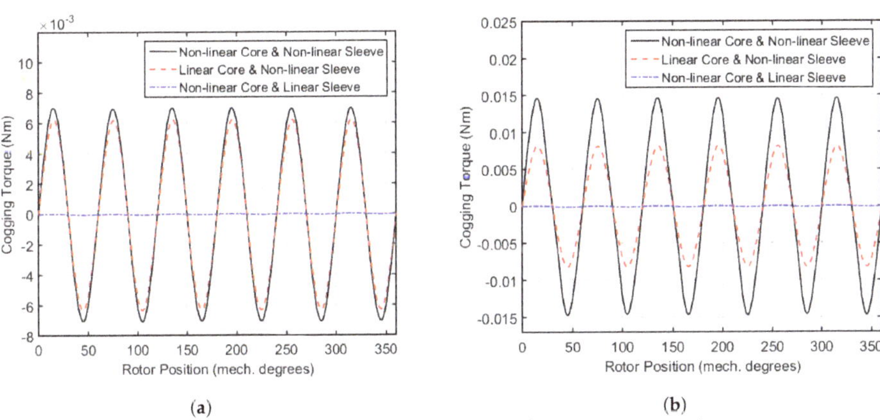

Figure 7. Cogging torque FEMM simulation with: (**a**) $\hat{B}_{rem} = 1.2$ T; and (**b**) $\hat{B}_{rem} = 1.3$ T.

4. Application Example Results and Validation

The validation was supported by problems simplification. Machine slotting effect was neglected and many magnetic parameters were approximated, as discussed in Section 2. This approach was chosen to help clarify some key aspects in the new formulation proposed. Furthermore, the relative permeability was defined only with a constant and one fundamental space-varying component, i.e., $\mu_r(\varphi) = \hat{\mu}_{r,0} + \hat{\mu}_{r,1}\cos(2\varphi_r)$. This approximation is justified here to help evaluate the interaction between space harmonic of different orders, showing that even low order spatial harmonics in the RS can significantly affect the air-gap density flux distribution. In addition, results from the analytical method were validated from FEM results.

The geometrical parameters used in the mathematical formulation are based on the PM machine structure illustrated in Figure 1 with values given in Table 3. The rotor type allows one more simplification without loss of generality. The rotor iron core region was neglected and the rotor structure is composed of full cylindrical PM and the RS, as illustrated in Figure 8. This reduces one boundary condition and, as a consequence, one linear equation (i.e., $D_k^{(1)} = 0$, from (3b)). In addition, the machine PM configuration simplifies the Poisson's equation since the curl of the parallel magnetization vector is null. Therefore, the radial and tangential components of the magnetization were defined only with a fundamental space harmonic, i.e., $\mathbf{B}_{rem} = \hat{B}_{rem}\cos(\varphi_r)\,\mathbf{a}_r - \hat{B}_{rem}\sin(\varphi_r)\,\mathbf{a}_\varphi$, where $\mathbf{a}_r$ and $\mathbf{a}_\varphi$ are the unit vectors in the radial and tangential direction.

Table 3. Parameters for analytical model of slotless geometry.

Parameters	Symbol	Value
Radius of magnet surface	r_m	9.8935 mm
Radius of retaining sleeve surface	r_{sl}	10.475 mm
Radius of stator inner surface	r_s	11.0 mm
Magnet remanence	$\hat{B}_{rem}$	1.2 T

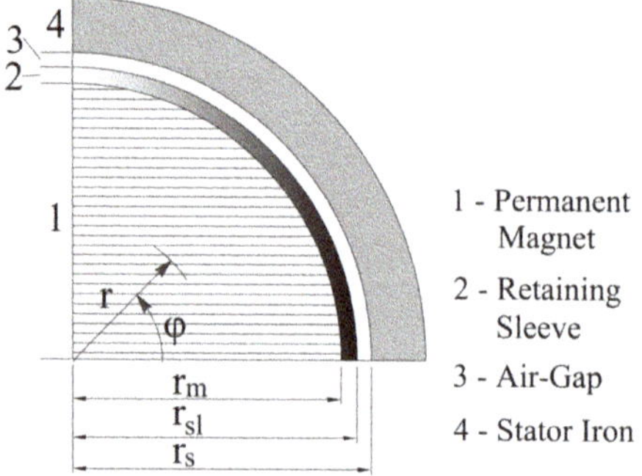

Figure 8. A quarter cross section of the PM synchronous machine with a segmented RS region for FEMM simulation.

Combining Equations (3a) and (3b) with the boundary conditions (4) and (5) establishes a set of linear equations. However, the RS boundary conditions were reevaluated with the procedure developed in Section 2. First, for the studied machine with only fundamental magnetization vector, the index l in (13) is set as equal to one. Table 4 presents the space harmonic indexes m and n, as defined in (12), which satisfies the conditions in Table 1. In addition, the superscripts in Table 4 relate the conditions in Table 1 that are satisfied.

Table 4. Space harmonic index relation for an evaluated PM machine.

Base Index	Relative Permeability Index	
	m = 0	m = 1
$k = 1$		$n^2 = 3; n^3 = 1$
$k = 3$	$n^{1,2} = k$	$n^1 = 1; n^2 = 5$
$k = 5$		$n^1 = 3; n^2 = 7$
$k = 7$		$n^1 = 5; n^2 = 9$

With the presented space harmonic indexes combinations, the system of linear equations can be defined. First, three boundary conditions used to derive the linear equation system remain unchanged. For the normal flux density continuity at $r = r_m$ and $r = r_{sl}$ and for the tangential field intensity at $r = r_s$, the expressions can be derived for any space harmonic indexes as follows:

$$r_s^{k-1}C_k^{(3)} - r_s^{-k-1}D_k^{(3)} = 0, \tag{16}$$

$$r_{sl}^{k-1}C_k^{(3)} + r_{sl}^{-k-1}D_k^{(3)} - r_{sl}^{k-1}C_k^{(2)} - r_{sl}^{-k-1}D_k^{(2)} = 0, \tag{17}$$

$$r_m^{k-1}C_k^{(2)} + r_m^{-k-1}D_k^{(2)} - r_m^{k-1}C_k^{(1)} = 0. \tag{18}$$

Furthermore, from the space harmonic relations defined in Table 4, the tangential field intensity continuity at $r = r_m$ and $r = r_{sl}$ can be expressed as

- For $k = 1$:

$$\begin{aligned}\left(\hat{\mu}_{r,0} - \frac{1}{2}\hat{\mu}_{r,1}\right) C_1^{(3)} - \left(\hat{\mu}_{r,0} - \frac{1}{2}\hat{\mu}_{r,1}\right) r_{sl}^{-2} D_1^{(3)} - C_1^{(2)} + r_{sl}^{-2} D_1^{(2)} \\ + \frac{3}{2}\hat{\mu}_{r,1} r_{sl}^{2} C_3^{(3)} - \frac{3}{2}\hat{\mu}_{r,1} r_{sl}^{-4} D_3^{(3)},\end{aligned} \tag{19}$$

$$\begin{aligned}C_1^{(2)} - r_m^{-2} D_1^{(2)} - \left(\hat{\mu}_{r,0} - \frac{1}{2}\hat{\mu}_{r,1}\right) C_1^{(1)} - \frac{3}{2}\hat{\mu}_{r,1} r_m^{2} C_3^{(1)} \\ = -\left(\hat{\mu}_{r,0} - \frac{1}{2}\hat{\mu}_{r,1}\right) \hat{B}_{rem}.\end{aligned} \tag{20}$$

- For $k = 3$:

$$\begin{aligned}\frac{1}{6}\hat{\mu}_{r,1} C_1^{(3)} - \frac{1}{6}\hat{\mu}_{r,1} r_{sl}^{-2} D_1^{(3)} + \hat{\mu}_{r,0} r_{sl}^{2} C_3^{(3)} - \hat{\mu}_{r,0} r_{sl}^{-4} D_3^{(3)} \\ - r_{sl}^{2} C_3^{(2)} + r_{sl}^{-4} D_3^{(2)} + \frac{5}{6}\hat{\mu}_{r,1} r_{sl}^{4} C_5^{(3)} - \frac{5}{6}\hat{\mu}_{r,1} r_{sl}^{-6} D_5^{(3)} = 0,\end{aligned} \tag{21}$$

$$\begin{aligned}-\frac{1}{6}\hat{\mu}_{r,1} C_1^{(1)} + r_m^{2} C_3^{(2)} - r_m^{-4} D_3^{(2)} - \hat{\mu}_{r,0} r_m^{2} C_3^{(1)} - \frac{5}{6}\hat{\mu}_{r,1} r_m^{4} C_5^{(1)} \\ = \frac{1}{6}\hat{\mu}_{r,1} \hat{B}_{rem}.\end{aligned} \tag{22}$$

- For $k \geq 5$:

$$\begin{aligned}\frac{k-2}{2k}\hat{\mu}_{r,1} r_{sl}^{k-3} C_{k-2}^{(3)} - \frac{k-2}{2k}\hat{\mu}_{r,1} r_{sl}^{-k+1} D_{k-2}^{(3)} \\ + \hat{\mu}_{r,0} r_{sl}^{k-1} C_k^{(3)} - \hat{\mu}_{r,0} r_{sl}^{-k-1} D_k^{(3)} - r_{sl}^{k-1} C_k^{(2)} + r_{sl}^{-k-1} C_k^{(2)} \\ + \frac{k+2}{2k}\hat{\mu}_{r,1} r_{sl}^{k+1} C_{k+2}^{(3)} - \frac{k+2}{2k}\hat{\mu}_{r,1} r_{sl}^{-k-3} D_{k+2}^{(3)} = 0,\end{aligned} \tag{23}$$

$$\begin{aligned}-\frac{k-2}{2k}\hat{\mu}_{r,1} r_m^{k-3} C_{k-2}^{(1)} + r_m^{k-1} C_k^{(2)} - r_m^{-k-1} D_k^{(2)} - \hat{\mu}_{r,0} r_m^{k-1} C_k^{(1)} - \\ \frac{k+2}{2k}\hat{\mu}_{r,1} r_m^{k+1} C_{k+2}^{(1)} = 0.\end{aligned} \tag{24}$$

The proposed formulation requires different space harmonics orders to be solved integrally, which differs from the usual procedure. In addition, the analytical method was validated based on results from numerical simulations. The FEM analysis was set with the same premises used in the analytical model. For the space-varying relative permeability in the RS, the region was discretized. To obtain the needed accuracy, 80 regions were defined to characterize this space-varying parameter in the RS, as illustrated in Figure 8.

The simulations were carried out with several values for the RS relative permeability, considering both homogeneous and space-varying parameter. Table 5 summarizes the obtained result where the RS was modeled based on the evaluated PM machine. The results show the radial component of the magnetic flux density in the air-gap. Moreover, to illustrate the analytical model application with different permeability values, the model validation was done with similar parameters but with different magnitude orders, as illustrated in Figure 9 with the results presented in Tables 6 and 7.

Table 5. Simulation result for $\hat{B}^{(3)}_{r,k}$ (T): relative permeability based on measurements.

Harmonic	Relative Permeability Model					
	$\mu_r = 300$		$\mu_r = 151 + 150\cos(2\varphi_r)$			
	FEM	Analytical	FEM	Analytical $k_{max} = 3$	Analytical $k_{max} = 7$	Analytical $k_{max} = 13$
$k = 1$	0.584	0.585	0.957	0.932	0.948	0.948
$k = 3$	0	0	−0.160	−0.118	−0.151	−0.156
$k = 5$	0	0	0.060	–	0.050	0.058
$k = 7$	0	0	−0.029	–	−0.017	−0.028

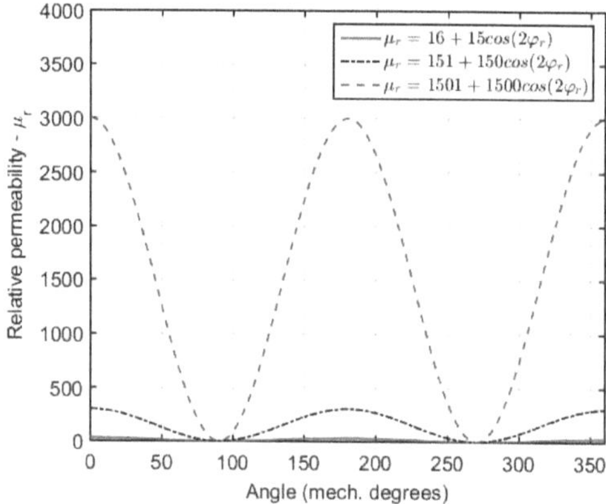

Figure 9. Different relative permeability profile for method validation.

Table 6. Simulation result for $\hat{B}^{(3)}_{r,k}$ (T): relative permeability based on measurements.

Harmonic	Relative Permeability Model					
	$\mu_r = 3000$		$\mu_r = 1501 + 1500\cos(2\varphi_r)$			
	FEM	Analytical	FEM	Analytical $k_{max} = 3$	Analytical $k_{max} = 15$	Analytical $k_{max} = 31$
$k = 1$	0.117	0.117	0.843	0.518	0.759	0.803
$k = 3$	0	0	−0.245	−0.084	−0.204	−0.227
$k = 5$	0	0	0.134	–	0.101	0.120
$k = 7$	0	0	−0.088	–	−0.059	−0.076

Table 7. Simulation result for $\hat{B}^{(3)}_{r,k}$ (T): relative permeability based on measurements.

Harmonic	Relative Permeability Model					
	$\mu_r = 30$		$\mu_r = 16 + 15\cos(2\varphi_r)$			
	FEM	Analytical	FEM	Analytical $k_{max} = 3$	Analytical $k_{max} = 7$	Analytical $k_{max} = 13$
$k = 1$	0.972	0.972	1.030	1.019	1.015	1.014
$k = 3$	0	0	−0.036	−0.037	−0.034	−0.033
$k = 5$	0	0	−0.001	–	0.002	0.000
$k = 7$	0	0	0.004	–	0.000	0.002

The results show a very good agreement with those obtained from the FEM simulations. As expected from the derived equations, the results show great interaction among space harmonics of different orders. Tables 5–7 present the results with different harmonic spectrum considered for the analytical method, ranging from $k = 1$ to k_{max}. In addition, the results are illustrated in Figure 10, which depicts the magnetic flux density waveforms from both methods in different regions (viz., the average radius of the air-gap, RS and PM). In these figures, results are for a maximum number of harmonics equal to $k_{max} = 100$.

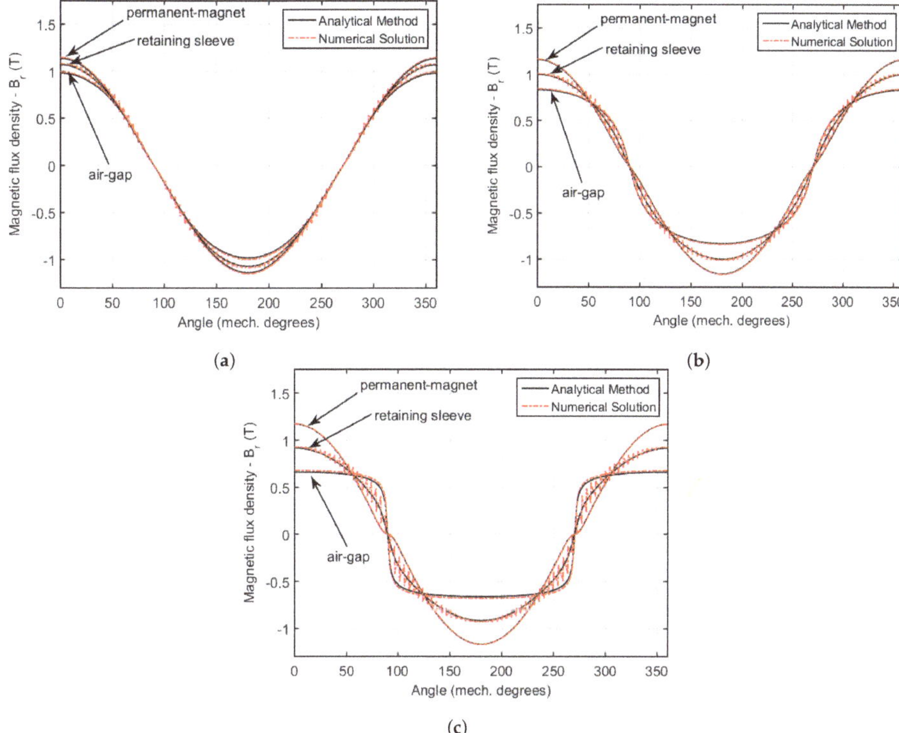

Figure 10. Analytical and numarical simulation comparison at different regions: (**a**) $\mu_r = 16 + 15\cos(2\varphi_r)$; (**b**) $\mu_r = 151 + 150\cos(2\varphi_r)$;and (**c**) $\mu_r = 1501 + 1500\cos(2\varphi_r)$.

As can be appreciated from the results in Tables 5–7, the analyital method accuracy is dependent on the maximum order of space harmonics, i.e., k_{max}. For a higher relative permeability case presented in Table 6 and in Figure 10c, the analytical method converged (with an error smaller than 5%) only for space harmonic content evaluated up to the thirtieth order. In addition, the relative permeability evaluated with only one harmonic content leads to a air-gap magnetic flux density with significant spacial harmonic content.

Furthermore, if only the fundamental component would exist in a system, with $\mu_{r,1} = 0$ in the evaluated PM machine, no harmonic content would be observed using either techniques, FEM or analytical method. This result is expected since the PM topology with parallel magnetization would result in only fundamental magnetic fields. In addition, as expected from the no-load voltage measurements in the prototype in comparison with numerical results, higher relative permeability in the RS region increases the pole-to-pole leakage flux significantly. This behavior would diminish air-gap magnetic fields and stator linkage flux if no saturation effect took place in the RS. Therefore, the space-varying relative permeability helps to maintain the magnetization level.

5. Conclusions

In this work, a new analytical model is presented that incorporates non-homogeneous magnetic parameters. The novel procedure evaluates an annular region defined with a space-varying relative permeability, where iterative methods could be used to study saturation phenomenon. This new solution is needed because experimental analysis pointed out a higher order space harmonics spectrum for the calculated magnetic field. In the assessed PM machine, linear models provide only fundamental components of the magnetic fields. Such assumption would lead to erroneous conclusions.

The novel analytical method was validated with results obtained from finite-element analysis. Seeking a clearer and direct way to analyze the derived results, the PM machine model and parameters not related to the proposed solution were simplified. In this case, linear system equations were derived illustrating the proposed methodology. Comparison with numerical solutions for magnetic flux density at different PM machine regions showed very good agreement for both amplitude and phase.

The proposed technique is derived from well-established analytical solution methodology, where other researched improvements can be extended with it. For example, evaluation of slotting effect can be easily carried out with a subdomain model. This will help suppress known analytical method drawbacks, such as the need to define linear material properties to solve the problem. Thus, improved accuracy in magnetic field calculation, aligned with fast computation time, can extend analytical method application capabilities throughout PM machine and drive design procedures.

Author Contributions: Supervision, B.J.C.F.; Writing—original draft, G.A.M.; Writing—review & editing, T.A.C.M.

Acknowledgments: This study was financed in part by the Coordenação de Aperfeiçoamento de Pessoal de Nível Superior - Brasil (CAPES) - Finance Code 001, the Conselho Nacional de Desenvolvimento Científico e Tecnológico (CNPq) and the Universidade Federal de Minas Gerais (UFMG).

Conflicts of Interest: The authors declare no conflict of interest.

References

1. Luise, F.; Tessarolo, A.; Pieri, S.; Raffin, P.; Chiara, M.D.; Agnolet, F.; Scalabrin, M. Design and technology solutions for high-efficiency high-speed motors. In Proceedings of the 2012 XXth International Conference on Electrical Machines, Marseille, France, 2–5 September 2012; pp. 157–163.
2. Borisavljević, A. Limits, Modeling and Design of High-Speed Permanent Magnet Machines. Ph.D. Thesis, Delft University of Technology, Delft, The Netherlands, 2011.
3. Holm, S.R. Modelling and Optimization of a Permanent Magnet Machine in a Flywheel. Ph.D. Thesis, Delft University of Technology, Delft, The Netherlands, 2003.
4. Roubache, L.; Boughrara, K.; Dubas, F.; Ibtiouen, R. New Subdomain Technique for Electromagnetic Performances Calculation in Radial-Flux Electrical Machines Considering Finite Soft-Magnetic Material Permeability. *IEEE Trans. Magn.* **2018**, *54*, 1–15. [CrossRef]
5. Qiu, H.; Duan, Q.; Yao, L.; Dong, Y.; Yi, R.; Cui, G.; Li, W. Analytical analysis of sleeve permeability for output performance of high speed permanent magnet generators driven by micro gas turbines. *Appl. Math. Model.* **2016**, *40*, 9017–9028. [CrossRef]
6. Polinder, H. On the Losses in a High-Speed Permanent-Magnet Generator with Rectifier. Ph.D. Thesis, Delft University of Technology, Delft, The Netherlands, 1998.
7. Boubaker, N.; Matt, D.; Enrici, P.; Nierlich, F.; Durand, G.; Orlandini, F.; Longère, X.; Aïgba, J. Study of eddy-current loss in the sleeves and Sm-Co magnets of a high-performance SMPM synchronous machine (10 kRPM, 60 kW). *Electr. Power Syst. Res.* **2017**, *142*, 20–28. [CrossRef]
8. Willerich, S.; Herzog, H.G. Prediction of the magnetic field in the air-gap of synchronous machines on a preliminary design level—Machine modelling and field calculation. In Proceedings of the 2015 IEEE International Electric Machines Drives Conference (IEMDC), Coeur d'Alene, ID, USA, 10–13 May 2015; pp. 1292–1298.
9. Holm, S.R.; Polinder, H.; Ferreira, J.A. Analytical Modeling of a Permanent-Magnet Synchronous Machine in a Flywheel. *IEEE Trans. Magn.* **2007**, *43*, 1955–1967. [CrossRef]

10. Zhu, Z.Q.; Howe, D.; Bolte, E.; Ackermann, B. Instantaneous magnetic field distribution in brushless permanent magnet DC motors. I. Open-circuit field. *IEEE Trans. Magn.* **1993**, *29*, 124–135. [CrossRef]
11. Ackermann, B.; Sottek, R. Analytical modeling of the cogging torque in permanent magnet motors. *Electr. Eng.* **1995**, *78*, 117–125. [CrossRef]
12. Boules, N. Prediction of No-Load Flux Density Distribution in Permanent Magnet Machines. *IEEE Trans. Ind. Appl.* **1985**, *IA-21*, 633–643. [CrossRef]
13. Ramakrishnan, K.; Curti, M.; Zarko, D.; Mastinu, G.; Paulides, J.J.H.; Lomonova, E.A. Comparative analysis of various methods for modelling surface permanent magnet machines. *IET Electr. Power Appl.* **2017**, *11*, 540–547. [CrossRef]
14. Hanic, A.; Zarko, D.; Kuhinek, D.; Hanic, Z. On-Load Analysis of Saturated Surface Permanent Magnet Machines Using Conformal Mapping and Magnetic Equivalent Circuits. *IEEE Trans. Energy Convers.* **2018**, *33*, 915–924. [CrossRef]
15. Shin, K.H.; Park, H.I.; Cho, H.W.; Choi, J.Y. Analytical prediction for electromagnetic performance of interior permanent magnet machines based on subdomain model. *AIP Adv.* **2017**, *7*, 056669. [CrossRef]
16. Tikellaline, A.; Boughrara, K.; Takorabet, N. Magnetic field analysis of double excited synchronous motor using numerical conformal mapping. In Proceedings of the 2017 5th International Conference on Electrical Engineering-Boumerdes (ICEE-B), Boumerdes, Algeria, 29–31 October 2017; pp. 1–6.
17. Ramakrishnan, K.; Zarko, D.; Hanic, A.; Mastinu, G. Improved method for field analysis of surface permanent magnet machines using Schwarz-Christoffel transformation. *IET Electr. Power Appl.* **2017**, *11*, 1067–1075. [CrossRef]
18. Hannon, B.; Sergeant, P.; Dupré, L. Study of the Effect of a Shielding Cylinder on the Torque in a Permanent-Magnet Synchronous Machine Considering Two Torque-Producing Mechanisms. *IEEE Trans. Magn.* **2017**, *53*, 1–8. [CrossRef]
19. Djelloul-Khedda, Z.; Boughrara, K.; Dubas, F.; Ibtiouen, R. Nonlinear Analytical Prediction of Magnetic Field and Electromagnetic Performances in Switched Reluctance Machines. *IEEE Trans. Magn.* **2017**, *53*, 1–11. [CrossRef]
20. Djelloul-Khedda, Z.; Boughrara, K.; Dubas, F.; Kechroud, A.; Souleyman, B. Semi-Analytical Magnetic Field Predicting in Many Structures of Permanent-Magnet Synchronous Machines Considering the Iron Permeability. *IEEE Trans. Magn.* **2018**, *54*, 1–21. [CrossRef]
21. Ben Yahia, M.; Boughrara, K.; Dubas, F.; Roubache, L.; Ibtiouen, R. Two-Dimensional Exact Subdomain Technique of Switched Reluctance Machines with Sinusoidal Current Excitation. *Math. Comput. Appl.* **2018**, *23*, 59.
22. Boughrara, K.; Dubas, F.; Ibtiouen, R. 2-D Exact Analytical Method for Steady-State Heat Transfer Prediction in Rotating Electrical Machines. *IEEE Trans. Magn.* **2018**, *54*, 1–19. [CrossRef]
23. Dubas, F.; Boughrara, K. New Scientific Contribution on the 2-D Subdomain Technique in Cartesian Coordinates: Taking into Account of Iron Parts. *Math. Comput. Appl.* **2017**, *22*, 17. [CrossRef]
24. Dubas, F.; Boughrara, K. New Scientific Contribution on the 2-D Subdomain Technique in Polar Coordinates: Taking into Account of Iron Parts. *Math. Comput. Appl.* **2017**, *22*, 42. [CrossRef]
25. Tiegna, H.; Amara, Y.; Barakat, G. Overview of analytical models of permanent magnet electrical machines for analysis and design purposes. *Math. Comput. Simul.* **2013**, *90*, 162–177. [CrossRef]
26. Abbaszadeh, K.; Alam, F.R. On-Load Field Component Separation in Surface-Mounted Permanent-Magnet Motors Using an Improved Conformal Mapping Method. *IEEE Trans. Magn.* **2016**, *52*, 1–12. [CrossRef]
27. Alam, F.R.; Abbaszadeh, K. Magnetic Field Analysis in Eccentric Surface-Mounted Permanent-Magnet Motors Using an Improved Conformal Mapping Method. *IEEE Trans. Energy Convers.* **2016**, *31*, 333–344. [CrossRef]
28. Oner, Y.; Zhu, Z.Q.; Wu, L.J.; Ge, X.; Zhan, H.; Chen, J.T. Analytical On-Load Subdomain Field Model of Permanent-Magnet Vernier Machines. *IEEE Trans. Ind. Electr.* **2016**, *63*, 4105–4117. [CrossRef]
29. Qian, H.; Guo, H.; Wu, Z.; Ding, X. Analytical Solution for Cogging Torque in Surface-Mounted Permanent-Magnet Motors With Magnet Imperfections and Rotor Eccentricity. *IEEE Trans. Magn.* **2014**, *50*, 1–15. [CrossRef]
30. Chebak, A.; Viarouge, P.; Cros, J. Improved Analytical Model for Predicting the Magnetic Field Distribution in High-Speed Slotless Permanent-Magnet Machines. *IEEE Trans. Magn.* **2015**, *51*, 1–4. [CrossRef]
31. Rahideh, A.; Korakianitis, T. Analytical calculation of open-circuit magnetic field distribution of slotless brushless PM machines. *Int. J. Electr. Power Energy Syst.* **2013**, *44*, 99–114. [CrossRef]

32. Ortega, A.J.P.; Paul, S.; Islam, R.; Xu, L. Analytical Model for Predicting Effects of Manufacturing Variations on Cogging Torque in Surface-Mounted Permanent Magnet Motors. *IEEE Trans. Ind. Appl.* **2016**, *52*, 3050–3061. [CrossRef]
33. Sprangers, R.L.J.; Paulides, J.J.H.; Gysen, B.L.J.; Lomonova, E.A. Magnetic Saturation in Semi-Analytical Harmonic Modeling for Electric Machine Analysis. *IEEE Trans. Magn.* **2016**, *52*, 1–10. [CrossRef]
34. Hannon, B.; Sergeant, P.; Dupré, L. Two-dimensional fourier-based modeling of electric machines. In Proceedings of the 2017 IEEE International Electric Machines and Drives Conference (IEMDC), Miami, FL, USA, 21–24 May 2017; pp. 1–8.
35. Pfister, P.; Yin, X.; Fang, Y. Slotted Permanent-Magnet Machines: General Analytical Model of Magnetic Fields, Torque, Eddy Currents, and Permanent-Magnet Power Losses Including the Diffusion Effect. *IEEE Trans. Magn.* **2016**, *52*, 1–13. [CrossRef]
36. Dubas, F.; Espanet, C. Analytical Solution of the Magnetic Field in Permanent-Magnet Motors Taking Into Account Slotting Effect: No-Load Vector Potential and Flux Density Calculation. *IEEE Trans. Magn.* **2009**, *45*, 2097–2109. [CrossRef]
37. Maia, T.A.; Faria, O.A.; Barros, J.E.; Porto, M.P.; Filho, B.J.C. Test and simulation of an electric generator driven by a micro-turbine. *Electr. Power Syst. Res.* **2017**, *147*, 224–232. [CrossRef]
38. Maia, T.A.; Barros, J.E.; Filho, B.J.C.; Porto, M.P. Experimental performance of a low cost micro-CAES generation system. *Appl. Energy* **2016**, *182*, 358–364. [CrossRef]
39. Meeker, D. *Finite Element Method Magnetics v4.2*; 2015. Available online: www.femm.info/wiki/HomePage (accessed on 9 November 2018).
40. Fofanov, D.; Riedner, S. Magnetic Properties of Stainless Steels: Applications, Opportunities and New Developments. In Proceedings of the 2011 World Stainless Steel Conference & Expo, Maastricht, The Netherlands, 29 November–1 December 2011; pp. 1–13.
41. Zarko, D.; Ban, D.; Lipo, T.A. Analytical Solution for Cogging Torque in Surface Permanent-Magnet Motors Using Conformal Mapping. *IEEE Trans. Magn.* **2008**, *44*, 52–65. [CrossRef]
42. Zarko, D.; Ban, D.; Lipo, T.A. Analytical calculation of magnetic field distribution in the slotted air gap of a surface permanent-magnet motor using complex relative air-gap permeance. *IEEE Trans. Magn.* **2006**, *42*, 1828–1837. [CrossRef]

Article

Memory Efficient Method for Electromagnetic Multi-Region Models Using Scattering Matrices

C. H. H. M. Custers *, J. W. Jansen and E. A. Lomonova

Department of Electrical Engineering, Electromechanics and Power Electronics, Eindhoven University of Technology, 5612 AZ Eindhoven, The Netherlands; j.w.jansen@tue.nl (J.W.J.); e.lomonova@tue.nl (E.A.L.)
* Correspondence: c.h.h.m.custers@tue.nl

Received: 25 October 2018; Accepted: 7 November 2018; Published: 9 November 2018

Abstract: This paper describes the scattering matrix approach to obtain the solution to electromagnetic field quantities in harmonic multi-layer models. Using this approach, the boundary conditions are solved in such way that the maximum size of any matrix used during the computations is independent of the number of regions defined in the problem. As a result, the method is more memory efficient than classical methods used to solve the boundary conditions. Because electromagnetic sources can be located inside the regions of a configuration, the scattering matrix formulation is developed to incorporate these sources into the solving process. The method is applied to a 3D electromagnetic configuration for verification.

Keywords: scattering matrix; Fourier analysis; permanent magnet machines; analytical modeling

1. Introduction

The design and optimization of electrical machines and other electromagnetic applications necessitate accurate models of electromagnetic fields to precisely predict their performance. As these systems and their structures get more complex, it is becoming more difficult to accurately model, e.g., eddy current effects [1,2] or hysteresis effects [3] in both the spatial and time domain. To model electromagnetic field distributions and the related forces and power losses, the finite element method (FEM) is often used because of its ability to produce accurate results when it is correctly utilized. However, the method can be demanding in terms of memory and relatively slow in terms of computation time. Therefore, semi-analytical models have been proposed over the years for increasingly complex structures in both 2D and 3D. One of the semi-analytical models is the harmonic modeling technique [4–7], which uses a Fourier basis to describe the solutions to the electromagnetic field quantities. By including position dependent material properties, such as the electrical conductivity [8] and the permeability [9–11], into the solutions, the spatial distribution of electromagnetic fields can also be accurately modeled for complex devices.

In many electromagnetic configurations, accurate results are obtained using a relatively low number of harmonics. However, for more complex structures, with detailed features compared to the device size, the number of harmonics has to be increased to retain accuracy. Computationally, this leads to an increase in the required memory. In addition, as an electromagnetic configuration has to be divided into regions (or layers), the number of regions influences the required memory. As a result, the number of harmonics that can be considered is limited when a relatively large number of regions has to be incorporated. Consequently, the advantage in terms of memory of the harmonic model in comparison to FEM is reducing. The scattering matrix approach [12–16] reformulates a multiple region electromagnetic problem in such a way that it can be expressed as a single region with in- and out-going fields. Consequently, the dependency of the computational memory on the number of regions is removed. All matrices that are manipulated with this method have a maximum size of $L \times L$, where L is the total number of harmonics. In [12], where the scattering matrices are used to solve a high-frequent electromagnetic problem, the source is an incident wave on one side of the problem. However, to apply the scattering formulation in low-frequent problems in which sources (currents or magnets) are present in regions, the scattering formulation has to be extended to incorporate the related source terms.

In this paper, the scattering matrix approach for harmonic models of electromagnetic configurations is described. Using this approach, the memory required to obtain the solutions of the model is significantly reduced. The method presented in [12] is extended to take electromagnetic sources inside regions into account. The method is verified by application to a 3D electromagnetic configuration presented in [10], where the permeability varies in both periodic directions as a function of position.

2. Model Assumptions

The mathematical models of electromagnetic configurations that can be solved with the method described in this paper are subject to a number of assumptions. The considered models are set up in the Cartesian coordinate system, where, for 2D configurations in one direction, periodicity is assumed. For 3D configurations, two directions are assumed periodic. Because of the periodicity, the solution in the periodic directions is expressed by Fourier series. Based upon the electromagnetic material properties, such as the permeability, permittivity and conductivity, and the electromagnetic field sources, the model is divided into regions in the non-periodic direction. In the non-periodic direction, the properties of a region may not vary, whereas, in the periodic direction, they are allowed to change as a function of position. Lastly, any configuration needs to have an air region at both sides, extending to plus and minus infinity in the non-periodic direction, for the applied method.

3. Scattering Matrix Formulation

For all problems at hand, the Fourier coefficients of describing the electromagnetic fields have the same form. For the remainder of the paper, the non-periodic direction is the z-direction. The z-dependent solution describing the coefficients $\mathbf{u}$ and $\mathbf{v}$ of particular field components of a region i is given by

$$\mathbf{u}_i = \mathbf{Q}_{\mathbf{u}i} \left(e^{\boldsymbol{\lambda}_i z} \mathbf{c}_i^+ + e^{-\boldsymbol{\lambda}_i z} \mathbf{c}_i^- \right) + \mathbf{p}_{u,i}, \tag{1}$$

$$\mathbf{v}_i = \mathbf{Q}_{\mathbf{v}i} \left(e^{\boldsymbol{\lambda}_i z} \mathbf{c}_i^+ - e^{-\boldsymbol{\lambda}_i z} \mathbf{c}_i^- \right) + \mathbf{p}_{v,i}, \tag{2}$$

where $\boldsymbol{\lambda}$ is a vector with eigenvalues and $\mathbf{Q}_\mathbf{u}$ and $\mathbf{Q}_\mathbf{v}$ are eigenvector matrices. Combined, $\boldsymbol{\lambda}$ and $\mathbf{Q}$ describe the propagation of field components inside a region. The particular solutions described by $\mathbf{p}_u$ and $\mathbf{p}_v$ relate source terms, such as coil currents and magnetization, which are known in advance, to the solutions $\mathbf{u}$ and $\mathbf{v}$. All these variables are determined by the properties of a region that is under consideration. The way to obtain these variables for a variety of problems has been described in e.g., [8–10]. The unknown coefficients $\mathbf{c}_i^+$ and $\mathbf{c}_i^-$ in (1) and (2) are determined by applying continuity boundary conditions between regions.

The continuity boundary conditions force both $\mathbf{u}$ and $\mathbf{v}$ to be continuous on an interface between adjacent regions. Often in harmonic models used for electrical machines, all boundary conditions are gathered in a large matrix which forms a system of linear equations,

$$\underbrace{\begin{bmatrix} \mathbf{Q}_{u1}e^{\lambda_1 h_1} & \mathbf{Q}_{u1}e^{-\lambda_1 h_1} & -\mathbf{Q}_{u2}e^{\lambda_2 h_1} & -\mathbf{Q}_{u2}e^{-\lambda_2 h_1} & 0 & \cdots & \cdots & \cdots \\ \mathbf{Q}_{v1}e^{\lambda_1 h_1} & -\mathbf{Q}_{v1}e^{-\lambda_1 h_1} & -\mathbf{Q}_{v2}e^{\lambda_2 h_1} & \mathbf{Q}_{v2}e^{-\lambda_2 h_1} & 0 & \cdots & \cdots & \cdots \\ 0 & 0 & \mathbf{Q}_{u2}e^{\lambda_2 h_2} & \mathbf{Q}_{u2}e^{-\lambda_2 h_2} & \mathbf{Q}_{u3}e^{\lambda_3 h_2} & \cdots & \cdots & \cdots \\ \vdots & \vdots & \vdots & \vdots & \vdots & \ddots & \vdots & \vdots \\ 0 & 0 & 0 & 0 & 0 & \cdots & -\mathbf{Q}_{ui}e^{-\lambda_i h_{i-1}} & -\mathbf{Q}_{ui}e^{\lambda_i h_{i-1}} \\ 0 & 0 & 0 & 0 & 0 & \cdots & -\mathbf{Q}_{vi}e^{-\lambda_i h_{i-1}} & \mathbf{Q}_{vi}e^{\lambda_i h_{i-1}} \end{bmatrix}}_{\mathbf{E}} \begin{bmatrix} \mathbf{c}_1^+ \\ \mathbf{c}_1^- \\ \mathbf{c}_2^+ \\ \mathbf{c}_2^- \\ \mathbf{c}_3^+ \\ \vdots \\ \mathbf{c}_i^+ \\ \mathbf{c}_i^- \end{bmatrix} = \begin{bmatrix} \mathbf{P}_{u,2} - \mathbf{P}_{u,1} \\ \mathbf{P}_{v,2} - \mathbf{P}_{v,1} \\ \mathbf{P}_{u,3} - \mathbf{P}_{u,2} \\ \vdots \\ \mathbf{P}_{u,i} - \mathbf{P}_{u,i-1} \\ \mathbf{P}_{v,i} - \mathbf{P}_{v,i-1} \end{bmatrix}, \quad (3)$$

where h specifies the height of a boundary in the z-direction. By solving the system of equations, all unknowns can be obtained. A drawback of obtaining the unknowns in this way, is that the size of the matrix **E** depends on the number of harmonics that is considered for the problem and, on the number of regions that build up the model. Hence, with a certain amount of available computational memory, the number of harmonics has to be compromised when regions are added to the problem.

To develop a more memory efficient manner of solving, the scattering matrix approach presented in [12] is applied. With this solving method, the unknowns of all regions between the top and bottom air region of the problem are eliminated. For the considered model formulations, the solution in the non-periodical direction is described by two waves, where one wave is traveling in the positive and the other in the negative z-direction, as presented in (1) and (2). The scattering matrix, visually represented in Figure 1, couples the incoming waves of a region to the outgoing waves of that region. The sub-matrices $\mathbf{S}_{11}$ and $\mathbf{S}_{22}$ represent the transfer, in negative and positive z-direction respectively, from incoming to outgoing waves and $\mathbf{S}_{12}$ and $\mathbf{S}_{21}$ represent the reflection at each side of the region. The method described in [12] is adapted, since sources can be located inside a region. The source terms in scattering vector form are represented by a vector **t** which describes the contribution of the sources to the outgoing waves of the region. By combining all scattering matrices of the regions in the model, a scattering matrix for electromagnetic configuration under consideration can be obtained.

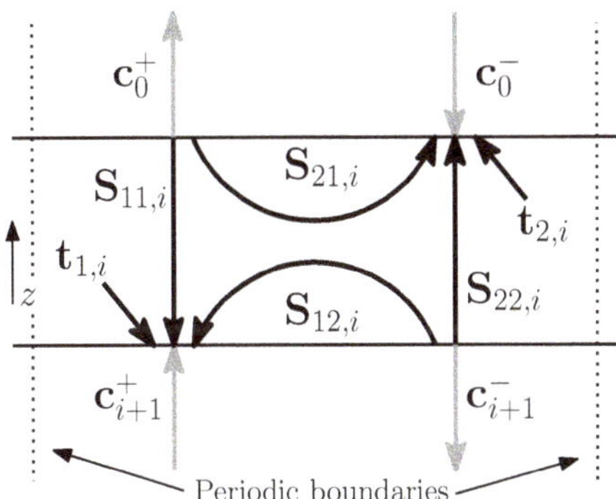

Figure 1. Visual representation of the scattering formulation for a single region.

To derive the sub-matrices of the scattering matrix, the continuity boundary conditions between a region with the index i and the index $i+1$ are written in matrix form

$$\begin{aligned}\begin{bmatrix}\mathbf{c}_i^+\\ \mathbf{c}_i^-\end{bmatrix} = \begin{bmatrix}\mathbf{E}_i\mathbf{A}_i & \mathbf{E}_i\mathbf{B}_i\mathbf{E}_{i+1}^{-1}\\ \mathbf{B}_i & \mathbf{A}_i\mathbf{E}_{i+1}^{-1}\end{bmatrix}\cdot\begin{bmatrix}\mathbf{c}_{i+1}^+\\ \mathbf{c}_{i+1}^-\end{bmatrix}\\ +\begin{bmatrix}\mathbf{E}_i & \mathbf{0}\\ \mathbf{0} & \mathbf{I}\end{bmatrix}\cdot\begin{bmatrix}\mathbf{d}_{1,i}\\ \mathbf{d}_{2,i}\end{bmatrix},\end{aligned} \tag{4}$$

where $\mathbf{I}$ is the identity matrix and

$$\mathbf{A}_i = \frac{1}{2}\left(\mathbf{Q_u}_i^{-1}\,\mathbf{Q_u}_{i+1} + \mathbf{Q_v}_i^{-1}\,\mathbf{Q_v}_{i+1}\right), \tag{5}$$

$$\mathbf{B}_i = \frac{1}{2}\left(\mathbf{Q_u}_i^{-1}\,\mathbf{Q_u}_{i+1} - \mathbf{Q_v}_i^{-1}\,\mathbf{Q_v}_{i+1}\right), \tag{6}$$

$$\mathbf{E}_i = e^{\lambda_i h_i}, \tag{7}$$

$$\mathbf{d}_{1,i} = \frac{1}{2}\begin{bmatrix}\mathbf{Q_u}_i^{-1} & \mathbf{Q_v}_i^{-1}\end{bmatrix}\cdot\begin{bmatrix}\mathbf{p}_{u,i+1}-\mathbf{p}_{u,i}\\ \mathbf{p}_{v,i+1}-\mathbf{p}_{v,i}\end{bmatrix}, \tag{8}$$

$$\mathbf{d}_{2,i} = \frac{1}{2}\begin{bmatrix}\mathbf{Q_u}_i^{-1} & -\mathbf{Q_v}_i^{-1}\end{bmatrix}\cdot\begin{bmatrix}\mathbf{p}_{u,i+1}-\mathbf{p}_{u,i}\\ \mathbf{p}_{v,i+1}-\mathbf{p}_{v,i}\end{bmatrix}, \tag{9}$$

where h_i is the height of region i in the z-direction. The scattering relation, denoted with index $i-1$, between the top region with index 0 and an arbitrary region i is given by

$$\begin{bmatrix}\mathbf{c}_i^+\\ \mathbf{c}_0^-\end{bmatrix} = \begin{bmatrix}\mathbf{S}_{11,i-1} & \mathbf{S}_{12,i-1}\\ \mathbf{S}_{21,i-1} & \mathbf{S}_{22,i-1}\end{bmatrix}\cdot\begin{bmatrix}\mathbf{c}_0^+\\ \mathbf{c}_i^-\end{bmatrix} + \begin{bmatrix}\mathbf{t}_{1,i-1}\\ \mathbf{t}_{2,i-1}\end{bmatrix}. \tag{10}$$

The scattering matrices, denoted by $\mathbf{S}_i$ and $\mathbf{t}_i$, for the scattering relation between region 0 and $i+1$ can now be defined, thereby removing the unknowns of region i from the problem as shown in Figure 1.

This relation is given by

$$\begin{bmatrix}\mathbf{c}_{i+1}^+\\ \mathbf{c}_0^-\end{bmatrix} = \begin{bmatrix}\mathbf{S}_{11,i} & \mathbf{S}_{12,i}\\ \mathbf{S}_{21,i} & \mathbf{S}_{22,i}\end{bmatrix}\cdot\begin{bmatrix}\mathbf{c}_0^+\\ \mathbf{c}_{i+1}^-\end{bmatrix} + \begin{bmatrix}\mathbf{t}_{1,i}\\ \mathbf{t}_{2,i}\end{bmatrix}, \tag{11}$$

where the sub-matrices of $\mathbf{S}_i$ and vectors $\mathbf{t}_{1,i}$ and $\mathbf{t}_{2,i}$ can be computed by

$$\begin{aligned}\mathbf{S}_{11,i} &= \left(\mathbf{A}_i\mathbf{E}_i^{-1}\mathbf{S}_{12,i-1}\mathbf{B}_i\right)^{-1}\left(\mathbf{E}_i^{-1}\mathbf{S}_{11,i-1}\right),\\ \mathbf{S}_{12,i} &= \left(\mathbf{A}_i\mathbf{E}_i^{-1}\mathbf{S}_{12,i-1}\mathbf{B}_i\right)^{-1}\\ &\quad\left(\mathbf{E}_i^{-1}\mathbf{S}_{12,i-1}\mathbf{A}_i\mathbf{E}_{i+1}^{-1} - \mathbf{B}_i\mathbf{E}_{i+1}^{-1}\right),\\ \mathbf{S}_{21,i} &= \mathbf{S}_{22,i-1}\mathbf{B}_i\mathbf{S}_{11,i} + \mathbf{S}_{21,i-1},\\ \mathbf{S}_{22,i} &= \mathbf{S}_{22,i-1}\mathbf{B}_i\mathbf{S}_{12,i} + \mathbf{S}_{22,i-1}\mathbf{A}_i\mathbf{E}_{i+1}^{-1},\end{aligned} \tag{12}$$

and

$$\begin{aligned}\mathbf{t}_{1,i} &= \left(\mathbf{A}_i\mathbf{E}_i^{-1}\mathbf{S}_{12,i-1}\mathbf{B}_i\right)^{-1}\\ &\quad\left(\mathbf{E}_i^{-1}\mathbf{S}_{12,i-1}\mathbf{d}_{2,i} + \mathbf{E}_i^{-1}\mathbf{t}_{1,i-1} - \mathbf{d}_{1,i}\right),\\ \mathbf{t}_{2,i} &= \mathbf{S}_{22,i-1}\mathbf{B}_i\mathbf{t}_{1,i} + \mathbf{S}_{22,i-1}\mathbf{d}_{2,i} + \mathbf{t}_{2,i-1}.\end{aligned} \tag{13}$$

In these equations, stability is ensured by the formulation where only decaying exponential functions are used. Starting from an initial scattering matrix $\mathbf{S}_0$, which is an identity matrix, and vector $\mathbf{t}_0$, which contains zeros, the scattering formulation of all regions can be found by repeating (12) and (13). Now, let the air region on the bottom of the problem be denoted with index I, so a total of $I+1$ regions is assumed. The unknowns of the top and bottom region are now related through

$$\begin{bmatrix} \mathbf{c}_I^+ \\ \mathbf{c}_0^- \end{bmatrix} = \mathbf{S}_{I-1} \cdot \begin{bmatrix} \mathbf{c}_0^+ \\ \mathbf{c}_I^- \end{bmatrix} + \mathbf{t}_{I-1}. \tag{14}$$

Because all field components of the bottom region have to disappear when z goes to minus infinity, the vector of unknowns $\mathbf{c}_I^-$ has to equal zero. When applying the same analysis on the top region, it is obtained that also $\mathbf{c}_0^+$ has to equal zero. The non-zero unknowns of the top and bottom air region are then, using (14), calculated through

$$\begin{bmatrix} \mathbf{c}_I^+ \\ \mathbf{c}_0^- \end{bmatrix} = \mathbf{t}_{I-1}. \tag{15}$$

The unknowns of all other regions can be computed from the unknowns $\mathbf{c}_0^-$ and $\mathbf{c}_I^+$, by consequently applying (4).

4. Application to an Electromagnetic Configuration

The scattering approach is applied to the electromagnetic configuration shown in Figure 2, where a rectangular slab of magnetic material is placed above three permanent magnets. The configuration is divided into five regions in the z-direction, as depicted in Figure 2. The model is periodic in both the x- and y- direction and the permeability inside Regions 2 and 4 varies as a function of position. The configuration was analyzed in [10], where also the expressions for the magnetic field components have been obtained and all geometric properties have been given. The magnetic properties are given in Table 1. In [10], however, the unknowns where obtained by solving a linear system of equations as presented in (3).

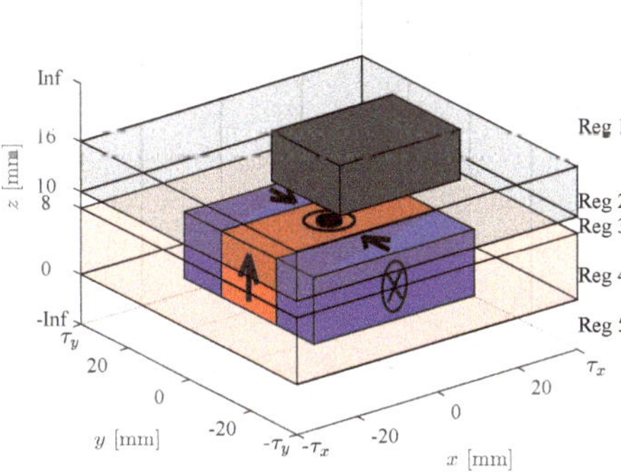

Figure 2. Electromagnetic configuration consisting of three magnets with a slab of magnetic material above [10].

Table 1. Model parameters.

Parameter	Value	Unit	Description
B_{rem}	1.3	T	Remanent flux density of the magnets
$\mu_{r_{mag}}$	1.2	-	Relative permeability of the magnets
μ_{r_p}	800	-	Relative permeability of the plate in Region 2

Because only the magneto-static fields have to be considered for the configuration, the magnetic scalar potential, ψ, is introduced to obtain the expressions for the magnetic field quantities. In the boundary conditions, the continuity of this scalar potential and the z-component of the magnetic flux density are forced. Therefore, $\mathbf{u} = \mathbf{b_z}$ and $\mathbf{v} = \psi$ in (1) and (2), respectively. The expressions for $\mathbf{b_z}$ and ψ are described in [10]. Note that, for a problem where also an electric field exists, **u** and **v** would consist of the tangential components of the $\vec{H}$- and $\vec{E}$-field, respectively, and their size would be twice that compared to the current magneto-static problem. Hence, the expressions for **u** and **v** and their size depend on the problem at hand.

The number of harmonics that is used to calculate the solution to the configuration of Figure 2 is denoted by L. The number of harmonics used in the x- and y-directions is denoted by N and M, respectively, and thus the total number of harmonics $L = NM$. With the solving method of (3), the matrix that has to be inverted has a size of $8L \times 8L$, while with the scattering approach the maximum size of any matrix that is inverted is equal to $L \times L$. This means that the matrix size is reduced by 87.5%. Moreover, with the classical solving method, adding a region to the problem would mean that the matrix **E** becomes larger by $2L$ in both directions. In the scattering approach, the maximum size of any matrix would be $L \times L$, independent of the number of regions that makes the method especially beneficiary when the number of regions in a model is relatively large. The solving method using (3) is performed in Matlab R2017a using the command 'A\b', which means that the method of inversion is automatically chosen. It should be noted that properties reducing the computational effort of the inversion of a matrix, such as zeros in the off-diagonal entries and matrix symmetry are limited in the boundary condition matrix of (3), in contrast to the matrices inverted in (5), (6), (8), (9), (12) and (13).

In Figure 3a, the magnetic field strength, computed with the scattering matrix approach with a total of 6561 harmonics, in the center of Region 2 is depicted. The difference with the result computed by the finite element method (FEM) is shown in Figure 3b. In the FEM analysis, a total of 2,384,628 second-order mesh elements is used. It has been verified that, with the applied mesh, the error of the FEM (compared to a FEM result obtained with a denser mesh) has converged to an error less than 1%. In [10], a total of 1089 harmonics was used to obtain the same result. The maximum relative error with respect to FEM with 6561 harmonics is equal to 9.2%, where, computed with 1089 harmonics, the error was equal to 17.4%. For problems where the electromagnetic properties of a region are varying with a high spatial frequencies, the addition of harmonics can be necessary to obtain accurate field results.

Using the hardware available (quadcore Intel Core i7-4790 with 32 GB RAM), the computation time for the 'classical' solving method for a total of 1089 harmonics is equal to 20.4 s. With the developed scattering matrix approach, the computation time, for the same number of harmonics, is equal to 12.6 s. Hence, the number of calculations is increased in the scattering matrix approach, however, because all manipulations are performed on matrices of size $L \times L$, the computation time is not significantly increased or even decreased compared to the classical solving method. On the same hardware, the FEM analysis takes around 125 s to mesh and solve the electromagnetic problem. The semi-analytical approach obtains the results for the electromagnetic configuration of Figure 2 faster than the FEM; however, the solving time of the semi-analytical approach depends on the number of harmonics and number of regions in the model. This means that, for models where the geometry contains high spatial frequencies (relatively small details with respect to the periodic width), and the number of harmonics needs to be large to accurately compute the fields, and the advantage over FEM regarding computation time will decrease.

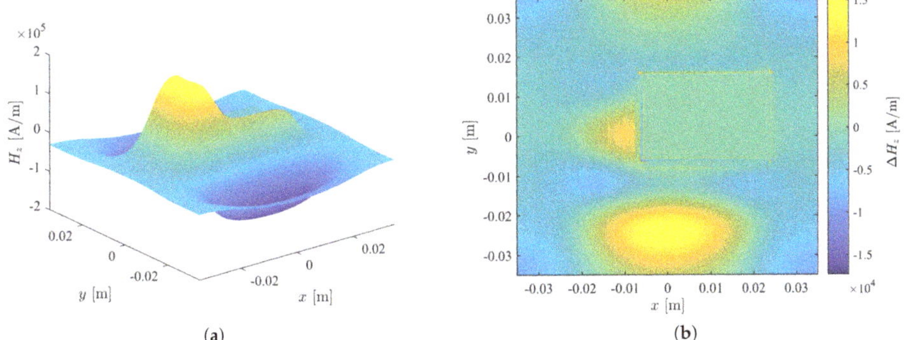

Figure 3. Magnetic field strength in the z-direction on an xy-plane in the center of the region with magnetic material (Region 2). (**a**) calculated H_z with the scattering matrix approach; (**b**) difference in calculated H_z between the scattering matrix approach and the finite element method.

5. Conclusions

The paper presents the scattering matrix approach to solve the boundary conditions of a multiple region electromagnetic problem, with Fourier based solutions. Compared to the classical solving method, where all boundary conditions are collected in a single matrix, the scattering approach is more memory efficient. The boundary conditions for each region are reformulated, so that all unknowns, except the ones from the top and bottom region, can be removed from the problem. As a result, the maximum size of any matrix that is used during calculations is determined by the number of harmonics and, more importantly, independent of the number of regions inside the problem. Therefore, the scattering approach is especially beneficial for problems with a relatively large number of regions. Sources located inside a certain region of the electromagnetic configuration are taken into account in the described scattering formulation. Furthermore, due to the general formulation used, the approach can be used for a variety of electromagnetic problems with continuity boundary conditions between regions.

Author Contributions: The theory presented in this paper has been developed by C.H.H.M.C. The analysis of the results has been performed in cooperation with J.W.J. and E.A.L. The paper was written by C.H.H.M.C. and contributions and improvements to the content have been made by J.W.J. and E.A.L.

Conflicts of Interest: The authors declare no conflict of interest.

References

1. Kawase, Y.; Yamaguchi, T.; Onogi, Y. Eddy current analysis of three-phase transformer using 3-D parallel finite element method. In Proceedings of the 2016 XXII International Conference on Electrical Machines (ICEM), Lausanne, Switzerland, 4–7 September 2016; pp. 2828–2832, doi:10.1109/ICELMACH.2016.7732923. [CrossRef]
2. Mirzaei, M.; Binder, A.; Funieru, B.; Susic, M. Analytical Calculations of Induced Eddy Currents Losses in the Magnets of Surface Mounted PM Machines With Consideration of Circumferential and Axial Segmentation Effects. *IEEE Trans. Magn.* **2012**, *48*, 4831–4841, doi:10.1109/TMAG.2012.2203607. [CrossRef]
3. Nasiri-Zarandi, R.; Mirsalim, M. Analysis and Torque Calculation of an Axial Flux Hysteresis Motor Based on Hyperbolic Model of Hysteresis Loop in Cartesian Coordinates. *IEEE Trans. Magn.* **2015**, *51*, 1–10, doi:10.1109/TMAG.2015.2396912. [CrossRef]
4. Lubin, T.; Mezani, S.; Rezzoug, A. 2-D Exact Analytical Model for Surface-Mounted Permanent-Magnet Motors With Semi-Closed Slots. *IEEE Trans. Magn.* **2011**, *47*, 479–492, doi:10.1109/TMAG.2010.2095874. [CrossRef]

5. Smeets, J.P.C.; Overboom, T.T.; Jansen, J.W.; Lomonova, E.A. Three-Dimensional Analytical Modeling Technique of Electromagnetic Fields of Air-Cored Coils Surrounded by Different Ferromagnetic Boundaries. *IEEE Trans. Magn.* **2013**, *49*, 5698–5708, doi:10.1109/TMAG.2013.2278528. [CrossRef]
6. Dwivedi, A.; Singh, S.K.; Srivastava, R.K. Analysis of Permanent Magnet Brushless AC Motor Using Fourier Transform Approach. *IET Electr. Power Appl.* **2016**, *10*, 539–547, doi:10.1049/iet-epa.2015.0420. [CrossRef]
7. Hannon, B.; Sergeant, P.; Dupré, L. 2-D Analytical Subdomain Model of a Slotted PMSM With Shielding Cylinder. *IEEE Trans. Magn.* **2014**, *50*, 1–10, doi:10.1109/TMAG.2014.2309325. [CrossRef]
8. Custers, C.H.H.M.; Jansen, J.W.; Lomonova, E.A. 2-D Semi-Analytical Modeling of Eddy Currents in Multiple Non-Connected Conducting Elements. *IEEE Trans. Magn.* **2017**, *53*, 1–6, doi:10.1109/TMAG.2017.2697946. [CrossRef]
9. Djelloul-Khedda, Z.; Boughrara, K.; Dubas, F.; Ibtiouen, R. Nonlinear Analytical Prediction of Magnetic Field and Electromagnetic Performances in Switched Reluctance Machines. *IEEE Trans. Magn.* **2017**, *53*, 1–11, doi:10.1109/TMAG.2017.2679686. [CrossRef]
10. Custers, C.H.H.M.; Jansen, J.W.; van Beurden, M.C.; Lomonova, E.A. 3D Harmonic Modeling of Magnetostatic Fields Including Rectangular Structures with Finite Permeability. In Proceedings of the 2016 XXII International Conference on Electrical Machines (ICEM), Lausanne, Switzerland, 4–7 September 2016; pp. 1230–1236, doi:10.1109/ICELMACH.2016.7732682. [CrossRef]
11. Djelloul-Khedda, Z.; Boughrara, K.; Dubas, F.; Kechroud, A.; Souleyman, B. Semi-Analytical Magnetic Field Predicting in Many Structures of Permanent-Magnet Synchronous Machines Considering the Iron Permeability. *IEEE Trans. Magn.* **2018**, *54*, doi:10.1109/TMAG.2018.2824278. [CrossRef]
12. Cotter, N.P.K.; Preist, T.W.; Sambles, J.R. Scattering-Matrix Approach to Multilayer Diffraction. *J. Opt. Soc. Am. A* **1995**, *12*, 1097–1103. [CrossRef]
13. Li, L. Formulation and Comparison of Two Recursive Matrix Algorithms for Modeling Layered Diffraction Gratings. *J. Opt. Soc. Am. A* **1996**, *13*, 1024–1035. [CrossRef]
14. Rumpf, R. Improved Formulation of Scattering Matrices for Semi-Analytical Methods that is Consistent with Convention. *Prog. Electromagn. Res. B* **2011**, *35*, 241–261. [CrossRef]
15. Onishi, M.; Crabtree, K.; Chipman, R.A. Formulation of Rigorous Coupled-Wave Theory for Gratings in Bianisotropic Media. *J. Opt. Soc. Am. A* **2011**, *28*, 1747–1758. [CrossRef] [PubMed]
16. Pisarenco, M.; Maubach, J.; Setija, I.; Mattheij, R. *Efficient Solution of Maxwell's Equations for Geometries with Repeating Patterns by an Exchange of Discretization Directions in the Aperiodic Fourier Modal Method*; CASA-Report; Technische Universiteit: Eindhoven, The Netherlands, 2012.

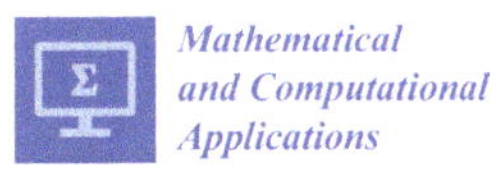

Article

Exact Two-Dimensional Analytical Calculations for Magnetic Field, Electromagnetic Torque, UMF, Back-EMF, and Inductance of Outer Rotor Surface Inset Permanent Magnet Machines

AmirAbbas Vahaj *, Akbar Rahideh, Hossein Moayed-Jahromi and AliReza Ghaffari

Department of Electrical and Electronics Engineering, Shiraz University of Technology, Shiraz 13876-71557, Iran; rahide@sutech.ac.ir (A.R.); h_moaied@yahoo.com (H.M.-J.); a.ghaffari@sutech.ac.ir (A.G.)

* Correspondence: a.vahaj@sutech.ac.ir

Received: 7 January 2019; Accepted: 13 February 2019; Published: 17 February 2019

Abstract: This paper presents a two-dimensional analytical model of outer rotor permanent magnet machines equipped with surface inset permanent magnets. To obtain the analytical model, the whole model is divided into the sub-domains, according to the magnetic properties and geometries. Maxwell equations in each sub-domain are expressed and analytically solved. By using the boundary/interface conditions between adjacent sub-regions, integral coefficients in the general solutions are obtained. At the end, the analytically calculated results of the air-gap magnetic flux density, electromagnetic torque, unbalanced magnetic force (UMF), back-electromotive force (EMF) and inductances are verified by comparing them with those obtained from finite element method (FEM). One of the merits of this method in comparison with the numerical model is the capability of rapid calculation with the highest precision, which made it suitable for optimization problems.

Keywords: analytical model; partial differential equations; separation of variable technique; electrical machines; surface inset permanent magnet

---

1. Introduction

The existence of different types of PM brushless machines (PMBLM) made them applicable for a wide range of applications. PMBLMs have superiorities in comparison with other rivals like induction machines or reluctance synchronous machines due to higher efficiency, high torque per volume, lower torque ripple, lower vibration, and lower acoustic noise.

PMBLM can be categorized in terms of various criteria such as the topology of PMs, the relative position of the rotor and stator, the slotted or slotless stator structure, etc.

Various PM topologies such as surface-mounted, surface-inset, and interior are used where each has its own advantages and disadvantages. Among these topologies, surface-inset can provide a compromise between the other two topologies.

Electric machines with single rotor and single stator can be either inner rotor or outer rotor. The outer rotor motors can develop more output torque than the inner ones for the same volume of the machine. Usually, inner rotor machines are used for applications, which need rapid acceleration and deceleration. Outer rotor machines usually are used for applications which need constant speed. Also, the mechanical robustness of the PMs in the outer rotor configuration is higher than the inner one.

In this paper because of the aforementioned advantages of the outer rotor machines and surface inset PMs, an exact two-dimensional electromagnetic model for this type of machines is extracted.

In the design procedure, the static model is normally considered. Numerous static models have been presented for electric machines, in which some of them are based on the analytical approaches [1–52], and the others are based on numerical methods [53–55].

The presented analytical model for electric machines are based on permeance model [1] or magnetic equivalent circuit, also known as (a.k.a.) 0-D analytic model [2], or resolution of the Maxwell's equations in 2-D plane (a.k.a. 2-D analytic model) [3–48] or 3-D plane (a.k.a. 3-D analytic model) [49,50]. Also, other methods based on mapping techniques such as Schwarz-Christoffel have been used to extract the model of some machine with the analytical approach [51,52]. The most accurate presented models among the analytical methods are 3-D analytic and 2-D analytic. The 2-D analytic method can be used instead of 3-D when the model is symmetric and has no skewing. Two-dimensional analytical models are not only capable of considering a high number of harmonics which elevate the precision of the models, but also have less computational time in comparison with the numerical methods and made them appropriate for optimal design problems.

Two-dimensional analytical models are presented for the slotted [3–17,19,20,29–38,41,42,44,45] and slotless [21–28] machines, equipped with surface mounted [3–6,10–12,14–20,23,26,32,34,37,41,52] and surface inset [9,13,21,22,24,25,35,36] or spoke type magnets [7,8], in order to obtain the important quantities like magnetic flux density, electromagnetic torque, unbalanced magnetic force, back-EMF, and inductances. Also, the 2-D analytical model is used to calculate the eddy current effect [11,18,22,31,32,49,50] in electrical machines. Most of the abovementioned publications are focused on the inner rotor structure [5–18,22–27,31–38,42,44–46] and only a few of them present the 2-D model for outer rotor machines with surface mounted PMs [4,28–30,41]. Therefore, to the best knowledge of the authors, it is for the first time that 2-D analytical model of brushless PM machines with outer rotor and surface inset PM is presented using the subdomain technique.

Most of the developed 2-D or 3-D analytic model are assumed with the infinite permeability of the iron parts. New techniques to account for finite soft-magnetic permeability have been recently developed, i.e., the multi-layer model using the Cauchy's product theorem is presented in [38], and the subdomain technique by applying the superposition principle in both directions is proposed in [39–46] which can be used to calculate the core losses and saturation phenomena.

An overview of the analytical models in the Maxwell-Fourier method with a global or local saturation effect has been realized in [40]. According to [48], Dubas' superposition technique [39,40] is very interesting since it enables the magnetic field calculation in the material of slotted geometries. This superposition technique has been implemented in radial-flux electrical machines with(out) PMs supplied by a direct or alternate current [44,45].

The presented technique in [39–46] is not only used to predict the magnetic field in all parts of the electrical machines, but also it is used to obtain a 2-D analytical model of the steady state heat transfer of the electrical machines by solving the heat equations [47].

The aim of this paper is to extract a 2-D analytical model of PM brushless outer rotor machines equipped with surface inset PMs. The model is used to analytically compute the electromagnetic torque, torque ripple, back-electromotive force (EMF), inductances and unbalanced magnetic forces (UMF).

Therefore, this paper is organized as follows. In Section 2 the procedure of extracting the 2-D model is explained. Section 3 is dedicated to the calculation of the electromagnetic quantities. In Section 4 the analytical results of the case study are presented and compared with those of the numerical method. In the final part this paper is concluded.

2. Extracting the Magnetic Model

2.1. Assumptions

Figure 1 shows the topology of an outer rotor surface inset brushless permanent magnet machine.

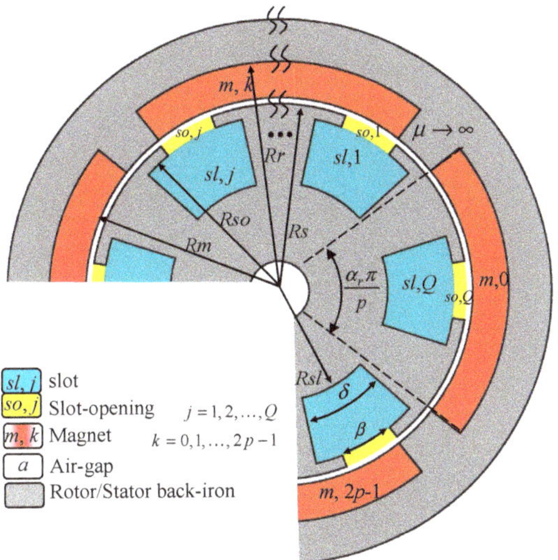

Figure 1. Outer rotor surface inset brushless permanent magnet machine illustration.

In order to make the problem solvable, below assumptions are made:

(a) According to the geometry and the absence of skewing, the problem is solved in 2-D polar coordinates which means the end effect is neglected.
(b) Magnetic vector potential has just axial component which is function of r and θ. Consequently, magnetic flux density has radial and tangential component, i.e., $\mathbf{A} = [0, 0, A_z]$, $\mathbf{B} = [B_r, B_\theta, 0]$.
(c) All materials are isotropic.
(d) Rotor and stator back iron have infinite permeability.
(e) The edges of the slots and slot-openings have radial direction.
(f) The eddy current effect is neglected.

2.2. Dividing Region into Sub-Regions

According to the shape and material characteristics, the whole domain is divided into a number of sub-domains. All the sub-domains are illustrated in Figure 1 and listed in Table 1 for a PMBLM with Q slots and p pole-pairs.

When the winding is single-layer or double-layer non-overlapping, as shown respectively in Figure 2a,b, each slot is a single subdomain; however, if the winding is two-layer overlapping, as shown in Figure 2c, each slot is divided into two subdomains.

Table 1. The sub-domains and related symbols.

Sub-Domain	Symbol	Number of Sub-Regions
Magnet	m	$1, 2, \ldots, 2p$
Air-gap	a	1
Slot-opening	so	$1, 2, \ldots, Q$
Slot	sl	$1, 2, \ldots, Q$

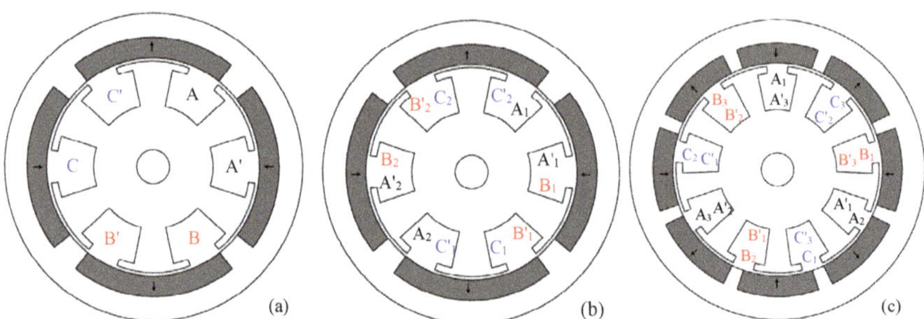

Figure 2. Winding topologies (**a**) single layer alternate teeth wound (**b**) double layer all teeth wound non-overlapping (**c**) double-layer overlapping.

2.3. Extracting the Magnetic Model

In this part, for each sub-domain a partial differential equation is extracted based on Maxwell's equations.

Maxwell's equations in quasi-static form are as follows:

$$\nabla \cdot \mathbf{B} = 0 \tag{1}$$

$$\nabla \times \mathbf{H} = \mathbf{J} + \hat{\mathbf{J}} \tag{2}$$

where **B** is the magnetic flux density vector. Ampere's law represents the relation between the magnetic field intensity vector (**H**), external current density vector (**J**), and current density vector in media ($\hat{\mathbf{J}}$). In this investigation the current density vector in media is assumed to be negligible, i.e., $\hat{\mathbf{J}} = 0$.

The relation between the magnetic flux density vector and the magnetic field intensity vector in permanent magnet media with linear demagnetizing curve is as follows:

$$\mathbf{B} = \mu_0 \mu_r \mathbf{H} + \mu_0 \mathbf{M} \tag{3}$$

where **M** is the magnetization vector.

Substituting (3) in (2) yields

$$\nabla \times \mathbf{B} = \mu_0 \mu_r \mathbf{J} + \mu_0 \nabla \times \mathbf{M} \tag{4}$$

The magnetic flux density vector can be represented as the curl of the magnetic vector potential (**A**):

$$\mathbf{B} = \nabla \times \mathbf{A} \tag{5}$$

Using (4) and (5) the following expression is obtained:

$$\nabla^2 \mathbf{A} = -\mu_0 \mu_r \mathbf{J} - \mu_0 \nabla \times \mathbf{M} \tag{6}$$

For each sub-domain, Equation (6) results in Poisson equations for the magnet and slot regions, and Laplace equations for the air-gap and slot-opening sub-domains, as represented below.

$$\nabla^2 \mathbf{A}^{sl,j} = -\mu_0 \mu_r \mathbf{J} \tag{7}$$

$$\nabla^2 \mathbf{A}^{m,k} = -\mu_0 \nabla \times \mathbf{M} \tag{8}$$

$$\nabla^2 \mathbf{A}^{\chi} = 0,\ \chi = \{(a), (so, j)\} \tag{9}$$

In 2-D polar coordinates, the magnetic vector potential and the current density vector have just a component along z, i.e., $\mathbf{A} = [0, 0, A_z(r,\theta)]$ and $\mathbf{J} = [0, 0, J_z(\theta, t)]$. Also, the magnetic flux density vector and the magnetization vector have radial and tangential components as below:

$$\mathbf{B} = [B_r(r,\theta), B_\theta(r,\theta), 0]$$

$$\mathbf{M} = [M_r(r,\theta), M_\theta(r,\theta), 0]$$

Therefore, Equations (7)–(9) are rewritten as

$$\frac{1}{r}\frac{\partial}{\partial r}\left(r\frac{\partial A_z^{sl,j}}{\partial r}\right) + \frac{1}{r^2}\frac{\partial^2 A_z^{sl,j}}{\partial \theta^2} = -\mu_0 J_z^{sl,j} \tag{10}$$

$$\frac{1}{r}\frac{\partial}{\partial r}\left(r\frac{\partial A_z^{m,k}}{\partial r}\right) + \frac{1}{r^2}\frac{\partial^2 A_z^{m,k}}{\partial \theta^2} = -\frac{\mu_0}{r}\left(M_\theta^k - \frac{\partial M_r^k}{\partial \theta}\right) \tag{11}$$

$$\frac{1}{r}\frac{\partial}{\partial r}\left(r\frac{\partial A_z^{\chi}}{\partial r}\right) + \frac{1}{r^2}\frac{\partial^2 A_z^{\chi}}{\partial \theta^2} = 0,\ \chi = \{(a), (so, j)\} \tag{12}$$

2.4. Boundary Conditions

The perpendicular magnetic flux density in two adjacent sub-domains must be equal as mathematically represented as follows:

$$\mathbf{n}.(\mathbf{B}^i - \mathbf{B}^{i+}) = 0 \tag{13}$$

In this equation, $\mathbf{B}^i$ is the magnetic flux density in sub-domain i, and $\mathbf{B}^{i+}$ is the magnetic flux density in the sub-domain $i+$.

Also, if there is no current between the two adjacent sub-domains, the tangential components of the magnetic field intensity at the boundary of the two sub-domains are equal; this expression is shown mathematically by Relation (14).

$$\mathbf{n} \times (\mathbf{H}^i - \mathbf{H}^{i+}) = 0 \tag{14}$$

In this equation $\mathbf{H}^i$ is the magnetic field intensity of the sub-domain i and $\mathbf{H}^{i+}$ is the magnetic field intensity of sub-domain $i+$.

In both (13) and (14), $\mathbf{n}$ is the perpendicular unit vector to the interface between two adjacent sub-domains.

According to Figure 1, all boundary/interface conditions between sub-domains have been shown from (15) to (25) where α_r, β, and δ are respectively the magnet arc per pole pitch ratio, the span angle of slot-openings, and the span angle of slots as shown in Figure 1.

Domain (i)	Domain (i+)	Equation	Border of the Interface	Limit	
Magnet	Rotor yoke	$H_\theta^{m,k}(r,\theta)=0$	$r=R_r$	$\lvert\theta-\alpha-k\pi/p\rvert\le\alpha_r\pi/2p$	(15)
Magnet	Iron next to the PM pole	$H_r^{m,k}(r,\theta)=0$	$\theta=\alpha+k\pi/p\pm\alpha_r\pi/2p$	$R_r\le r\le R_m$	(16)
Air-gap	Magnet	$B_r^a(r,\theta)=B_r^{m,k}(r,\theta)$	$r=R_m$	$\lvert\theta-\alpha-k\pi/p\rvert\le\alpha_r\pi/2p$	(17)
Air-gap	Magnet	$H_\theta^a(r,\theta)=\begin{cases}\sum_{k=0}^{p-1}H_\theta^{m,k}(r,\theta)\\0\end{cases}$	$r=R_m$	$\begin{cases}\lvert\theta-\alpha-k\pi/p\rvert\le\alpha_r\pi/2p\\\text{elsewhere}\end{cases}$	(18)
Slot-opening	Edges of slot-opening	$H_r^{so,j}(r,\theta)=0$	$\theta=\theta_j\pm\beta/2$	$R_s\le r\le R_{so}$	(19)
Air-gap	Slot-opening	$B_r^a(r,\theta)=B_r^{so,j}(r,\theta)$	$r=R_s$	$\lvert\theta-\theta_j\rvert\le\frac{\beta}{2}$	(20)
Air-gap	Slot-opening	$H_\theta^a(r,\theta)=\begin{cases}\sum_{j=1}^{Q}H_\theta^{so,j}(r,\theta)\\0\end{cases}$	$r=R_s$	$\begin{cases}\lvert\theta-\theta_j\rvert\le\frac{\beta}{2}\\\text{elsewhere}\end{cases}$	(21)
Slot	Slot-opening	$B_r^{sl,j}(r,\theta)=B_r^{so,j}(r,\theta)$	$r=R_{so}$	$\lvert\theta-\theta_j\rvert\le\frac{\beta}{2}$	(22)
Slot	Edge of tooth Slot-opening Edge of tooth	$H_\theta^{sl,j}(r,\theta)=\begin{cases}0\\H_\theta^{so,j}(r,\theta)\\0\end{cases}$	$r=R_{so}$	$\begin{cases}\theta_j-\frac{\delta}{2}\le\theta<\theta_j-\frac{\beta}{2}\\\theta_j-\frac{\beta}{2}\le\theta\le\theta_j+\frac{\beta}{2}\\\theta_j+\frac{\beta}{2}<\theta\le\theta_j+\frac{\delta}{2}\end{cases}$	(23)
Slot	Edges of the slot	$H_r^{sl,j}(r,\theta)=0$	$\theta=\theta_j\pm\delta/2$	$R_{so}\le r\le R_{sl}$	(24)
Slot	Stator yoke	$H_\theta^{sl,j}(r,\theta)=0$	$r=R_{sl}$	$\lvert\theta-\theta_j\rvert\le\frac{\delta}{2}$	(25)

2.5. Extracting the Fourier Series of the Armature Reaction

To consider the effect of the armature reaction, the current density of each slot should be represented as Fourier series. The current of each phase varies with time and can be represented as Equation (26).

$$I_k(t)=\sum_v I_v\sin[v(p\omega t-\gamma_k)+\theta_v],\ k=1,2,\cdots,q \tag{26}$$

where q is the number of the phases, v shows the order of the harmonics, ω is the angular velocity of the rotor, p is the number of the pole-pairs, $\gamma_k=2\pi(k-1)/q$ is the time offset of the phase kth respect to the first phase. Also I_v and θ_v are the magnitude and phase offset of vth harmonic, respectively.

It is obvious that the relation between the current density in each slot and phase current is dependent on the winding configuration. If the winding configuration is like Figure 2a,b, each slot is considered as one sub-region, like Figure 3a,b. But, if the configuration is similar to Figure 2c, each slot consists of two sub-regions (upper and lower sub-regions), as represented in Figure 3c.

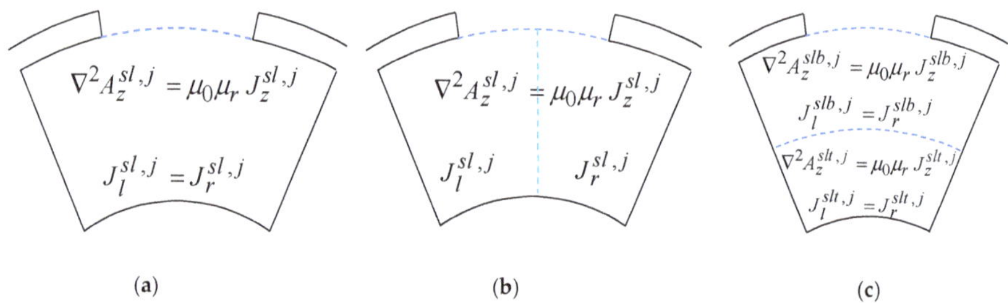

Figure 3. Sub-region division according to the winding configuration: (**a**) whole slot is considered as one region and belongs to one coil, (**b**) whole slot is considered as one region and left side and right side of the slot belongs to different coils, (**c**) whole slot divided into upper and lower sub-regions and each part belongs to one coil.

The current density in each sub-region of a slot can be represented in Fourier series form as in Equation (27).

$$J_z{}^j(\theta,t) = J_0^j(t) + \sum_{v=1}^{\infty} J_v^j(t)\cos\left(\tfrac{\pi v}{\delta}(\theta-\theta_j+\delta/2)\right) \quad \theta_j-\delta/2 \le \theta \le \theta_j+\delta/2. \tag{27}$$

where $J_0^j(t)$ and $J_v^j(t)$ are as follows:

$$J_0^j(t) = \frac{J_\ell^j(t)+J_r^j(t)}{2} \tag{28}$$

$$J_v^j(t) = \frac{J_\ell^j(t)-J_r^j(t)}{\pi v/2}\sin(\pi v/2) \tag{29}$$

In order to complete the Fourier series of each sub-region of a slot, it is necessary to obtain the current density of phases in each sub-region of a slot at a specific time by Equations (30) and (31):

$$J = \frac{I}{K_f A_{slot}} \tag{30}$$

$$J = \frac{I}{K_f A_{slot}/2} \tag{31}$$

For instance, the current density in each sub-domain of a slot could be as Figure 4a,b.

If the winding configuration is as shown in Figure 2a,c, the figure of the current density in each sub-region of slot, at a time instant will be as Figure 4a, and if the winding configuration is as shown in Figure 2b, the figure of the current density in a time instant will be as Figure 4b. Also if the configuration of the winding is similar to Figure 2c, the current density in each sub-region of the slot will be as Figure 4a.

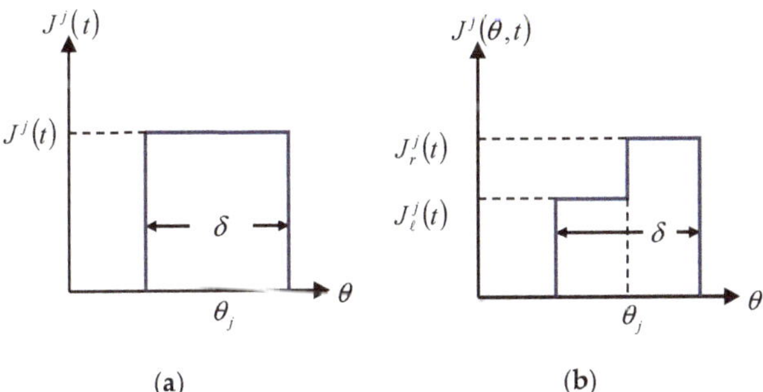

Figure 4. Value of the current density in each sub-region of slot, according to represented winding configurations.

2.6. Extracting the Fourier Series of the Magnetization

In the 2-D polar coordinate system, the magnetization vector only has the radial and tangential components as Equation (32).

$$\mathbf{M} = M_r\mathbf{r} + M_\theta\boldsymbol{\theta} \tag{32}$$

where $\mathbf{r}$ and $\boldsymbol{\theta}$ are the radial and tangential unit vectors. M_r and M_θ are the components of magnetization vector which can be represented as Fourier series expansion of Equations (33) and (34).

$$M_r(\theta_r) = \sum_{n=1,3,5,\ldots}^{\infty} M_{rn}\cos(np\theta_r) \tag{33}$$

$$M_\theta(\theta_r) = \sum_{n=1,3,5,\ldots}^{\infty} M_{\theta n} \sin(np\theta_r) \tag{34}$$

where M_{rn} and $M_{\theta n}$ respectively are the radial and tangential coefficient of the Fourier series and will be determined according to the magnetization pattern (Table 2). In this paper, only the radial magnetization pattern has been used and represented in Figure 5.

Table 2. Radial magnetization pattern and its Fourier series components.

Magnetization Pattern	Illustration	Radial Waveform Component	Tangential Waveform Component	Coefficient of the Radial Component $M_{rw}^{k} = (-1)^{\frac{w-1}{2}+k} \frac{B_{rem}}{\mu_0} \times$	Coefficient of the Tangential Component $M_{\theta w}^{k} = (-1)^{\frac{w-1}{2}+k} \frac{B_{rem}}{\mu_0} \times$
Radial Magnetization	$\theta=0$, $\alpha_p\pi/p$	M_r, $\alpha_p\pi/p$	M_θ, θ	$\frac{4}{w\pi}\sin\left(\frac{w\pi\alpha_p}{2\alpha_r}\right)$	0

2.7. Finding the General Solution

The overall format for the general solution in all sub-domains can be represented as Equation (35):

$$A(r,\theta) = \sum_{n'=1}^{\infty} \left(A_{n'}r^{n'} + B_{n'}r^{n'}\right).(C_{n'} \cos n'\theta + D_{n'} \sin n'\theta) + (A_0 \ln r + B_0)(C_0\theta + D_0) \tag{35}$$

The general solution not only has the capability to satisfy the related PDE, but must satisfy the boundary conditions of the related sub-domain, especially Equations (16), (19), and (24). So the general solutions for sub-domains are as Equations (36)–(40).

The general solution of Poisson equation in slot sub-domain will be as follows:

$$\begin{aligned} A_z^{sl,j}(r,\theta) &= b_0^{sl,j} \ln r + A_p^{sl,j}(r,\theta) + \sum_{v=1}^{\infty} \left[a_v^{sl,j} \left(\frac{r}{R_{so}}\right)^{\frac{\pi v}{\delta}} + b_v^{sl,j} \left(\frac{R_{sl}}{r}\right)^{\frac{\pi v}{\delta}} \right] \\ &\quad \times \cos\left(\frac{\pi v}{\delta}\left(\theta - \theta_j + \frac{\delta}{2}\right)\right) \end{aligned} \tag{36}$$

The particular solution is as follows:

$$A_p^{sl,j}(r,\theta) = -\frac{\mu_0}{4} J_0^j r^2 + \sum_{v=1}^{\infty} \frac{\mu_0 J_v^j r^2}{\left(\frac{\pi v}{\delta}\right)^2 - 4} \cos\left(\frac{\pi v}{\delta}\left(\theta - \theta_j + \frac{\delta}{2}\right)\right) \tag{37}$$

Also the general solution for the slot-opening sub-domain is

$$\begin{aligned} A_z^{so,j}(r,\theta) &= b_0^{so,j} \ln r + \sum_{u=1}^{\infty} \left[a_u^{so,j} \left(\frac{r}{R_s}\right)^{\frac{\pi u}{\beta}} + b_u^{so,j} \left(\frac{R_{so}}{r}\right)^{\frac{\pi u}{\beta}} \right] \\ &\quad \times \cos\left(\frac{\pi u}{\beta}\left(\theta - \theta_j + \frac{\beta}{2}\right)\right) \end{aligned} \tag{38}$$

The general solution for the air-gap sub-domain is

$$\begin{aligned} A_z^{so,j}(r,\theta) &= b_0^{so,j} \ln r + \sum_{u=1}^{\infty} \left[a_u^{so,j} \left(\frac{r}{R_s}\right)^{\frac{\pi u}{\beta}} + b_u^{so,j} \left(\frac{R_{so}}{r}\right)^{\frac{\pi u}{\beta}} \right] \\ &\quad \times \cos\left(\frac{\pi u}{\beta}\left(\theta - \theta_j + \frac{\beta}{2}\right)\right) \end{aligned} \tag{39}$$

And finally, Poisson equation in the PMs sub-domain has the general solution as Equation (40).

$$\begin{aligned} A_z^{m,k}(r,\theta) &= b_0^{m,k}\ln r + \sum_{w=1}^{\infty}\left[a_w^{m,k}\left(\frac{r}{R_r}\right)^{\frac{wp}{\alpha_r}} + b_w^{m,k}\left(\frac{R_m}{r}\right)^{\frac{wp}{\alpha_r}} + k_w^k r\right] \\ &\times \cos\left(\frac{wp}{\alpha_r}\left(\theta - \alpha - \frac{k\pi}{p} + \frac{\alpha_r\pi}{2p}\right)\right) \end{aligned} \tag{40}$$

where

$$k_w^k = -\mu_0\chi w\begin{cases} \dfrac{\frac{wp}{\alpha_r}M_{rw}^k - M_{\theta w}^k}{\left(\frac{wp}{\alpha_r}\right)^2 - 1} & wp \neq \alpha_r \\ \dfrac{M_{rw}^k - M_{\theta w}^k}{2}\ln r & wp = \alpha_r \end{cases} \tag{41}$$

$$\chi w = \frac{1-(-1)^w}{2} \tag{42}$$

In order to simplify the general solution in PM sub-domain, boundary condition (15) has been implemented.

$$\begin{aligned} A_z^{m,k}(r,\theta) &= \sum_{w=1}^{\infty}\left\{b_w^{m,k}\left[\left(\frac{R_m}{R_r}\right)^{\frac{wp}{\alpha_r}}\left(\frac{r}{R_r}\right)^{\frac{wp}{\alpha_r}} + \left(\frac{R_m}{r}\right)^{\frac{wp}{\alpha_r}}\right]\right. \\ &\left. + R_r\zeta_{w1}^k\left(\frac{r}{R_r}\right)^{\frac{wp}{\alpha_r}} + k_w^k r\right\} \times \cos\left(\frac{wp}{\alpha_r}\left(\theta - \alpha - \frac{k\pi}{p} + \frac{\alpha_r\pi}{2p}\right)\right) \end{aligned} \tag{43}$$

where

$$\begin{aligned} \zeta_{w1}^k &= \frac{\alpha_r}{pw}\left(\left.\frac{dk_w^k r}{dr}\right|_{r=R_r} + \mu_0\,\chi w\,M_{\theta w}^k\right) = \\ -\mu_0\chi w &\begin{cases} \dfrac{M_{rw}^k - \frac{wp}{\alpha_r}M_{\theta w}^k}{\left(\frac{wp}{\alpha_r}\right)^2 - 1} & wp \neq \alpha_r \\ \dfrac{M_{rw}^k - M_{\theta w}^k}{2}(1+\ln R_r) - M_{\theta w}^k & wp = \alpha_r\,. \end{cases} \end{aligned} \tag{44}$$

Also, by implementing boundary condition (25), the general solution in slot sub-domain will be simplified as in Equation (45).

$$\begin{aligned} A_z^{sl,j}(r,\theta) &= \frac{\mu_0}{4}J_0^{sl,j}\left(2R_{sl}^2\ln(r) - r^2\right) + \sum_{v=1}^{V}\left\{b_v^{sl,j}\left[\left(\frac{r}{R_{so}}\right)^{\frac{\pi v}{\delta}} + \left(\frac{R_{sl}}{r}\right)^{\frac{\pi v}{\delta}}\left(\frac{R_{sl}}{R_{so}}\right)^{\frac{\pi v}{\delta}}\right]\right. \\ &\left. + \frac{\mu_0 J_v^j}{\left(\frac{\pi v}{\delta}\right)^2 - 4}\left[r^2 - \frac{2R_{sl}}{\frac{\pi v}{\delta}}\left(\frac{R_{sl}}{r}\right)^{\frac{\pi v}{\delta}}\right]\right\} \times \cos\left(\frac{\pi v}{\delta}\left(\theta - \theta_j + \frac{\delta}{2}\right)\right) \end{aligned} \tag{45}$$

2.8. Obtaining Integral Coefficients

For implementing boundary condition (17), the correlation technique must be used [31].

By multiplying Equation (17) in $\frac{2p}{\alpha_r\pi}\sin\left(\frac{wp}{\alpha_r}(\theta - \alpha - \frac{k\pi}{p} + \frac{\alpha_r\pi}{2p})\right)$ and integration over $\left[\alpha + \frac{k\pi}{p} - \frac{\pi\alpha_r}{2p}, \alpha + \frac{k\pi}{p} + \frac{\pi\alpha_r}{2p}\right]$, Equation (46) will be obtained.

$$\begin{aligned} &\frac{2p}{\alpha_r\pi}\int_{\alpha+k\pi/p-\alpha_r\pi/2p}^{\alpha+k\pi/p+\alpha_r\pi/2p} B_r^a\Big|_{r=R_m}\sin\left(\frac{wp}{\alpha_r}\left(\theta - \alpha - \frac{k\pi}{p} + \frac{\alpha_r\pi}{2p}\right)\right)d\theta \\ &= \frac{2p}{\alpha_r\pi}\int_{\alpha+k\pi/p-\alpha_r\pi/2p}^{\alpha+k\pi/p+\alpha_r\pi/2p} B_r^{m,k}\Big|_{r=R_m}\sin\left(\frac{wp}{\alpha_r}\left(\theta - \alpha - \frac{k\pi}{p} + \frac{\alpha_r\pi}{2p}\right)\right)d\theta \end{aligned} \tag{46}$$

From Equation (46), Equation (47) will be deduced.

$$\begin{aligned} &\frac{wp}{\alpha_r}\left[1 + \left(\frac{R_m}{R_r}\right)^{\frac{2wp}{\alpha_r}}\right]a_w^{m,k} - \sum_{n=1}^{N} n\left\{\left[a_n^a\left(\frac{R_s}{R_m}\right)^n + b_n^a\right]\sigma_s(n,w,k)\right. \\ &\left. - \left[c_n^a\left(\frac{R_s}{R_m}\right)^n + d_n^a\right]\sigma_c(n,w,k)\right\} = -R_m\frac{wp}{\alpha_r}\left[\zeta_{w1}^k\left(\frac{R_m}{R_r}\right)^{\frac{wp}{\alpha_r}+1} + \zeta_{w2}^k\right] \end{aligned} \tag{47}$$

where

$$\zeta_{w2}^{k} = k_w^k\Big|_{r=R_m} = -\mu_0\,\chi_w \begin{cases} \frac{\frac{wp}{\alpha_r}M_{rw}^k - M_{\theta w}^k}{\left(\frac{wp}{\alpha_r}\right)^2-1} & wp \neq \alpha_r \\ \frac{M_{rw}^k - M_{\theta w}^k}{2}\ln R_m & wp = \alpha_r \end{cases} \tag{48}$$

$$\sigma_c(n,w,k) = \frac{2p}{\alpha_r\pi}\int_{\alpha+k\pi/p-\alpha_r\pi/2p}^{\alpha+k\pi/p+\alpha_r\pi/2p} \cos(n\theta)\,\sin\left(\frac{wp}{\alpha_r}\left(\theta-\alpha-\frac{k\pi}{p}+\frac{\alpha_r\pi}{2p}\right)\right)d\theta \tag{49}$$

$$\sigma_s(n,w,k) = \frac{2p}{\alpha_r\pi}\int_{\alpha+k\pi/p-\alpha_r\pi/2p}^{\alpha+k\pi/p+\alpha_r\pi/2p} \sin(n\theta)\,\sin\left(\frac{wp}{\alpha_r}\left(\theta-\alpha-\frac{k\pi}{p}+\frac{\alpha_r\pi}{2p}\right)\right)d\theta \tag{50}$$

The solutions of the integrals have been given in the Appendix A.

By using correlation technique, Relation (18) must be multiplied by $\frac{1}{\pi}\sin(n\theta)$ and integration on the interval $[\alpha-\pi, \alpha+\pi]$, Equation (51) will be obtained.

$$\frac{1}{\pi}\int_{\alpha-\pi}^{\alpha+\pi} H_\theta^a|_{r=R_m}\sin(n\theta)\,d\theta = \frac{1}{\pi}\sum_{k=0}^{2p-1}\int_{\alpha+k\pi/p-\alpha_r\pi/2p}^{\alpha+k\pi/p+\alpha_r\pi/2p} H_\theta^{m,k}\Big|_{r=R_m}\sin(n\theta)\,d\theta \tag{51}$$

Equation (51) results in Equation (52):

$$\begin{aligned} &\sum_{k=0}^{2p-1}\sum_{w=1}^{W}\frac{wp}{\mu_r\alpha_r}\left[\left(\frac{R_m}{R_r}\right)^{\frac{2wp}{\alpha_r}}-1\right]\rho_s(n,w,k)\,a_w^{m,k} + n\left[c_n^a\left(\frac{R_s}{R_m}\right)^n - d_n^a\right] \\ &= \sum_{k=0}^{2p-1}\sum_{w=1}^{W}\frac{wpR_m}{\mu_r\alpha_r}\left[-\zeta_{w1}^k\left(\frac{R_m}{R_r}\right)^{\frac{wp}{\alpha_r}-1} + \zeta_{w3}^k\right]\rho_s(n,w,k) \end{aligned} \tag{52}$$

where

$$\begin{aligned} \zeta_{w3}^k = {} & \frac{\alpha_r}{wp}\left(\frac{dk_w^k r}{dr}\Big|_{r=R_m} + \mu_0\,\chi_w M_{\theta w}^k\right) = \\ & -\mu_0\,\chi_w \begin{cases} \frac{M_{rw}^k - \frac{wp}{\alpha_r}M_{\theta w}^k}{\left(\frac{wp}{\alpha_r}\right)^2-1} & wp \neq \alpha_r \\ \frac{M_{rw}^k - M_{\theta w}^k}{2}(1+\ln R_m) - M_{\theta w}^k & wp = \alpha_r \end{cases} \end{aligned} \tag{53}$$

$$\rho_s(n,w,k) = \frac{1}{\pi}\int_{\alpha+k\pi/p-\alpha_r\pi/2p}^{\alpha+k\pi/p+\alpha_r\pi/2p} \cos\left(\frac{wp}{\alpha_r}\left(\theta-\alpha-\frac{k\pi}{p}+\frac{\alpha_r\pi}{2p}\right)\right)\sin(n\theta)\,d\theta \tag{54}$$

The solution of the integral has been given in the Appendix A.

Again Equation (18) must be multiplied by $\frac{1}{\pi}\cos(n\theta)$, then integration on the interval $[\alpha-\pi, \alpha+\pi]$ causes:

$$\frac{1}{\pi}\int_{\alpha-\pi}^{\alpha+\pi} H_\theta^a|_{r=R_m}\cos(n\theta)\,d\theta = \frac{1}{\pi}\sum_{k=0}^{2p-1}\int_{\alpha+k\pi/p-\alpha_r\pi/2p}^{\alpha+k\pi/p+\alpha_r\pi/2p} H_\theta^{m,k}\Big|_{r=R_m}\cos(n\theta)\,d\theta \tag{55}$$

Equation (55) results in Equation (56).

$$\begin{aligned} &\sum_{k=0}^{2p-1}\sum_{w=1}^{W}\frac{wp}{\mu_r\alpha_r}\left[\left(\frac{R_m}{R_r}\right)^{\frac{2wp}{\alpha_r}}-1\right]\rho_c(n,w,k)\,a_w^{m,k} + n\left[a_n^a\left(\frac{R_s}{R_m}\right)^n - b_n^a\right] \\ &= \sum_{k=0}^{2p-1}\sum_{w=1}^{W}\frac{wp\,R_m}{\mu_r\alpha_r}\left[-\zeta_{w1}^k\left(\frac{R_m}{R_r}\right)^{\frac{wp}{\alpha_r}-1} + \zeta_{w3}^k\right]\rho_c(n,w,k) \end{aligned} \tag{56}$$

where

$$\rho_c(n,w,k) = \frac{1}{\pi}\int_{\alpha+k\pi/p-\alpha_r\pi/2p}^{\alpha+k\pi/p+\alpha_r\pi/2p} \cos\left(\frac{wp}{\alpha_r}\left(\theta-\alpha-\frac{k\pi}{p}+\frac{\alpha_r\pi}{2p}\right)\right)\cos(n\theta)\,d\theta \tag{57}$$

The solution of the integral has been given in the Appendix A.

For implementing boundary condition (20), multiplying $\frac{2}{\beta}\sin\left(\frac{\pi u}{\beta}\left(\theta-\theta_j+\frac{\beta}{2}\right)\right)$ to Equation (20) and integration over $\left[\theta_j-\frac{\beta}{2},\,\theta_j+\frac{\beta}{2}\right]$ yields

$$\begin{aligned}&\frac{2}{\beta}\int\limits_{\theta_j-\beta/2}^{\theta_j+\beta/2} B_r^{so,j}(r,\theta)\Big|_{r=R_s}\sin\left(\frac{\pi u}{\beta}\left(\theta-\theta_j+\frac{\beta}{2}\right)\right)d\theta=\\&\frac{2}{\beta}\int\limits_{\theta_j-\beta/2}^{\theta_j+\beta/2} B_r^{a}(r,\theta)\Big|_{r=R_s}\sin\left(\frac{\pi u}{\beta}\left(\theta-\theta_j+\frac{\beta}{2}\right)\right)d\theta\end{aligned} \tag{58}$$

Equation (59) is obtained via the simplification of Equation (58).

$$\begin{aligned}&-\sum_{n=1}^{N} n\left\{\left[a_n^a+b_n^a\left(\frac{R_s}{R_m}\right)^n\right]\varepsilon_s(n,u,j)\;-\;\left[c_n^a+d_n^a\left(\frac{R_s}{R_m}\right)^n\right]\varepsilon_c(n,u,j)\right\}\\&+\frac{\pi u}{\beta}\left(\frac{R_{so}}{R_s}\right)^{\frac{\pi u}{\beta}}a_u^{so,j}+\frac{\pi u}{\beta}b_u^{so,j}=0\end{aligned} \tag{59}$$

where

$$\varepsilon_s(n,u,j)=\frac{2}{\beta}\int\limits_{\theta_j-\beta/2}^{\theta_j+\beta/2}\sin(n\theta)\sin\left(\frac{\pi u}{\beta}\left(\theta-\theta_j+\frac{\beta}{2}\right)\right)d\theta \tag{60}$$

$$\varepsilon_c(n,u,j)=\frac{2}{\beta}\int\limits_{\theta_j-\beta/2}^{\theta_j+\beta/2}\cos(n\theta)\sin\left(\frac{\pi u}{\beta}\left(\theta-\theta_j+\frac{\beta}{2}\right)\right)d\theta \tag{61}$$

The solution of the integral has been given in the Appendix A.

The correlation technique is used for boundary condition (21) and $\frac{1}{\pi}\cos(n\theta)$ is multiplied to Equation (21) and integration is taken over $[-\pi,\,\pi]$.

$$\frac{1}{\pi}\int\limits_{-\pi}^{\pi}H_\theta^a(r,\theta)|_{r=R_s}\cos(n\theta)\,d\theta=\frac{1}{\pi}\sum_{j=1}^{Q}\int\limits_{\theta_j-\beta/2}^{\theta_j+\beta/2}H_\theta^{so,j}(r,\theta)\Big|_{r=R_s}\cos(n\theta)\,d\theta \tag{62}$$

Simplifying Equation (62) yields (63).

$$\begin{aligned}&n\left[a_n^a-b_n^a\left(\frac{R_s}{R_m}\right)^n\right]-\sum_{j=1}^{Q}\sum_{u=1}^{U}\frac{\pi u}{\beta}\left[\left(\frac{R_{so}}{R_s}\right)^{\frac{\pi u}{\beta}}a_u^{so,j}-b_u^{so,j}\right]\eta_c(n,u,j)\\&-\sum_{j=1}^{Q}\eta_c(n,0,j)\,b_0^{so,j}=0\end{aligned} \tag{63}$$

where

$$\eta_c(n,u,j)=\frac{1}{\pi}\int\limits_{\theta_j-\beta/2}^{\theta_j+\beta/2}\cos(n\theta)\cos\left(\frac{\pi u}{\beta}\left(\theta-\theta_j+\frac{\beta}{2}\right)\right)d\theta \tag{64}$$

Also, by multiplying Equation (21) to $\frac{1}{\pi}\sin(n\theta)$ and integration over interval $[-\pi,\,\pi]$ the following expression is obtained:

$$\frac{1}{\pi}\int\limits_{-\pi}^{\pi}H_\theta^a(r,\theta)|_{r=R_s}\sin(n\theta)\,d\theta=\frac{1}{\pi}\sum_{j=1}^{Q}\int\limits_{\theta_j-\beta/2}^{\theta_j+\beta/2}H_\theta^{so,j}(r,\theta)\Big|_{r=R_s}\sin(n\theta)\,d\theta \tag{65}$$

Simplification of Equation (65) causes the formation of Equation (66).

$$n\left[c_n^a-\left(\frac{R_s}{R_m}\right)^n d_n^a\right]-\sum_{j=1}^{Q}\sum_{u=1}^{U}\frac{\pi u}{\beta}\left[\left(\frac{R_{so}}{R_s}\right)^{\frac{\pi u}{\beta}}a_u^{so,j}-b_u^{so,j}\right]\eta_s(n,u,j)$$
$$-\sum_{j=1}^{Q}\eta_s(n,0,j)\,b_0^{so,j}=0 \tag{66}$$

where

$$\eta_s(n,u,j)=\frac{1}{\pi}\int_{\theta_j-\beta/2}^{\theta_j+\beta/2}\sin(n\theta)\cos\left(\frac{\pi u}{\beta}\left(\theta-\theta_j+\frac{\beta}{2}\right)\right)d\theta \tag{67}$$

The correlation technique is used and boundary condition (22) is multiplied by $\frac{2}{\beta}\sin\left(\frac{\pi u}{\beta}\left(\theta-\theta_j+\frac{\beta}{2}\right)\right)$. Integration over interval $\left[\theta_j-\frac{\beta}{2},\theta_j+\frac{\beta}{2}\right]$ results in Equation (68).

$$\frac{2}{\beta}\int_{\theta_j-\beta/2}^{\theta_j+\beta/2}B_r^{sl,j}(r,\theta)\Big|_{r=R_{so}}\sin\left(\frac{\pi u}{\beta}\left(\theta-\theta_j+\frac{\beta}{2}\right)\right)d\theta=$$
$$\frac{2}{\beta}\int_{\theta_j-\beta/2}^{\theta_j+\beta/2}B_r^{so,j}(r,\theta)\Big|_{r=R_{so}}\sin\left(\frac{\pi u}{\beta}\left(\theta-\theta_j+\frac{\beta}{2}\right)\right)d\theta \tag{68}$$

Simplifying (68) results in (69).

$$\frac{\pi u}{\beta}\left[a_u^{so,j}+\left(\frac{R_{so}}{R_s}\right)^{\frac{\pi u}{\beta}}b_u^{so,j}\right]-\sum_{v=1}^{V}\frac{\pi v}{\delta}\left[\left(\frac{R_{sl}}{R_{so}}\right)^{\frac{2\pi v}{\delta}}+1\right]\gamma_s(u,v)\,b_v^{sl,j}$$
$$=\sum_{v=1}^{V}\frac{\mu_0 J_v^j}{\left(\frac{\pi v}{\delta}\right)^2-4}\left[\frac{\pi v}{\delta}R_{sl}^2-2R_{so}^2\left(\frac{R_{sl}}{R_{so}}\right)^{\frac{\pi v}{\delta}}\right]\gamma_s(u,v) \tag{69}$$

where

$$\gamma_s(u,v)=\frac{2}{\beta}\int_{\theta_j-\beta/2}^{\theta_j+\beta/2}\sin\left(\frac{\pi v}{\delta}\left(\theta-\theta_j+\frac{\delta}{2}\right)\right)\sin\left(\frac{\pi u}{\beta}\left(\theta-\theta_j+\frac{\beta}{2}\right)\right)d\theta \tag{70}$$

Solution of the integral are given in the Appendix A.

Correlation technique is implemented for boundary condition (23) and it is multiplied by $\frac{2}{\delta}\cos\left(\frac{\pi v}{\delta}\left(\theta-\theta_j+\frac{\delta}{2}\right)\right)$. Integration over interval $\left[\theta_j-\frac{\delta}{2},\theta_j+\frac{\delta}{2}\right]$ yields Equation (71).

$$\frac{2}{\delta}\int_{\theta_j-\delta/2}^{\theta_j+\delta/2}H_\theta^{sl,j}(r,\theta)\Big|_{r=R_{so}}\cos\left(\frac{\pi v}{\delta}\left(\theta-\theta_j+\frac{\delta}{2}\right)\right)d\theta=$$
$$\frac{2}{\delta}\int_{\theta_j-\beta/2}^{\theta_j+\beta/2}H_\theta^{so,j}(r,\theta)\Big|_{r=R_{so}}\cos\left(\frac{\pi v}{\delta}\left(\theta-\theta_j+\frac{\delta}{2}\right)\right)d\theta \tag{71}$$

Simplifying Equation (71) results in Equation (72).

$$\sum_{u=1}^{U}\frac{\pi u}{\beta}\left[-a_u^{so,j}+\left(\frac{R_{so}}{R_s}\right)^{\frac{\pi u}{\beta}}b_u^{so,j}\right]\gamma_c(u,v)-\gamma_c(0,v)\,b_0^{so,j}+\frac{\pi v}{\delta}\left[\left(\frac{R_{sl}}{R_{so}}\right)^{\frac{2\pi v}{\delta}}-1\right]b_v^{sl,j}$$
$$=\frac{-2\mu_0 J_v^j}{\left(\frac{\pi v}{\delta}\right)^2-4}\left[R_{sl}^2-R_{so}^2\left(\frac{R_{sl}}{R_{so}}\right)^{\frac{\pi v}{\delta}}\right] \tag{72}$$

where

$$\gamma_c(u,v) = \frac{2}{\delta}\int_{\theta_j-\beta/2}^{\theta_j+\beta/2} \cos\left(\frac{\pi v}{\delta}\left(\theta-\theta_j+\frac{\delta}{2}\right)\right)\cos\left(\frac{\pi u}{\beta}\left(\theta-\theta_j+\frac{\beta}{2}\right)\right)d\theta \tag{73}$$

Integration over interval $\left[\theta_j-\frac{\delta}{2},\,\theta_j+\frac{\delta}{2}\right]$ on boundary condition (23) causes Equation (74).

$$\int_{\theta_j-\delta/2}^{\theta_j+\delta/2} H_\theta^{sl,j}(r,\theta)\Big|_{r=R_{so}}d\theta = \int_{\theta_j-\beta/2}^{\theta_j+\beta/2} H_\theta^{so,j}(r,\theta)\Big|_{r=R_{so}}d\theta \tag{74}$$

From Equation (74), Equation (75) will be deduced.

$$b_0^{so,j} = \frac{\mu_0 J_0^j}{2}\left(R_{so}^2-R_{sl}^2\right)\frac{\delta}{\beta} \tag{75}$$

2.9. Overlapping Winding

In overlapping winding, each slot is divided into two sub-domains as represented in Figure 3c. The upper and lower sub-domains respectively are indicated with *slb* and *slt* indices. Hence Equation (69) will be as follows:

$$\frac{\pi u}{\beta}\left[a_u^{so,j}+\left(\frac{R_{so}}{R_s}\right)^{\frac{\pi u}{\beta}}b_u^{so,j}\right]-\sum_{v=1}^{V}\frac{\pi v}{\delta}\left[\frac{(R_{sl}/R_{so})^{\frac{2\pi v}{\delta}}+1}{(R_{slm}/R_{so})^{\frac{\pi v}{\delta}}}\right]\gamma_s(u,v)b_v^{slb,j}=0 \tag{76}$$

Also, Equations (72) and (75) will be modified as (77) and (78) respectively.

$$\begin{aligned}&\sum_{u=1}^{U}\frac{\pi u}{\beta}\left[-a_u^{so,j}+\left(\frac{R_{so}}{R_s}\right)^{\frac{\pi u}{\beta}}b_u^{so,j}\right]\gamma_c(u,v)+\frac{\pi v}{\delta}\left[\frac{(R_{sl}/R_{so})^{\frac{2\pi v}{\delta}}-1}{(R_{slm}/R_{so})^{\frac{\pi v}{\delta}}}\right]b_v^{slb,j}\\&-\gamma_c(0,v)\,b_0^{so,j}=0\end{aligned} \tag{77}$$

$$b_0^{so,j} = \frac{\mu_0}{2}\left[J_{b0}^j\left(R_{slm}^2-R_{sl}^2\right)+J_{t0}^j\left(R_{so}^2-R_{slm}^2\right)\right]\frac{\delta}{\beta} \tag{78}$$

where $R_{slm}=\sqrt{(R_{sl}^2+R_{so}^2)/2}$ is the radii of middle of the slot which divides it into two equal areas. J_{b0}^j and J_{t0}^j are current densities in the lower and upper sub-domains in a slot.

3. Quantities

3.1. Flux Density

The air-gap flux density vector is one of the most important quantities required for the calculation of other quantities. For obtaining the air-gap flux density, Equation (5) is expanded and Relations (78) and (79) in 2-D polar coordinates are deduced.

$$\begin{aligned}B_r^a(r,\theta) = \frac{1}{r}\frac{\partial A_z^a}{\partial\theta} = -\sum_{n=1}^{N} n\Bigg\{&\left[\frac{a_n^a}{R_m}\left(\frac{r}{R_m}\right)^{n-1}+\frac{b_n^a}{R_s}\left(\frac{R_s}{r}\right)^{n+1}\right]\sin(n\theta)\\&-\left[\frac{c_n^a}{R_m}\left(\frac{r}{R_m}\right)^{n-1}+\frac{d_n^a}{R_s}\left(\frac{R_s}{r}\right)^{n+1}\right]\cos(n\theta)\Bigg\}\end{aligned} \tag{79}$$

$$\begin{aligned}B_\theta^a(r,\theta) = -\frac{\partial A_z^a}{\partial r} = -\sum_{n=1}^{N} n\Bigg\{&\left[\left[\frac{a_n^a}{R_m}\left(\frac{r}{R_m}\right)^{n-1}-\frac{b_n^a}{R_s}\left(\frac{R_s}{r}\right)^{n+1}\right]\right]\cos(n\theta)\\&+\left[\frac{c_n^a}{R_m}\left(\frac{r}{R_m}\right)^{n-1}-\frac{d_n^a}{R_s}\left(\frac{R_s}{r}\right)^{n+1}\right]\sin(n\theta)\Bigg\}\end{aligned} \tag{80}$$

3.2. Inductances

For calculation of the inductances, just the flux produced by armature is considered. Inductances between phase k and k' are obtained as follows:

$$L_{k,k'} = \sum_{j \in k \,\&\, j' \in k'} \frac{\lambda_{j,j'}}{i_{j'}} \tag{81}$$

where $j = 1, 2, \ldots, Q$ and $j' = 1, 2, \ldots, Q$ are the indices of the coils. $i_{j'}$ is the current of phase k' and $\lambda_{j,j'}$ is the flux linked by coil j, which is produced by coil j'. If $k = k'$, self-inductance of the phase is calculated.

3.3. Back-EMF

In order to calculate the no-load back-EMF of a phase, the permanent magnet flux linked by coils must be calculated by Equation (82).

$$\varphi = \int \mathbf{B} \cdot d\mathbf{S} \tag{82}$$

According to Faraday's law, induced voltage in jth coil obtain by Equation (83).

$$E_j = -N_t \omega \frac{d\varphi_j}{d\alpha} \tag{83}$$

where N_t is the number of turns of the coil, ω and α are the angular velocity and rotor position respectively. The total back-EMF of a phase depends on coils connections.

3.4. Instantaneous Electromagnetic Torque

Instantaneous torque consists of cogging torque (T_{cog}), electromagnetic torque (T_{em}) and reluctance torque (T_{rel}).

$$T(t) = T_{cog}(t) + T_{em}(t) + T_{rel}(t) \tag{84}$$

By using Maxwell stress tensor, the instantaneous electromagnetic torque can be obtained as follows:

$$T(t) = \int\int \frac{1}{\mu_0} B_r B_\theta \, ds \tag{85}$$

By expanding Equation (85), Relations (86) and (87) will be obtained.

$$T(t) = L_s \int_{-\pi}^{\pi} \frac{1}{\mu_0} \left(B^a_{r,PM} + B^a_{r,AR}\right)\left(B^a_{\theta,PM} + B^a_{\theta,AR}\right)\big|_{r=R_c} R_c^2 \, d\theta \tag{86}$$

$$T(t) = \frac{L_s R_c^2}{\mu_0} \int_{-\pi}^{\pi} \left(B^a_{r,PM} B^a_{\theta,PM} + B^a_{r,AR} B^a_{\theta,PM} + B^a_{r,PM} B^a_{\theta,AR} + B^a_{r,AR} B^a_{\theta,AR}\right)\big|_{r=R_c} d\theta \tag{87}$$

where $B^a_{r,PM}$ and $B^a_{\theta,PM}$ respectively are the radial and tangential flux density components in the air-gap, due to the PMs. Also $B^a_{r,AR}$ and $B^a_{\theta,AR}$ are the radial and tangential magnetic flux density components due to the armature reaction in the air-gap. The parameter R_c is the radius of the middle of inner air-gap.

3.5. Unbalanced Magnetic Force

The radial and tangential components of the local traction exerted on the rotor surface can be obtained by Maxwell stress tensor as follows:

$$f_r = \frac{1}{2\mu_0}\left(B_r^2 - B_\theta^2\right) \tag{88}$$

$$f_\theta = \frac{1}{\mu_0} B_r B_\theta \tag{89}$$

By transforming these local tractions to the Cartesian plane, and summation of the same directions, Equations (90)–(93) will be obtained.

$$f_x = f_r \cos\theta - f_\theta \sin\theta \tag{90}$$

$$f_y = f_r \sin\theta + f_\theta \cos\theta \tag{91}$$

$$F_x(t) = \int_{-L/2}^{L/2}\int_{-\pi}^{\pi} f_x\, r\, d\theta\, dz = L\int_{-\pi}^{\pi} f_x\, r\, d\theta \tag{92}$$

$$F_y(t) = \int_{-L/2}^{L/2}\int_{-\pi}^{\pi} f_y\, r\, d\theta\, dz = L\int_{-\pi}^{\pi} f_y\, r\, d\theta \tag{93}$$

Finally, the amplitude of the unbalanced magnetic force can be obtained by Equation (94).

$$F_r = |F(t)| = \sqrt{F_x^2(t) + F_y^2(t)} \tag{94}$$

4. Results

In order to investigate the efficacy of the model, a case study with the parameters listed in Table 3 has been used.

Also, the winding configuration for this case study has been shown in Figure 5.

Table 3. Parameters used in the case study.

Parameters	Unit	Symbol	Value
Number of phases		q	3
Number of the pole-pair		p	4
Number of slots		Q	9
Outer radius of the slots	(mm)	R_{sl}	18
Outer radius of the slot-opening	(mm)	R_{so}	29
Stator radius	(mm)	R_s	31
Magnet radius	(mm)	R_m	32
Radius of the rotor back iron	(mm)	R_r	38
Axial length	(m)	L_s	0.1
Span of the slot	(rad)	δ	0.6
Span of the slot-opening	(rad)	β	0.3
Pole arc to pole pith of the magnet		α_p	0.85
Ratio of the rotor back iron to the pole pitch		α_r	0.85
Remanence of magnet	(T)	B_{rem}	1
Relative permeability of the magnet		μ_r	1.05
Number of harmonics in each sub-domain		N, U, V, W	100

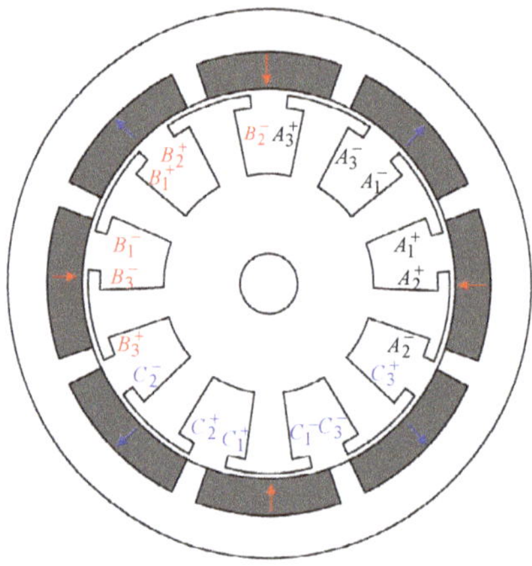

Figure 5. The machine topology and winding configuration.

4.1. Flux Density

The radial and tangential magnetic flux density components due to the open-circuit and armature reaction are respectively depicted in Figures 6–9. The numerical results obtained from FEM are shown respectively in each figure and confirm the accuracy of the proposed model.

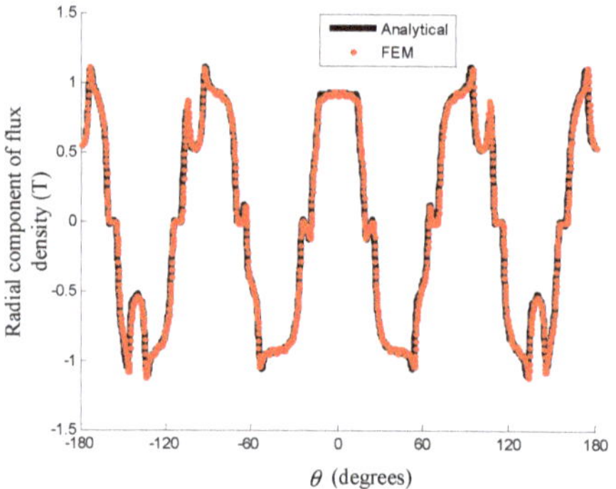

Figure 6. Radial magnetic flux density due to just PMs in the middle of the air-gap, when the rotor position is set to zero.

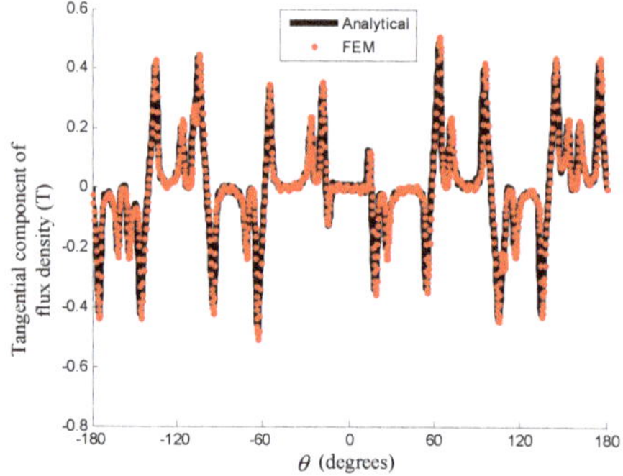

Figure 7. Tangential magnetic flux density due to just PMs in the middle of the air-gap, when the rotor position is set to zero.

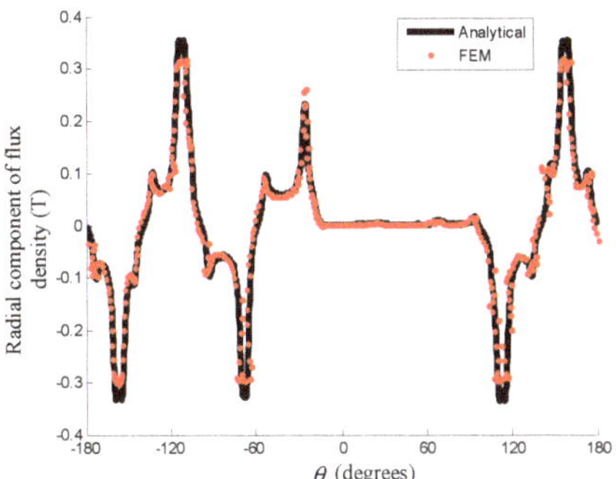

Figure 8. Radial magnetic flux density due to just armature winding in the middle of the air-gap, when the current density of phase A is zero, current density of phase B is 4.33 A/mm^2 and phase C is 4.33 A/mm^2.

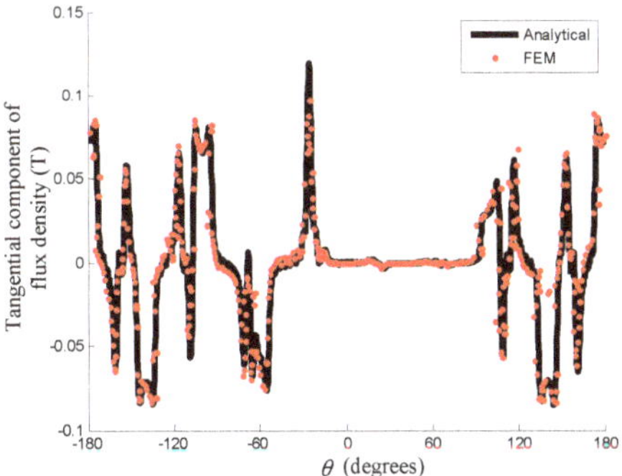

Figure 9. Tangential magnetic flux density due to just armature winding in the middle of the air-gap, when the current density of phase A is zero, current density of phase B is 4.33 A/mm^2 and phase C is 4.33 A/mm^2.

4.2. Torque

Instantaneous electromagnetic torque, reluctance torque and cogging torque of the machine have been depicted in Figures 10–12. Both analytic and numeric methods show good agreement which confirms the efficacy of the proposed model.

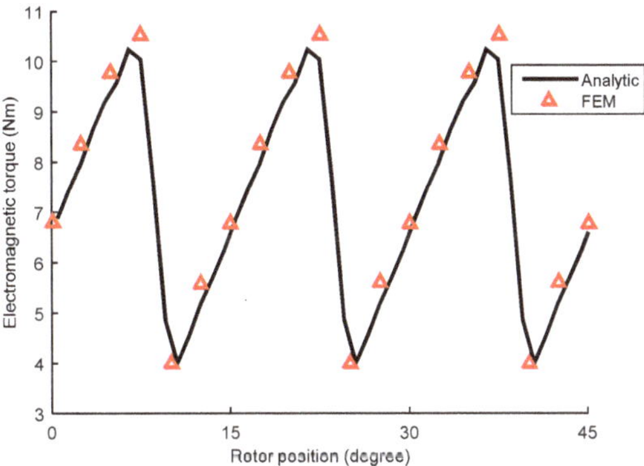

Figure 10. Electromagnetic torque vs. rotor position.

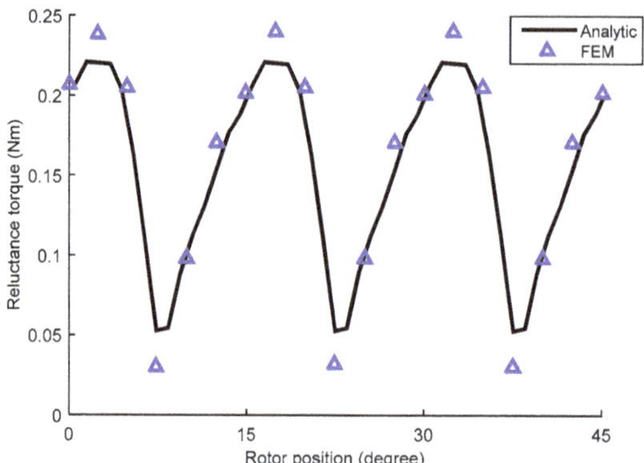

Figure 11. Reluctance torque vs. rotor position.

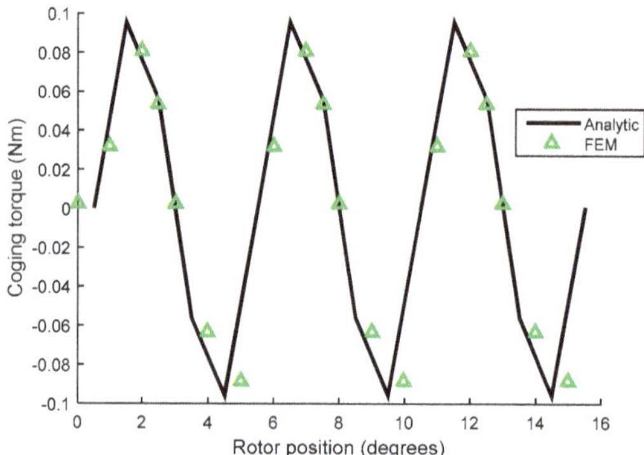

Figure 12. Cogging torque vs. rotor position.

4.3. Back-EMF and Inductance

Results of the phase back-EMF and line back-EMF are shown in Figure 13. Again it is shown that both analytic and numeric results have good conformity.

Also, the self and mutual inductances have been depicted in Figure 14.

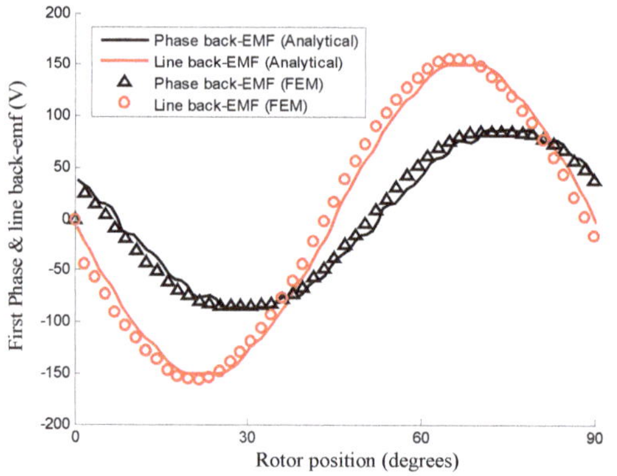

Figure 13. Back-electromotive force (EMF) of the first phase and first line.

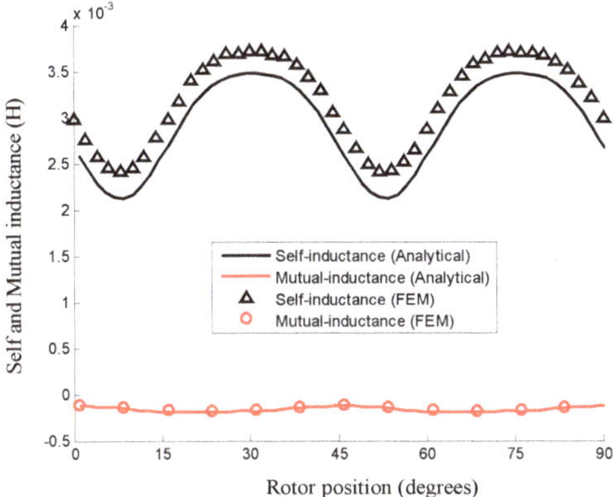

Figure 14. Self and mutual inductance of a phase.

4.4. Unbalanced Magnetic Force (UMF)

Unbalanced magnetic forces due to the open-circuit, armature reaction, and both of them have been depicted in Figures 15–17. As evident from these figures, unbalanced forces due to the armature reaction exert considerable forces compared to those of the open circuit.

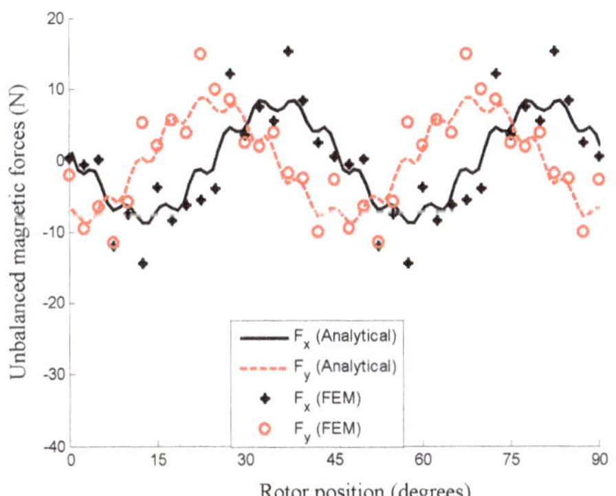

Figure 15. Unbalanced magnetic forces just due to the PMs.

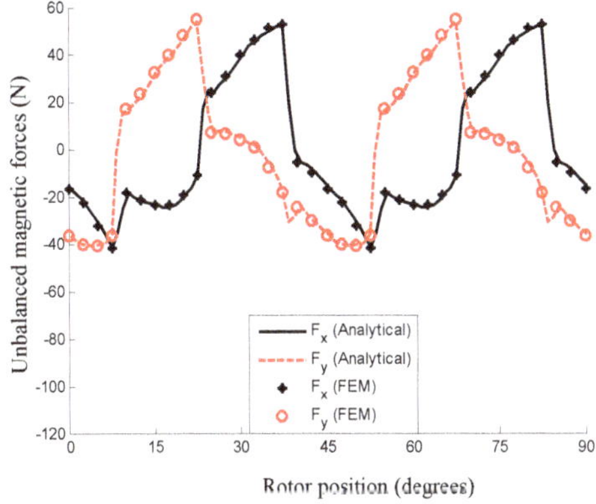

Figure 16. Unbalanced magnetic forces just due to the armature reaction.

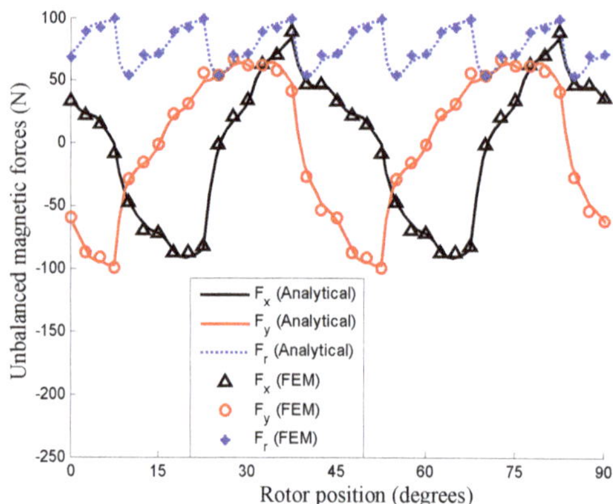

Figure 17. Total unbalanced magnetic forces exerted on the rotor.

Both analytic and numeric results are shown in Figures 15–17 and confirm the correctness of the proposed model.

5. Conclusions

A 2-D analytical magnetic model is presented for brushless synchronous outer rotor machines with surface inset PMs. For this purpose, Maxwell's equations in the form of the Laplace and Poisson equations are solved in predefined sub-domains of the 2-D polar coordinates. The general and particular solutions for each sub-region are presented so that they have the capability to satisfy the governing PDE and related boundary conditions. Finally, by imposing boundary conditions and solving simultaneous linear algebraic equations, all important quantities such as magnetic flux density, electromagnetic torque, UMF, back-EMF, and inductances are calculated and validated by those obtained by FEM. The results of the analytical model show the efficacy of the proposed approach.

Author Contributions: A.V. has done formal analysis, investigation, methodology, writing and editing the original draft. A.R. has done supervision, conceptualization, project administration and reviewing the original draft. H.M.-J. has done formal analysis, investigation, methodology, software, and validation. A.G. has done review and editing.

Conflicts of Interest: The authors declare no conflict of interest.

Appendix A

To define the left and right side current density in each slot for the two layers non-overlapping concentrated winding topology, we have:

$$J_r = \frac{i(t)}{K_f A_w/2} \mathbf{C}_r \quad \text{(A1)}$$

$$J_l = \frac{i(t)}{K_f A_w/2} \mathbf{C}_l \quad \text{(A2)}$$

where $J_r = [J_r^1 \ldots J_r^j \ldots J_r^Q]$, $J_l = [J_l^1 \ldots J_l^j \ldots J_l^Q]$, $i(t) = [i_a\, i_b\, i_c]$, K_f is the filling factor and A_w is the slot area. On the other hand, $\mathbf{C}_r$ and $\mathbf{C}_l$ are as follows:

$$\mathbf{C}_r(i,j) = \begin{cases} \frac{\mathbf{C}(i,j)+|\mathbf{C}(i,j)|}{2} & if \quad -1 < \mathbf{C}(i,j) < 1 \\ 1 & if \quad \mathbf{C}(i,j) = 2 \\ -1 & if \quad \mathbf{C}(i,j) = -2 \end{cases} \quad \text{(A3)}$$

$$\mathbf{C}_l(i,j) = \begin{cases} \frac{\mathbf{C}(i,j) - |\mathbf{C}(i,j)|}{2} & if \quad -1 < \mathbf{C}(i,j) < 1 \\ 1 & if \quad \mathbf{C}(i,j) = 2 \\ -1 & if \quad \mathbf{C}(i,j) = -2 \end{cases} \tag{A4}$$

According to the winding topology, **C** is as follows:

$$\mathbf{C} = \begin{bmatrix} 2 & -2 & 1 & 0 & 0 & 0 & 0 & 0 & -1 \\ 0 & 0 & -1 & 2 & -2 & 1 & 0 & 0 & 0 \\ 0 & 0 & 0 & 0 & 0 & -1 & 2 & -2 & 1 \end{bmatrix}_{3\times Q=3\times 9} \tag{A5}$$

where $\mathbf{C}(i,j) = 2$ or -2 means slot j accommodates two sides of two coils of phase i, which respectively carry positive or negative current. $\mathbf{C}(i,j) = 1$ or -1 means slot j accommodates one side of one coil of phase i, which respectively carry positive or negative current. Also, 0 means there is no coil of the phase i th in the slot.

For $\alpha_r n \neq wp$:

$$\rho_s(n, w, k) = \frac{\alpha_r}{2\pi}\left\{\frac{-\cos\left(w\pi + \frac{n\pi\alpha_r}{2p} + n\alpha + \frac{kn\pi}{p}\right) + \cos\left(\frac{n\pi\alpha_r}{2p} - n\alpha - \frac{kn\pi}{p}\right)}{\alpha_r n + wp} - \frac{\cos\left(w\pi - \frac{n\pi\alpha_r}{2p} - n\alpha - \frac{kn\pi}{p}\right) - \cos\left(\frac{n\pi\alpha_r}{2p} - n\alpha - \frac{kn\pi}{p}\right)}{\alpha_r n - wp}\right\} \tag{A6}$$

$$\rho_c(n, w, k) = \frac{\alpha_r}{2\pi}\left\{\frac{\sin\left(w\pi + \frac{n\pi\alpha_r}{2p} + n\alpha + \frac{kn\pi}{p}\right) + \sin\left(\frac{n\pi\alpha_r}{2p} - n\alpha - \frac{kn\pi}{p}\right)}{\alpha_r n + wp} - \frac{\sin\left(w\pi - \frac{n\pi\alpha_r}{2p} - n\alpha - \frac{kn\pi}{p}\right) - \sin\left(\frac{n\pi\alpha_r}{2p} - n\alpha - \frac{kn\pi}{p}\right)}{\alpha_r n - wp}\right\} \tag{A7}$$

$$\sigma_s(n, w, k) = \frac{p}{\pi}\left\{\frac{-\sin\left(w\pi + \frac{n\pi\alpha_r}{2p} + n\alpha + \frac{kn\pi}{p}\right) - \sin\left(\frac{n\pi\alpha_r}{2p} - n\alpha - \frac{kn\pi}{p}\right)}{\alpha_r n + wp} - \frac{\sin\left(w\pi - \frac{n\pi\alpha_r}{2p} - n\alpha - \frac{kn\pi}{p}\right) - \sin\left(\frac{n\pi\alpha_r}{2p} - n\alpha - \frac{kn\pi}{p}\right)}{\alpha_r n - wp}\right\} \tag{A8}$$

$$\sigma_c(n, w, k) = \frac{p}{\pi}\left\{\frac{-\cos\left(w\pi + \frac{n\pi\alpha_r}{2p} + n\alpha + \frac{kn\pi}{p}\right) + \cos\left(\frac{n\pi\alpha_r}{2p} - n\alpha - \frac{kn\pi}{p}\right)}{\alpha_r n + wp} + \frac{\cos\left(w\pi - \frac{n\pi\alpha_r}{2p} - n\alpha - \frac{kn\pi}{p}\right) - \cos\left(\frac{n\pi\alpha_r}{2p} - n\alpha - \frac{kn\pi}{p}\right)}{\alpha_r n - wp}\right\} \tag{A9}$$

For $\alpha_r n = wp$:

$$\rho_s(n, w, k) = \frac{-1}{4n\pi}\left[\cos\left(3w\pi/2 + n\alpha + \frac{kn\pi}{p}\right) - \cos\left(w\pi/2 - n\alpha - \frac{kn\pi}{p}\right)\right] - \frac{\alpha_r}{2p}\sin\left(w\pi/2 - n\alpha - \frac{kn\pi}{p}\right) \tag{A10}$$

$$\rho_c(n, w, k) = \frac{1}{4n\pi}\left[\sin\left(3w\pi/2 + n\alpha + \frac{kn\pi}{p}\right) + \sin\left(w\pi/2 - n\alpha - \frac{kn\pi}{p}\right)\right] + \frac{\alpha_r}{2p}\cos\left(w\pi/2 - n\alpha - \frac{kn\pi}{p}\right) \tag{A11}$$

$$\sigma_s(n, w, k) = \frac{-1}{2w\pi}\left[\sin\left(3w\pi/2 + n\alpha + \frac{kn\pi}{p}\right) + \sin\left(w\pi/2 - n\alpha - \frac{kn\pi}{p}\right)\right] + \cos\left(w\pi/2 - n\alpha - \frac{kn\pi}{p}\right) \tag{A12}$$

$$\sigma_c(n, w, k) = \frac{-1}{2w\pi}\left[\cos\left(3w\pi/2 + n\alpha + \frac{kn\pi}{p}\right) - \cos\left(w\pi/2 - n\alpha - \frac{kn\pi}{p}\right)\right] + \sin\left(w\pi/2 - n\alpha - \frac{kn\pi}{p}\right) \tag{A13}$$

For $\pi u \neq \beta n$:

$$\varepsilon_s(n,u,j) = 2\pi u \frac{(-1)^{u+1}\sin\left(n\left(\theta_j + \frac{\beta}{2}\right)\right) + \sin\left(n\left(\theta_j - \frac{\beta}{2}\right)\right)}{\pi^2 u^2 - \beta^2 n^2} \tag{A14}$$

$$\varepsilon_c(n,u,j) = 2\pi u \frac{(-1)^{u+1}\cos\left(n\left(\theta_j + \frac{\beta}{2}\right)\right) + \cos\left(n\left(\theta_j - \frac{\beta}{2}\right)\right)}{\pi^2 u^2 - \beta^2 n^2} \tag{A15}$$

$$\eta_s(n,u,j) = \frac{\beta^2 n}{\pi} \frac{(-1)^{u}\cos\left(n\left(\theta_j + \frac{\beta}{2}\right)\right) - \cos\left(n\left(\theta_j - \frac{\beta}{2}\right)\right)}{\pi^2 u^2 - \beta^2 n^2} \tag{A16}$$

$$\eta_c(n,u,j) = \frac{\beta^2 n}{\pi} \frac{(-1)^{u+1}\sin\left(n\left(\theta_j + \frac{\beta}{2}\right)\right) + \sin\left(n\left(\theta_j - \frac{\beta}{2}\right)\right)}{\pi^2 u^2 - \beta^2 n^2} \tag{A17}$$

For $\pi u = \beta n$:

$$\varepsilon_s(n,u,j) = \cos\left(n\left(\theta_j - \frac{\beta}{2}\right)\right) - \frac{\sin\left(n\left(\theta_j + \frac{3\beta}{2}\right)\right) - \sin\left(n\left(\theta_j - \frac{\beta}{2}\right)\right)}{2n\beta} \tag{A18}$$

$$\varepsilon_c(n,u,j) = -\sin\left(n\left(\theta_j - \frac{\beta}{2}\right)\right) - \frac{\cos\left(n\left(\theta_j + \frac{3\beta}{2}\right)\right) - \cos\left(n\left(\theta_j - \frac{\beta}{2}\right)\right)}{2n\beta} \tag{A19}$$

$$\eta_s(n,u,j) = \frac{\sin\left(n\left(\theta_j - \frac{\beta}{2}\right)\right)}{2\pi/\beta} - \frac{\cos\left(n\left(\theta_j + \frac{3\beta}{2}\right)\right) - \cos\left(n\left(\theta_j - \frac{\beta}{2}\right)\right)}{4n\pi} \tag{A20}$$

$$\eta_c(n,u,j) = \frac{\cos\left(n\left(\theta_j - \frac{\beta}{2}\right)\right)}{2\pi/\beta} + \frac{\sin\left(n\left(\theta_j + \frac{3\beta}{2}\right)\right) - \sin\left(n\left(\theta_j - \frac{\beta}{2}\right)\right)}{4n\pi} \tag{A21}$$

For $\delta u \neq \beta v$:

$$\gamma_s(u,v) = \frac{2\delta^2 u}{\pi} \frac{(-1)^{u+1}\sin\left(\frac{\pi v}{2\delta}(\delta+\beta)\right) + \sin\left(\frac{\pi v}{2\delta}(\delta-\beta)\right)}{\delta^2 u^2 - \beta^2 v^2} \tag{A22}$$

$$\gamma_c(u,v) = \frac{2\beta^2 v}{\pi} \frac{(-1)^{u+1}\sin\left(\frac{\pi v}{2\delta}(\delta+\beta)\right) + \sin\left(\frac{\pi v}{2\delta}(\delta-\beta)\right)}{\delta^2 u^2 - \beta^2 v^2} \tag{A23}$$

For $\delta u = \beta v$:

$$\gamma_s(u,v) = \frac{2\pi u \cos\left(\frac{\pi}{2}(u-v)\right) - \sin\left(\frac{\pi}{2}(3u+v)\right) - \sin\left(\frac{\pi}{2}(u-v)\right)}{2\pi u} \tag{A24}$$

$$\gamma_c(u,v) = \frac{2\pi u \cos\left(\frac{\pi}{2}(u-v)\right) + \sin\left(\frac{\pi}{2}(3u+v)\right) + \sin\left(\frac{\pi}{2}(u-v)\right)}{2\pi v} \tag{A25}$$

References

1. Zarko, D.; Ban, D.; Lipo, T.A. Analytical calculation of magnetic field distribution in the slotted air gap of a surface permanent-magnet motor using complex relative air-gap permeance. *IEEE Trans. Magn.* **2006**, *42*, 1828–1837. [CrossRef]
2. Hur, J.; Yoon, S.; Hwang, D.; Hyun, D. Analysis of PMLSM using three dimensional equivalent magnetic circuit network method. *IEEE Trans. Magn.* **1997**, *33*, 4143–4145.
3. Rahideh, A.; Vahaj, A.A.; Mardaneh, M.; Lubin, T. Two-Dimensional Analytical Investigation of the Parameters and the Effects of Magnetisation Patterns on the Performance of Coaxial Magnetic Gears. *IET Electr. Syst. Transp.* **2017**, *7*, 230–245. [CrossRef]

4. Moayed-Jahromi, H.; Rahideh, A.; Mardaneh, M. 2-D Analytical Model for External Rotor Brushless PM Machines. *IEEE Trans. Energy Convers.* **2016**, *31*, 1100–1109. [CrossRef]
5. Wu, L.J.; Zhu, Z.Q.; Staton, D.; Popescu, M.; Hawkins, D. An improved subdomain model for predicting magnetic field of surface-mounted permanent-magnet machines accounting for toothtips. *IEEE Trans. Magn.* **2011**, *47*, 1693–1704. [CrossRef]
6. Rahideh, A.; Korakianitis, T. Subdomain analytical magnetic field prediction of slotted brushless machines with surface mounted magnets. *Int. Rev. Electr. Eng.* **2012**, *7*, 3891–3909.
7. Pourahmadi-Nakhli, M.; Rahideh, A.; Mardaneh, M. Analytical 2-D model of slotted brushless machines with cubic spoke-type permanent magnets. *IEEE Trans. Energy Convers.* **2018**, *33*, 373–382. [CrossRef]
8. Boughrara, K.; Ibtiouen, R.; Lubin, T. Analytical Prediction of Magnetic Field in Parallel Double Excitation and Spoke-Type Permanent-Magnet Machines Accounting for Tooth-Tips and Shape of Polar Pieces. *IEEE Trans. Magn.* **2012**, *48*, 2121–2137. [CrossRef]
9. Rahideh, A.; Korakianitis, T. Analytical magnetic field calculation of slotted brushless PM machines with surface inset magnets. *IEEE Trans. Magn.* **2012**, *48*, 2633–2649. [CrossRef]
10. Dubas, F.; Sari, A.; Kauffmann, J.M.; Espanet, C. Cogging torque evaluation through a magnetic field analytical computation in permanent magnet motors. In Proceedings of the 2009 International Conference on Electrical Machines and Systems, Tokyo, Japan, 15–18 November 2009; pp. 1–5.
11. Dubas, F.; Espanet, C. Semi-analytical Solution of 2-D rotor eddy-current losses due to the slotting effect in SMPMM. In Proceedings of the 17th Conference on the Computation of Electromagnetic Fields COMPUMAG 2009, Florianopolis, Brasil, 22–26 November 2009; pp. 20–25.
12. Lubin, T.; Mezani, S.; Rezzoug, A. 2-D exact analytical model for surface-mounted permanent magnet motors with semi-closed slots. *IEEE Trans. Magn.* **2011**, *47*, 479–492. [CrossRef]
13. Lubin, T.; Mezani, S.; Rezzoug, A. Two-dimensional analytical calculation of magnetic field and electromagnetic torque for surface-inset permanent magnet motors. *IEEE Trans. Magn.* **2012**, *48*, 2080–2091. [CrossRef]
14. Dubas, F.; Espanet, C. Analytical solution of the magnetic field in permanent-magnet motors taking into account slotting effect: No-load vector potential and flux density calculation. *IEEE Trans. Magn.* **2009**, *45*, 2097–2109. [CrossRef]
15. Zhu, Z.Q.; Wu, L.J.; Xia, Z.P. An accurate subdomain model for magnetic field computation in slotted surface mounted permanent-magnet machines. *IEEE Trans. Magn.* **2010**, *46*, 1100–1115. [CrossRef]
16. Boughrara, K.; Chikouche, B.L.; Ibtiouen, R.; Zarko, D.; Touhami, O. Analytical model of slotted air-gap surface mounted permanent-magnet synchronous motor with magnet bars magnetized in the shifting direction. *IEEE Trans. Magn.* **2009**, *45*, 747–758. [CrossRef]
17. Bellara, A.; Amara, Y.; Barakat, G.; Dakyo, B. Two-dimensional exact analytical solution of armature reaction field in slotted surface mounted PM radial flux synchronous machines. *IEEE Trans. Magn.* **2009**, *45*, 4534–4538. [CrossRef]
18. Dubas, F.; Espanet, C.; Miraoui, A. Field diffusion equation in high-speed surface mounted permanent magnet motors, parasitic eddy-current losses. In Proceedings of the 6th International Symposium on Advanced Electromechanical Motion Systems, Lausanne, Switzerland, 27–29 September 2005; pp. 1–6.
19. Liu, Z.J.; Li, J.T. Analytical solution of air-gap field in permanent-magnet motors taking into account the effect of pole transition over slots. *IEEE Trans. Magn.* **2007**, *43*, 3872–3883. [CrossRef]
20. Zhu, Z.Q.; Howe, D.; Chan, C.C. Improved analytical model for predicting the magnetic field distribution in brushless permanent-magnet machines. *IEEE Trans. Magn.* **2002**, *38*, 229–238. [CrossRef]
21. Rahideh, A.; Korakianitis, T. Analytical magnetic field distribution of slotless brushless machines with inset permanent magnets. *IEEE Trans. Magn.* **2011**, *47*, 1763–1774. [CrossRef]
22. Dubas, F.; Rahideh, A. 2-D analytical PM eddy-current loss calculations in slotless PMSM equipped with surface-inset magnets. *IEEE Trans. Magn.* **2014**, *50*, 6300320. [CrossRef]
23. Rahideh, A.; Korakianitis, T. Analytical open-circuit magnetic field distribution of slotless brushless permanent magnet machines with rotor eccentricity. *IEEE Trans. Magn.* **2011**, *47*, 4791–4808. [CrossRef]
24. Rahideh, A.; Mardaneh, M.; Korakianitis, T. Analytical 2-D calculations of torque, inductance, and back-EMF for brushless slotless machines with surface inset magnets. *IEEE Trans. Magn.* **2013**, *49*, 4873–4884. [CrossRef]
25. Rahideh, A.; Korakianitis, T. Analytical, armature reaction field distribution of slotless brushless machines with inset permanent magnets. *IEEE Trans. Magn.* **2012**, *48*, 2178–2191. [CrossRef]

26. Atallah, K.; Zhu, Z.Q.; Howe, D.; Birch, T.S. Armature reaction field and winding inductances of slotless permanent-magnet brushless machines. *IEEE Trans. Magn.* **1998**, *34*, 3737–3744. [CrossRef]
27. Pfister, P.D.; Perriard, Y. Slotless permanent-magnet machines: General analytical magnetic field calculation. *IEEE Trans. Magn.* **2011**, *47*, 1739–1752. [CrossRef]
28. Holm, S.R.; Polinder, H.; Ferreira, J.A. Analytical modeling of a permanent-magnet synchronous machine in a flywheel. *IEEE Trans. Magn.* **2007**, *43*, 1955–1967. [CrossRef]
29. Liu, Z.J.; Li, J.T. Accurate prediction of magnetic field and magnetic forces in permanent magnet motors using an analytical solution. *IEEE Trans. Energy Convers.* **2008**, *23*, 717–726. [CrossRef]
30. Liu, Z.J.; Li, J.T.; Jiang, Q. An improved analytical solution for predicting magnetic forces in permanent magnet motors. *J. Appl. Phys.* **2008**, *103*, 07F135. [CrossRef]
31. Amara, Y.; Reghem, P.; Barakat, G. Analytical prediction of eddy-current loss in armature windings of permanent magnet brushless AC machines. *IEEE Trans. Magn.* **2010**, *46*, 3481–3484. [CrossRef]
32. Pfister, P.; Yin, X.; Fang, Y. Slotted Permanent-Magnet Machines: General Analytical Model of Magnetic Fields, Torque, Eddy Currents, and Permanent-Magnet Power Losses Including the Diffusion Effect. *IEEE Trans. Magn.* **2016**, *52*, 1–13. [CrossRef]
33. Wu, L.J.; Zhu, Z.Q.; Staton, D.; Popescu, M.; Hawkins, D. Subdomain model for predicting armature reaction field of surface-mounted permanent-magnet machines accounting for tooth-tips. *IEEE Trans. Magn.* **2011**, *47*, 812–822. [CrossRef]
34. Gysen, B.L.J.; Meessen, K.J.; Paulides, J.J.H.; Lomonova, E.A. General formulation of the electromagnetic field distribution in machines and devices using fourier analysis. *IEEE Trans. Magn.* **2010**, *46*, 39–52. [CrossRef]
35. Zhu, Z.Q.; Howe, D.; Xia, Z.P. Prediction of open-circuit airgap field distribution in brushless machines having an inset permanent magnet rotor topology. *IEEE Trans. Magn.* **1994**, *30*, 98–107. [CrossRef]
36. Jian, L.; Chau, K.T.; Gong, Y.; Yu, C.; Li, W. Analytical calculation of magnetic field in surface-inset permanent-magnet motors. *IEEE Trans. Magn.* **2009**, *45*, 4688–4691. [CrossRef]
37. Zhu, Z.Q.; Ishak, D.; Howe, D.; Chen, J. Unbalanced magnetic forces in permanent-magnet brushless machines with diametrically asymmetric phase windings. *IEEE Trans. Ind. Appl.* **2007**, *43*, 1544–1553. [CrossRef]
38. Sprangers, R.L.J.; Paulides, J.J.H.; Gysen, B.L.J.; Lomonova, E.A. Magnetic Saturation in Semi-Analytical Harmonic Modeling for Electric Machine Analysis. *IEEE Trans. Magn.* **2016**, *52*, 1–10. [CrossRef]
39. Dubas, F.; Boughrara, K. New scientific contribution on the 2-D subdomain technique in polar coordinates: Taking into account of iron parts. *Math. Comput. Appl.* **2017**, *22*, 42. [CrossRef]
40. Dubas, F.; Boughrara, K. New scientific contribution on the 2-D subdomain technique in Cartesian coordinates: Taking into account of iron parts. *Math. Comput. Appl.* **2017**, *22*, 17. [CrossRef]
41. Djelloul-Khedda, Z.; Boughrara, K.; Dubas, F.; Kechroud, A.; Tikellaline, A. Analytical Prediction of Iron-Core Losses in Flux-Modulated Permanent-Magnet Synchronous Machines. *IEEE Trans. Magn.* **2019**, *55*, 1–12. [CrossRef]
42. Djelloul-Khedda, Z.; Boughrara, K.; Dubas, F.; Ibtiouen, R. Nonlinear Analytical Prediction of Magnetic Field and Electromagnetic Performances in Switched Reluctance Machines. *IEEE Trans. Magn.* **2017**, *53*, 1–11. [CrossRef]
43. Djelloul-Khedda, Z.; Boughrara, K.; Dubas, F.; Kechroud, A.; Souleyman, B. Semi-Analytical Magnetic Field Predicting in Many Structures of Permanent-Magnet Synchronous Machines Considering the Iron Permeability. *IEEE Trans. Magn.* **2018**, *54*, 1–21. [CrossRef]
44. Roubache, L.; Boughrara, K.; Dubas, F.; Ibtiouen, R. New Subdomain Technique for Electromagnetic Performances Calculation in Radial-Flux Electrical Machines Considering Finite Soft-Magnetic Material Permeability. *IEEE Trans. Magn.* **2018**, *54*, 1–15. [CrossRef]
45. Ben Yahia, M.; Boughrara, K.; Dubas, F.; Roubache, L.; Ibtiouen, R. Two-Dimensional Exact Subdomain Technique of Switched Reluctance Machines with Sinusoidal Current Excitation. *Math. Comput. Appl.* **2018**, *23*, 59.
46. Roubache, L.; Boughrara, K.; Dubas, F.; Ibtiouen, R. Elementary subdomain technique for magnetic field calculation in rotating electrical machines with local saturation effect. *Int. J. Comput. Math. Electr. Electron. Eng.* **2018**. [CrossRef]
47. Boughrara, K.; Dubas, F.; Ibtiouen, R. 2-D Exact Analytical Method for Steady-State Heat Transfer Prediction in Rotating Electrical Machines. *IEEE Trans. Magn.* **2018**, *54*, 1–19. [CrossRef]

48. Hannon, B.; Sergeant, P.; Dupré, L. Two-dimensional Fourier-based modeling of electric machines. In Proceedings of the 2017 IEEE International Electric Machines and Drives Conference, Miami, FL, USA, 21–24 May 2017; pp. 1–8.
49. Lubin, T.; Rezzoug, A. 3-D Analytical Model for Axial-Flux Eddy-Current Couplings and Brakes Under Steady-State Conditions. *IEEE Trans. Magn.* **2015**, *51*, 1–12. [CrossRef]
50. Lubin, T.; Rezzoug, A. Improved 3-D Analytical Model for Axial-Flux Eddy-Current Couplings with Curvature Effects. *IEEE Trans. Magn.* **2017**, *53*, 1–9. [CrossRef]
51. Zarko, D.; Ban, D.; Lipo, T.A. Analytical solution for cogging torque in surface permanent magnet motors using conformal mapping. *IEEE Trans. Magn.* **2008**, *44*, 52–65. [CrossRef]
52. Boughrara, K.; Zarko, D.; Ibtiouen, R.; Touhami, O.; Rezzoug, A. Magnetic field analysis of inset and surface-mounted permanent-magnet synchronous motors using Schwarz–Christoffel transformation. *IEEE Trans. Magn.* **2009**, *45*, 3166–3178. [CrossRef]
53. Clemens, M.; Lang, J.; Teleaga, D.; Wimmer, G. Transient 3D magnetic field simulations with combined space and time mesh adaptivity for lowest order Whitney finite element formulations. *IET Sci. Meas. Technol.* **2009**, *3*, 377–383. [CrossRef]
54. Liang, Y.; Bian, X.; Yang, L.; Wu, L. Numerical calculation of circulating current losses in stator transposition bar of large hydro-generator. *IET Sci. Meas. Technol.* **2015**, *9*, 485–491. [CrossRef]
55. Mohammed, O.A.; Liu, S.; Liu, Z. FE-based physical phase variable model of PM synchronous machines under stator winding short circuit faults. *IET Sci. Meas. Technol.* **2007**, *1*, 12–16. [CrossRef]

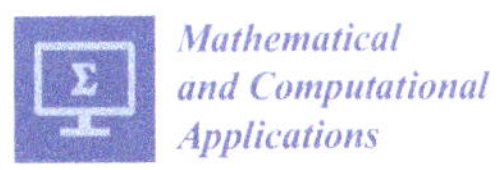

Article

Study of a Hybrid Excitation Synchronous Machine: Modeling and Experimental Validation

Salim Asfirane [1,2,*], Sami Hlioui [3], Yacine Amara [1] and Mohamed Gabsi [2]

1 GREAH, EA 3220, Université Le Havre Normandie, 25 Rue Philippe Lebon, 76600 Le Havre, France; Yacine.amara@univ-lehavre.fr
2 SATIE, CNRS, Ecole Normale Supérieure Paris-Saclay, 61 Avenue du Président Wilson, 94230 Cachan, France; Mohamed.gabsi@satie.ens-cachan.fr
3 SATIE, CNRS, Conservatoire National des Arts et Métiers (CNAM), 292 Rue Saint-Martin, F-75141 Paris CEDEX 03, France; Sami.hlioui@satie.ens-cachan.fr
* Correspondence: salim.asfirane@univ-lehavre.fr or salim.asfirane@satie.ens-paris-saclay.fr

Received: 1 February 2019; Accepted: 21 March 2019; Published: 27 March 2019

Abstract: This paper deals with a parallel hybrid excitation synchronous machine (HESM). First, an expanded literature review of hybrid/double excitation machines is provided. Then, the structural topology and principles of operation of the hybrid excitation machine are examined. With the aim of validating the double excitation principle of the topology studied in this paper, the construction of a prototype is presented. In addition, both the 3D finite element method (FEM) and 3D magnetic equivalent circuit (MEC) model are used to model the machine. The flux control capability in the open-circuit condition and results of the developed models are validated by comparison with experimental measurements. The reluctance network model is created from a mesh of the studied domain. The meshing technique aims to combine advantages of finite element modeling, i.e., genericity and expert magnetic equivalent circuit models, i.e., reduced computation time. It also allows taking the non-linear characteristics of ferromagnetic materials into consideration. The machine prototype is tested to validate the predicted results. By confronting results from both modeling techniques and measurements, it is shown that the magnetic equivalent circuit model exhibits fairly accurate results when compared to the 3D finite element method with a gain in computation time.

Keywords: electric machines; permanent magnet motor; rotating machines; hybrid excitation; permanent magnet machines; magnetic equivalent circuits; 3D finite element method

---

1. Introduction

Hybrid excitation synchronous machines (HESMs) are electric machines that use two excitation flux sources: Permanent magnets (PMs) and field coil excitation sources. The association of both excitation sources aims to combine advantages of PM machines and wound field synchronous machines [1]. The good performances of hybrid excitation machines, such as better flux-weakening capability and efficiency, is encouraging an increasing interest for their study. For the generator operating mode, hybrid excitation machines used together in a connection to a diode rectifier constitute an interesting alternative to permanent magnet alternators associated to an active power converter [2,3]. When operating in motor mode, the hybrid excitation principle permits an easier high-speed operation while the use of permanent magnets helps increase the energy efficiency [1]. It is also possible to use the hybrid excitation principle to reduce PM volumes and save material cost. Some comprehensive reviews on hybrid excited topologies can be found in [1] and [3–9]. An alternative and updated review will be provided in this paper. In addition, in this paper, a parallel hybrid excited machine topology is examined. A 3D magnetic equivalent circuit (MEC) and 3D finite element analysis (FEA) are used to model the HESM and thus, a review on MEC modeling will also be provided in this paper.

This machine has permanent magnets on the rotor and field coils in the stator. In order to improve the flux control capability, both the stator and the rotor contain laminated and massive ferromagnetic parts.

The structure and operating principle of this machine are described. In order to validate the hybrid excitation principle a prototype has been built based on requirements provided by a car manufacturer. A magnetic equivalent circuit (MEC) model, based on an original approach, is developed to predict the open circuit flux control capability of the studied machine [10–13]. This characteristic constitutes a good indicator of the ability of HESM to operate over a large speed range, in particular in the field weakening region [14–19]. In addition, MEC models are still widely used for the modeling of electric machines. MEC is suited for pre-design and optimization of electromagnetic devices [20–22]. Indeed, this technique helps get simple relations between geometric dimensions, physical properties of materials and machines performance. The goal, in this paper, is to evaluate the use of a 3D MEC model. Based on the approach of reluctance networks modeling [23–26], the 3D MEC is developed and adapted to the modeling of the studied HESM. The MEC model is generated from a mesh of the studied domain. This technique combines advantages of finite element method (genericity) and expert lumped parameter MEC models, i.e., reduced computation time, while considering non-linear characteristics of ferromagnetic materials. On another hand, the 3D finite element method is also to model the HESM and to predict its performance, while the prototype machine is tested to validate the predicted results.

2. State of The Art of Hybrid Excitation Machines

In scientific and technical literature, several terms are used to qualify electrical machines that use two excitation flux sources:

- Hybrid excitation synchronous machines;
- Double excitation synchronous machines;
- Dual excitation synchronous machines;
- Combined excitation synchronous machines;
- Permanent magnet synchronous machines with auxiliary exciting windings.

Before presenting the operating principles of the studied machine, criteria used for the classification of dual excitation machines are first discussed and an updated review of recently developed hybrid excitation machines will be provided in this section.

2.1. Classification Criteria of Hybrid Excitation Synchronous Machines

A large number of machine topologies structures can be realized when applying the double excitation principle. Therefore, a variety of criteria can be adopted for the classification of double excitation machines. A classical classification criteria used for other types of electric machines can be applied; such as magnetic flux paths as in 2D and 3D structures, linear [27,28] and rotating machines, axial field [18,29,30] and radial field structures. However, with regards to the structural particularity of double excitation machines, i.e., dual excitation flux sources, two criteria seem more appropriate for their classification [6]:

The first criterion is relative to where the excitation sources are located in the machine: Both sources in the stator, both sources in the rotor and mixed localization.

It is meant by mixed localization that one of the sources (excitation coils or permanent magnets) is located in the rotor and the other source in the stator or vice-versa. Having excitation coils in the stator is favored though to avoid sliding contacts [6].

The second is based on the analogy with electrical circuits. From the way the two excitation flux sources are combined, the criterion will be: Series and parallel double excitation machines [1].

It should be highlighted that HESMs are used in a large variety of applications. In [27], authors presented the design of a hybrid excitation linear eddy current brake which could be used in different applications, i.e., vibration suppression, vehicle suspension systems, high-speed train braking systems, transmission systems, etc. In [28], authors presented the design of a hybrid excited linear machine for oceanic wave power generation. While HESM has been first largely studied for transportation applications [1–3,16,31–34], many researchers are exploring the use of these machines in renewable energy applications [28,35–40].

2.2. Review of Recent Literature

A non-exhaustive review of recent literature, dedicated to hybrid excitation machines, covering the last few years, is presented in this section. One of the topologies that attracted considerable research efforts is the hybrid excited flux-switching machine. Several hybrid excited flux-switching topologies have been investigated in the last years [16,41–45]. The magnetic flux in flux-switching machines is of a 2D nature. In addition, all magnetic field sources (permanent magnets, armature windings and excitation coils) are located in the stator. This implies a completely passive rotor. These reasons make the hybrid excited flux-switching machine suitable for many different applications (hybrid/electrical vehicle [16,42], more electrical aircraft [31]).

Figure 1a presents a hybrid excited flux-switching structure which has been investigated in [31,41,42]. This topology has its field coils placed above the PMs and thus a magnetic bridge is present in the stator back iron. Figure 1b shows a similar hybrid excited flux-switching structure but without iron flux bridges with field coils placed below the PMs. This topology has been investigated in [44]. The structure shown in Figure 2 has been investigated in [16]. Figure 3 shows a doubly salient hybrid excited structure where both magnetic excitation field sources are located in the stator as flux-switching machines. This structure has been investigated in [46]. It has PMs placed in the stator yoke.

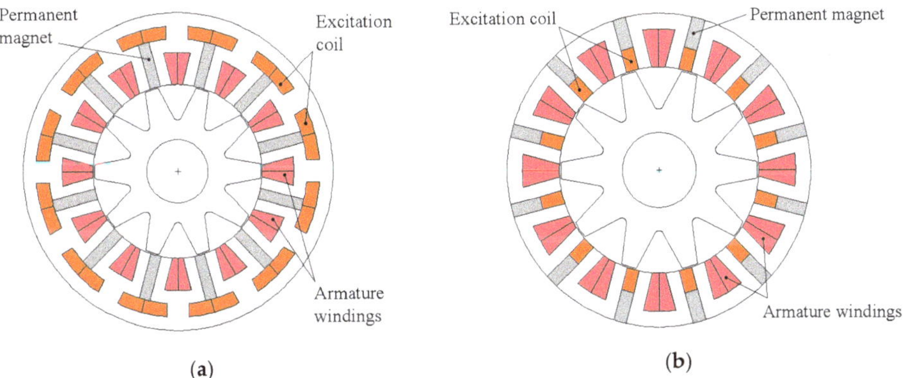

Figure 1. Hybrid excited flux-switching structures. (**a**) With iron flux bridges [31,41,42]; (**b**) without iron flux bridges [44].

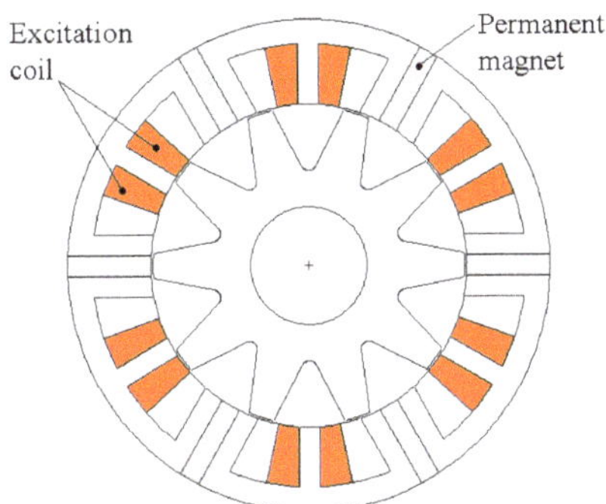

Figure 2. E-core hybrid excited flux-switching structure [16].

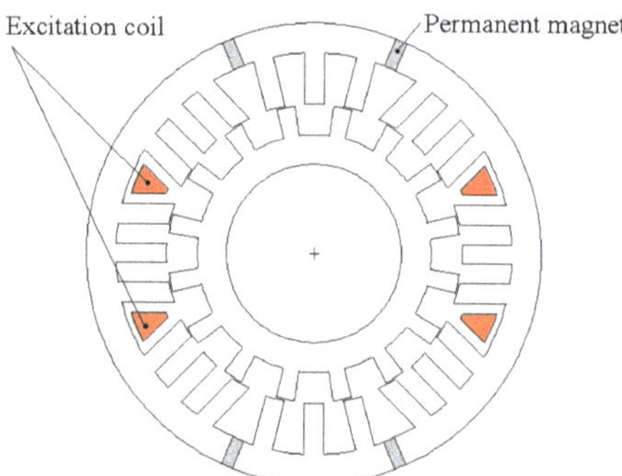

Figure 3. Doubly salient hybrid excited structure (inner rotor) [46].

Figure 4 shows a new hybrid excited structure where the electromagnetic field sources (permanent magnets, excitation windings and armature windings) are all located in static parts, but armature windings and excitation field sources (permanent magnets and excitation windings) are placed in separate stators. The rotor is completely passive. This structure helps overcome one of the drawbacks of previous flux-switching machines by a better space utilization. Magnetic saturation should appear for higher values of current densities, which should help improve the torque density [19,47].

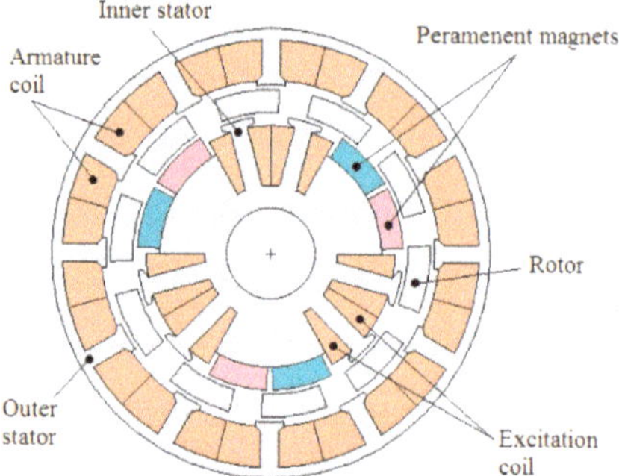

Figure 4. Separate stators hybrid excitation machine.

Figure 5 shows a hybrid excitation structure recently studied in [37]. Excitation coils are located in the stator avoiding the sliding contacts. Permanent magnets are present in both the stator and rotor armatures. All magnets are polarized in the same direction. Another original hybrid excited structure is presented in Figure 6. In this structure, the hybrid excitation principle is used within an electrical variable transmission. More details about the operation of this machine could be found in reference [32,33]. The structure shown in Figure 7 has been investigated in [6]. In this topology and that of Figure 3, the field created by magnets is in series with the dc excitation field. This limits the flux-adjusting capability because of the low-permeability of magnets. The location of excitation coils in the moving part will be an additional drawback. Other 2D structures have also been studied in [48–51]. Interested readers may consult these references. All structures presented previously are structures in which magnetic fluxes have a 2D nature. Even if 3D structures are relatively more difficult to analyze and manufacture than 2D ones, research on hybrid excitation machines having 3D structures is still relatively important. Figure 8a shows a hybrid excitation structure that can be considered as the combination of two synchronous structures, a classical permanent magnet structure in the middle and two homopolar inductor structures at both ends [52]. The basic operating principle of this kind of hybrid excitation structure has been previously described in [48].

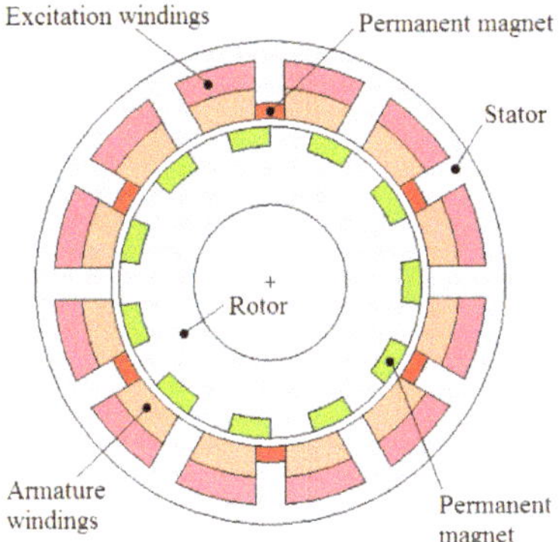

Figure 5. Hybrid excited dual-permanent-magnet (PM) machine [37].

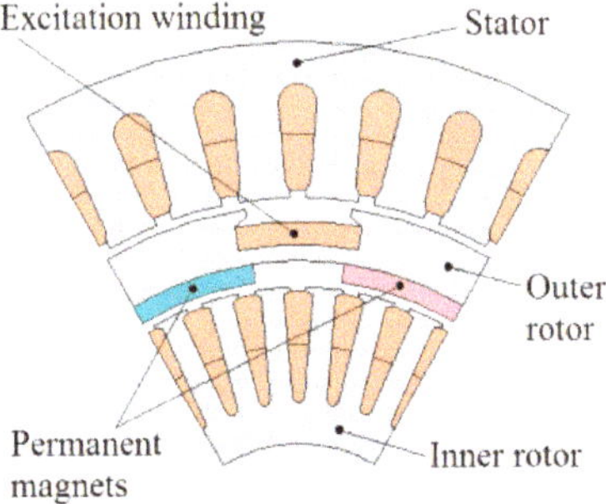

Figure 6. Electrical variable transmission with hybrid excitation [32,33].

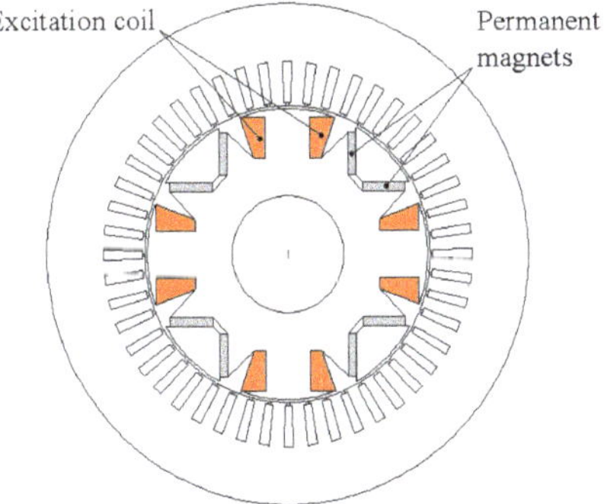

Figure 7. Series hybrid excited structure [6].

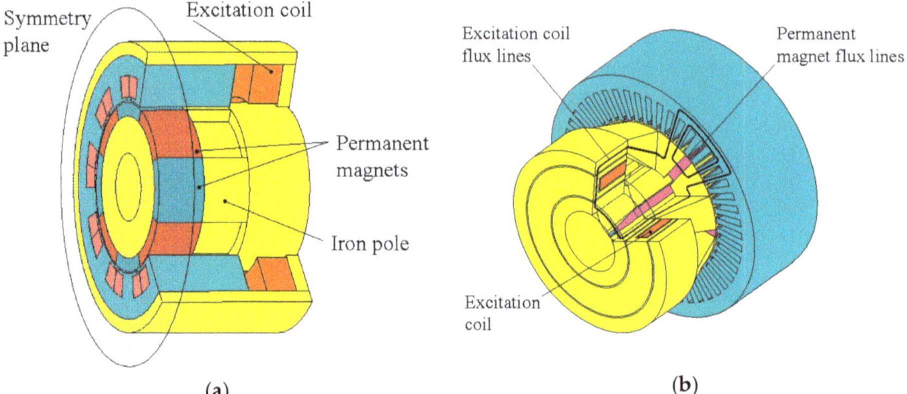

Figure 8. 3D parallel hybrid excited structures. (**a**) 3D juxtaposed structure [52]; (**b**) 3D imbricated structure [49,50].

Many 3D hybrid excited structures based on this principle have been recently studied [12,13,36,38,51,53]. Figure 8b illustrates another 3D hybrid excitation structure [10,34,49,54,55]. To a certain extent the basic operating principle of this structure is similar to that of the structure described in [1]. For the structure of Figure 8b, excitation coils are deported to a static part located at an axial end of the machine. This static part could be surrounded by the rotor's flux collectors (radial auxiliary air-gaps) [10,34,49,54], or in front of these flux collectors (axial auxiliary air-gaps) [55]. The advantage of deporting the excitation windings to this location is the reduction of copper volume and as a consequence the excitation Joule loss and the total machine volume and weight. However, an increase of the machine's axial length can be feared, and the adopted solution implies the presence of additional air-gaps in the flux path of dc excitation. It should be noted that many hybrid excitation structures have been reported in patent applications [48,56–75]. While the first applications were from European countries, Japan and USA, there is a significant increase of patent applications from China [57–59,67–75].

3. State of the Art of MEC Modeling

The magnetic equivalent circuits (MECs) modeling approach has been introduced in the late nineteen-sixties [76,77] and early nineteen-seventies [78]. More lately, the MEC modeling started to regain popularity among machine designers but MEC methods lack the genericity when compared to FEA. From the commercial software side, the MEC software is far less widespread than the FEA software. Two types of MEC approaches are mainly employed: Expert reluctance network (also called in literature lumped parameter MEC models) on one side and mesh-based reluctance network (MbRN) on the other side. Lumped parameter models are specifically developed for a dedicated topology and are based on the expertise of the designer. These models often need a prior knowledge of flux paths in the studied topology as shown in the works of Liu et al. [79] and Tang et al. [80]. On the other side, MbRN as a more generic approach is based on the space discretization of the studied domain with multi-directional reluctance block elements. Bidirectional blocks are used in 2D models [81,82] and axial reluctance branches are added to complete the third direction in 3D models [83,84]. Figure 9 shows an example of 2D and 3D reluctance elements.

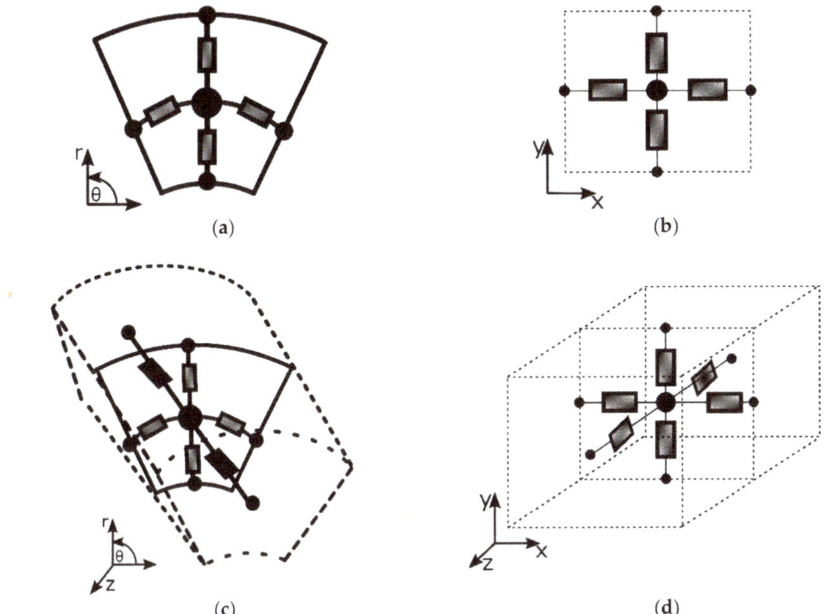

Figure 9. 2D and 3D reluctance block elements. (**a**) Rectangular bidirectional reluctances; (**b**) cylindrical bidirectional reluctances; (**c**) parallelepiped 3D reluctance block element; (**d**) cylindrical tridimensional reluctance block element.

MECs are also often used in hybrid modeling techniques combining either the analytical formulation, FEA or boundary element method. To model flux-switching machines, the analytical formal solution has been used in combination with MbRN by Laoubi et al. [85] and with lumped parameter MEC by Ilhan et al. in [86]. In [87], Pluk et al. have used 3D MbRN combined with 3D Fourier modeling to model a linear and planar actuator. The boundary element method combined with MEC has been used by Martins Araujo et al. in [88] to model a linear actuator. The FEA-MEC combination has been applied to model a permanent magnet machine by Philips in [89]. Regardless, a little number of computer-aided design (CAD) software exploiting RN modeling has appeared. In this section also, existing tools based on RN modeling are overviewed. On the academic side and in literature, we can mention Turbo-TCM [90] dedicated to small-power turbo-alternator modeling. On the commercial software side, RMxprt® in the ANSYS® Electromagnetic package [91] includes pre-defined designs of stator and rotor topologies that can be combined into one whole machine model for performance assessment but very few information on its working principle are available. SPEED [92] developed by Speed Laboratories (University of Glasgow), uses various analytical formulations as complementary to FEA but again with pre-defined geometries. In another approach, Reluctool® developed by G2ELab (Grenoble, France) is based on lumped parameter MEC for the modeling of electromagnetic devices and includes an optimization module for pre-design purposes [93,94]. Reluctool® models are intimately linked to a given topology, and the reluctance network needs to be built based on the expertise of the designer. All the previously mentioned software come with a graphical interface that allows interactions with the user/designer but none of them allows the automated processing of an arbitrary geometry. On this aspect, for a given structure, a dedicated MEC model needs to be developed. This makes model development duration longer for MEC methods as compared to FEA. Furthermore, if geometry parameters vary in a large scale, the model will no longer be valid and will have to be readjusted. The MEC modeling approach proposed in this study can be referred to as MbRN. It has been developed by many researchers [23,24,77,83,95,96]. The goal of this approach, as indicated earlier, is to overcome the genericity limitation of the classical MEC approach. Developing an analysis and design tool, which can compete with the finite element method in term of precision/computing time ratio, motivated researchers that have studied this technique. Even with a more generic technique, most works using the MEC modeling method have been dedicated to specific topologies as induction machines by Perho [25] or more recently, PM flux-switching machine by Benhamida et al. [97]. The 3D MEC modeling has been used to model a homopolar hybrid excitation synchronous machine with distributed windings and interiors permanent magnets in [98] and lumped parameter MEC for flux concentrating hybrid excitation machine in [99]. This technique consists of meshing the studied object, areas or volumes, using 2D or 3D reluctance block elements, respectively (see Figure 9) [24,82,84,100,101]. The FEM and meshed-based MEC methods share some common meshing rules, i.e., some areas of the studied object have to be more finely meshed than others (air-gap in electric machines). Different aspects related to the mesh-based generated MEC method could be found in [23–26,82,84,102,103]. A comprehensive review of 3D MEC modeling can be found in [84]. More details on the mesh-based reluctance network model of the hybrid excitation machine studied in this paper will be given in Section 4.2.

4. Hybrid Excitation Topology

Figure 10 shows a 3D cut view of the studied machine. It combines a wound field excitation with a permanent magnet's excitation. To avoid sliding contacts, excitation windings are located on top of armature end-windings in the stator part of the machine. The basic operating principles of this topology is similar to a structure studied in [50]. Nevertheless, there are a few differences between both topologies. These differences will be highlighted in Section 4.1.

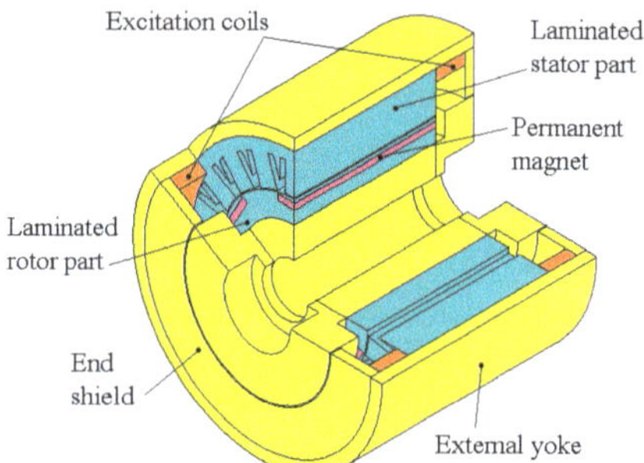

Figure 10. 3D cut view of the hybrid excitation machine.

4.1. Configuration and Operating Principle

The stator of the studied machine is composed of a laminated core, solid iron yoke and end-shields, AC three-phase windings, with concentrated coils, and two excitation annular coils. Solid iron components (external yoke and end-shields) provide a low reluctance path for wound field excitation flux. The rotor is, amongst other things, composed of two solid iron collectors located at both axial ends. Between the two rotoric flux collectors, a solid iron cylinder, located in the axial active length, is connected to both of them. A laminated cylinder in which six permanent magnets are embedded surrounds this massive cylinder. The six permanent magnets create the same type of magnetic poles, either North or South. Between two magnets, there is a laminated iron pole. As for solid iron parts of the stator, the two rotoric flux collectors and the massive cylinder offer a low reluctance path for wound field excitation flux. As for the machine studied in this paper, the structure presented in [50] have two annular excitation coils located in the stator (Figure 11). The machine studied in this paper is illustrated in Figure 11a. Annular excitation coils are placed above the armature end-windings. Another disadvantage affecting the efficiency of flux control using excitation windings is the flux cross-section that is located at the air-gaps between inner radii of end-shields and rotoric flux collectors. The flux cross-section will be further reduced when the annular excitation windings are placed under the armature end-windings. This matter is illustrated in Figure 11b where R_{min} is the radius corresponding to the smaller wound field flux cross-section.

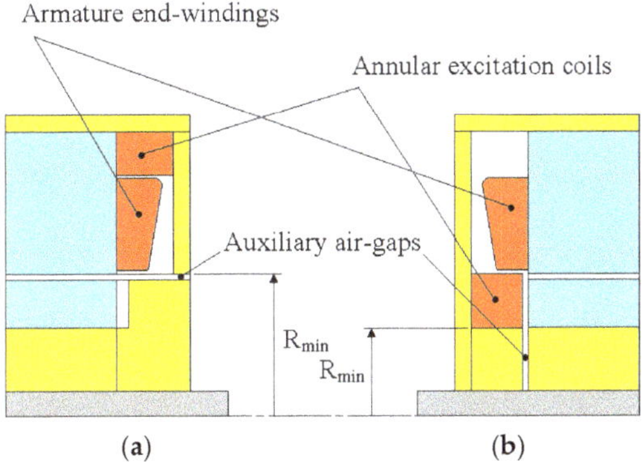

Figure 11. Annular excitation coils location. (**a**) Above armature end-windings; (**b**) below armature end-windings.

This implies that magnetic saturation may affect wound field excitation for lower values of excitation current and consequently reduce the flux control capability. Furthermore, while in [50], authors used soft magnetic composite (SMC) material to provide a low reluctance path for wound field excitation flux; massive iron is used in the machine studied in this paper. In addition, the machine presented in [50] is completely enclosed. This constitutes a drawback from the thermal point of view. Figure 12 shows principal flux trajectories of PM excitation flux. Flux trajectories can be divided into two categories: Bipolar flux lines and homopolar flux lines. The presence of a homopolar flux trajectory implies the presence of a DC component in armature flux linkage. Figure 12 also shows wound field excitation flux trajectories. Both annular excitation coils create a magnetic flux having a homopolar trajectory. They both create the same kind of magnetic poles, either North or South, depending on the circulation direction of excitation current in the excitation coils. Since the permanent magnets relative permeability is close to that of air, the flux created by the excitation coils will mainly circulate through the laminated iron pole located between the magnets. The excitation flux control is achieved by acting on the peak-to-peak amplitude of excitation flux using the excitation coils. If the excitation coils create magnetic poles with a reverse polarity as compared to magnets poles, the peak-to-peak amplitude of excitation flux linkage will be enhanced. Otherwise, the peak-to-peak value will be weakened.

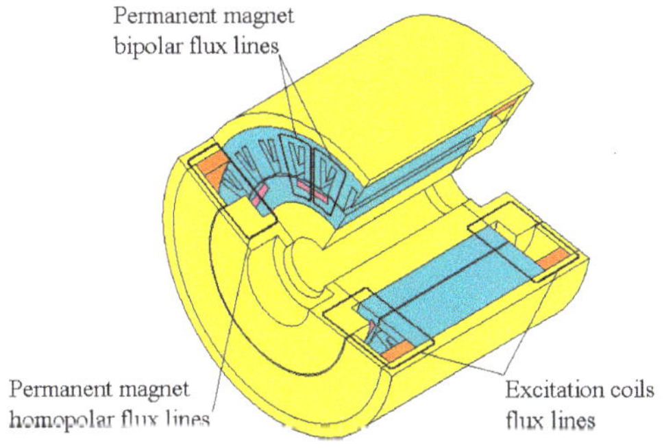

Figure 12. Main magnetic flux paths of the hybrid excitation machine.

4.2. Prototype Construction

In order to verify the operating principle and to assess the flux control capability of the proposed field sources combination, a prototype has been designed and constructed. The prototype has been designed via a parametric study, by use of the finite element method. The initial design parameters of this machine have been derived from a simple analytical model based on a simple reluctance network. For the double excitation circuit's design, the principle of equalization of flux cross-sections has been used [5]. The design constraints to be satisfied are given in Table 1. Figure 13a,b shows respectively, longitudinal cut view, and stator and rotor laminations of designed machine with main geometric dimensions. Values of these parameters are given in Table 2. Instead of being perfect cylinders, massive rotoric parts, are hollow cylinders with conical shapes in order to reduce rotor inertia.

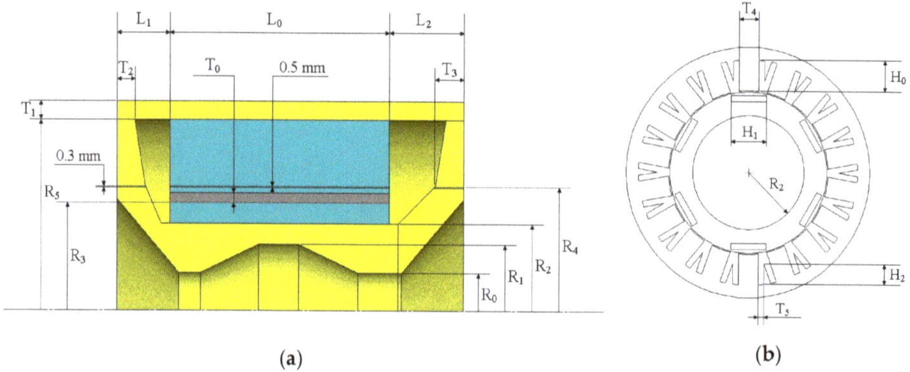

Figure 13. Main design dimensions. (**a**) Longitudinal cut view; (**b**) stator and rotor laminations.

The end-shields thickness at inner radius is greater than its thickness at outer radius, as can be seen in Figure 13a. It should be noted that air-gap thickness in the machine's active part (0.5 mm) is slightly greater than the air-gap thickness between end-shields and rotoric flux collectors at both axial ends (0.3 mm). The number of armature windings turns is equal to three. The length between the stator laminated stack axial end, at one side, and the machine axial end, at the same side, is greater at one side as compared to the other side ($L_2 > L_1$). Figure 13b shows the stator and rotor lamination sheets (M270-35A). Massive parts are made of solid iron XC18 (see Table 3).

Table 1. Design constraints.

Design Constraints Parameters	Values
Overall dimensions (Diameter × Length)	200 mm × 200 mm
Nominal torque T_n	80 N·m
Overload torque T_{max}	160 N·m
Base speed Ω_b	2 000 rpm
Maximum speed Ω_{max}	12 000 rpm
Maximum battery DC voltage	300 V

Table 2. Machine main geometric dimensions.

Geometric Dimensions	Values (mm)
R_0, R_1, R_2, R_3, R_4 and R_5	20, 36, 46.5, 55.4, 62.3 and 91.5
T_0, T_1, T_2, T_3, T_4 and T_5	4, 9, 9, 16, 15.5 and 5
L_0, L_1 and L_2	125, 35 and 42
H_0, H_1 and H_2	20, 29.5 and 16

Table 3. Double excitation synchronous machine data.

Parameter	Value/Designation
Stator outer diameter	201 mm
Machine's axial length	197 mm
Lamination material	M270-35A
Massive parts material	XC18
Magnet type	NdFeB 35EH (Br = 1.2 T)
Number of poles	12
Number of phases	3
Number of turns of each excitation coil	200
Excitation round wire dimension	∅ 1 mm
Number of turns of armature windings	3
Rectangular wire dimensions	5 mm × 1.12 mm
Armature phase resistance	16 mΩ
Excitation coils total resistance	4.73 Ω

Armature windings are constituted by connecting non-overlapping concentrated windings realised using rectangular section wires (see Figure 14). The triangular shape of lamination in each stator slot helps improve the heat transfer. The use of this kind of wire helps to improve the stacking factor and reduce armature end-windings volume. As can be seen from Figure 14, the armature end-windings volume is quite small. However, additional AC Joule loss can be feared due to the large wire section. Nevertheless, stranding armature windings conductors can reduce these losses. Figure 15 shows respectively the rotor and the machine during assembling.

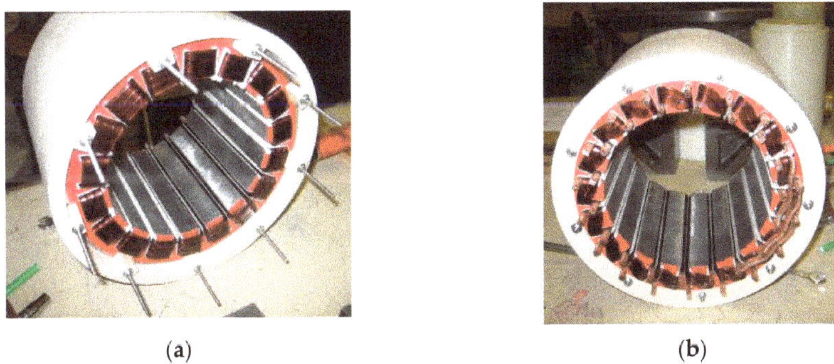

Figure 14. Stator's active part assembling. (**a**) Front view; (**b**) back view.

Figure 15. Machine during assembling. (**a**) Rotor assembling; (**b**) machine assembling.

5. Modeling of the Hybrid Excitation Synchronous Machine

Electric machines models are established for analysis and design purposes. Two methods are used, in this study, for the modeling of the double excitation machine: Finite element method (FEM), and magnetic equivalent circuits method (MEC). The finite element method, which is a numerical analysis method, is time consuming, in particular for 3D problems and especially at first design stages. This is why the MEC method, which is a semi-analytical method, is also used. The MEC method presents a good compromise between accuracy and computation time. However, the MEC method is not as generic as the finite element method. To overcome this, an improved and more generic MEC modeling approach is used [23–26,84]. The modeling study using the two methods is presented in this section.

5.1. 3D Finite Element Method

The structure of the studied machine requires the use of the 3D finite element multi-static analysis. Figure 16 shows the 3D finite element mesh of the studied machine (133,958 nodes). The non-linearity of B-H curves of the different parts of the machine is considered.

The laminated parts are modeled using anisotropic material characteristics. Due to symmetry consideration, only 1/6 of the machine is modeled (one unique pole pair is considered). The magnetic scalar potential formulation is used. The mesh of only 1/12 of the machine is shown to highlight the smoothness of the mesh.

The finite element calculations are done considering two air volumes at axial ends of the machine [5]. The Dirichlet boundary condition is applied to the bounding surface in both axial limits of the finite element model and for R_{ext} (the machine's external radius). The developed model takes into account the rotor motion.

The air-gap is divided into two parts; a part is linked to the rotor and the other part to the stator. Motion consideration has already been described in [5]. The lamination effect is considered in the finite element calculations via an anisotropic material property for laminated parts. Computation of relative permeability in the perpendicular direction to the lamination is described in [5,104].

The FEA is used for the analysis of flux control capability of the double excitation machine. To do so, flux variation with the rotor position is first calculated for different values of excitation current. The flux linkage in the three phases is calculated by getting flux density distributions in the three teeth. The EMF is obtained by differentiating the flux linkage.

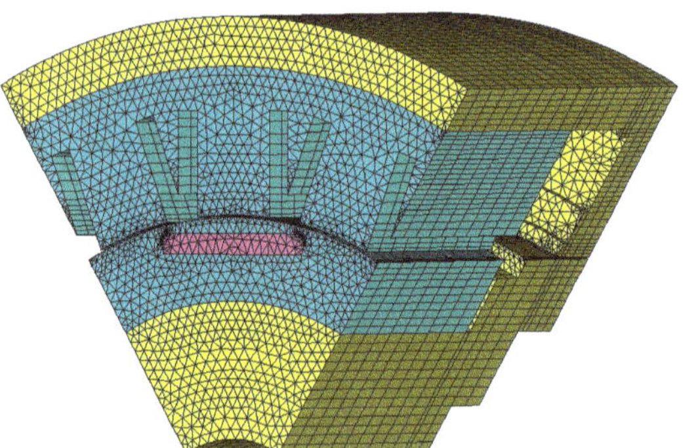

Figure 16. 3D finite element mesh.

5.2. Magnetic Equivalent Circuit Model

The used MEC modeling technique applied to the HESM consists of meshing the studied object, areas or volumes, using 2D or 3D reluctance elements, respectively. The elementary reluctance blocks used for 2D and 3D problems, concur to the geometries of flux tubes that appear most often in electromagnetic devices [82,86,102,105].

5.2.1. Mesh Generation Algorithm and Modeling of Motion

Figure 17a shows a parallelepiped flux path region and its corresponding passive 2D element. The values of permeances P_v and P_w are given by Equation (1) where l and h are respectively the element dimensions in the v and w directions. Accordingly, Figure 17b shows a cylindrical bidirectional reluctance flux path region also for a passive 2D element. The values of permeances P_{r1} and P_{r2} in the radial direction and permeance P_θ in the θ directions are given by Equation (2) where r_1, r_2 and r_3 are respectively the lower, the mid and the higher radius delimiting the reluctance block element and Δθ its opening angle.

$$\begin{cases} P_v = \mu_0 \mu_r \frac{h}{l} \\ P_w = \mu_0 \mu_r \frac{l}{h} \end{cases} \quad (H/m) \tag{1}$$

$$\begin{cases} P_{r1} = \mu_0 \mu_r \frac{\Delta\theta}{\ln\left(\frac{r_2}{r_1}\right)} \\ P_{r2} = \mu_0 \mu_r \frac{\Delta\theta}{\ln\left(\frac{r_3}{r_2}\right)} \\ P_\theta = \mu_0 \mu_r \frac{\ln\left(\frac{r_3}{r_1}\right)}{\Delta\theta} \end{cases} \quad (H/m) \tag{2}$$

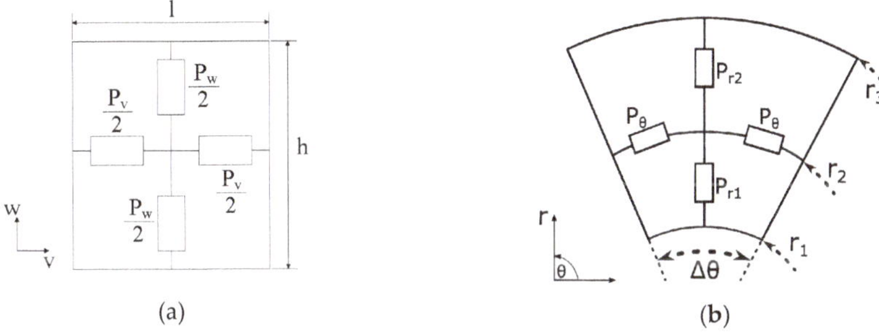

Figure 17. Flux paths regions and corresponding 2D elementary reluctance block elements. (**a**) Parallelepiped bidirectional reluctance block element; (**b**) cylindrical bidirectional reluctance block element.

Figure 18a,b show some elements used for the mesh of the studied machine. Figure 18c shows the different components contained in each branch of those elements. For a completely passive element, the MMF sources $Fs_{ei} = 0$ A and the flux sources $\Phi s_{ei} = 0$ Wb (i = 1, 2, 3, 4, 5, 6 in Figure 18a), or (i = 1, 2, 3, 4, 5 in Figure 18b). Figure 19 illustrates how the value of relative permeability in z direction (axial direction) is estimated for laminated machine parts. Laminated parts are considered as a succession of ferromagnetic and non-magnetic (lamination insulation and parasitic air-gaps) materials. A packing factor k_f, defined as the total length of ferromagnetic steel parts divided by total laminated pack length (active length), is set to 97%. Equation (3) gives then the value of the equivalent relative permeability in axial direction. μ_r is the relative permeability of ferromagnetic parts.

$$\mu_{rz} = \frac{\mu_r}{k_f + \mu_r \cdot (1 - k_f)} \tag{3}$$

As for the finite element computations, two air volumes at axial ends of the machine are considered in the MEC model. The lamination effect is also taken into account in the same way as the finite element model. The nodal method is used to formulate the MEC equations system. The unknowns for the generated circuit equations system are the magnetic scalar potentials at each node. The equations system referred to in Equation (4) is solved using the MATLAB software. For that purpose, it is expressed using the matrix formulation where, [P] is the permeance matrix, [U] is the magnetic scalar potential vector, [Φ] is the flux source vector and n is the total number of nodes. For each moving armature/stator relative position a new equations system is established. It should be noticed that only the air-gap region has to be remeshed for each position, and that the reluctances connecting the nodes located at the stator/air-gap and moving armature/air-gap interfaces have to be recalculated.

$$[P]_{n\times n}[U]_{n\times 1} = [\Phi]_{n\times 1} \tag{4}$$

Figure 20 summarizes the mesh-based generated MEC method incorporating magnetic saturation consideration and motion. The first step is to mesh the different regions of the studied object. Then, comes the node numbering before calculating the matrices [P] and [Φ]. After the solving of the algebraic system, the permeances of the block elements modeling the ferromagnetic parts are recalculated by adapting their permeabilities via the iterative process till convergence towards a magnetic equilibrium state.

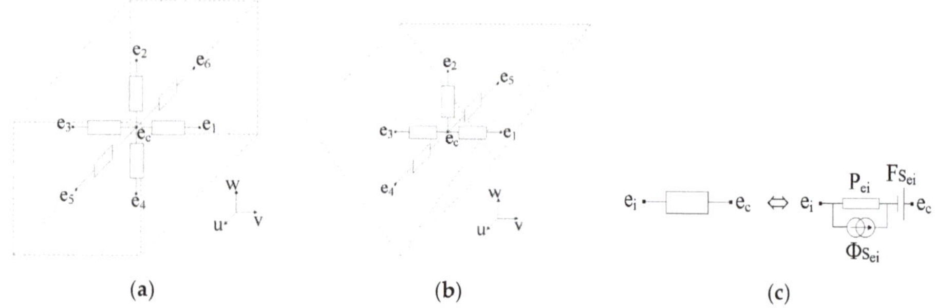

Figure 18. 3D elementary reluctance blocks. (**a**) 7-node block element; (**b**) 6-node block element; (**c**) branch components.

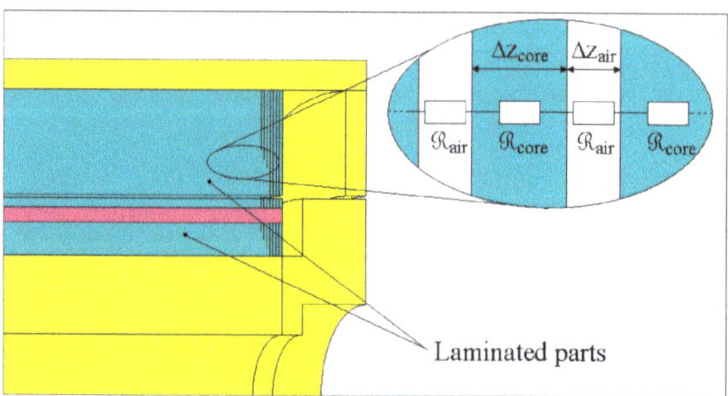

Figure 19. Lamination effect modeling.

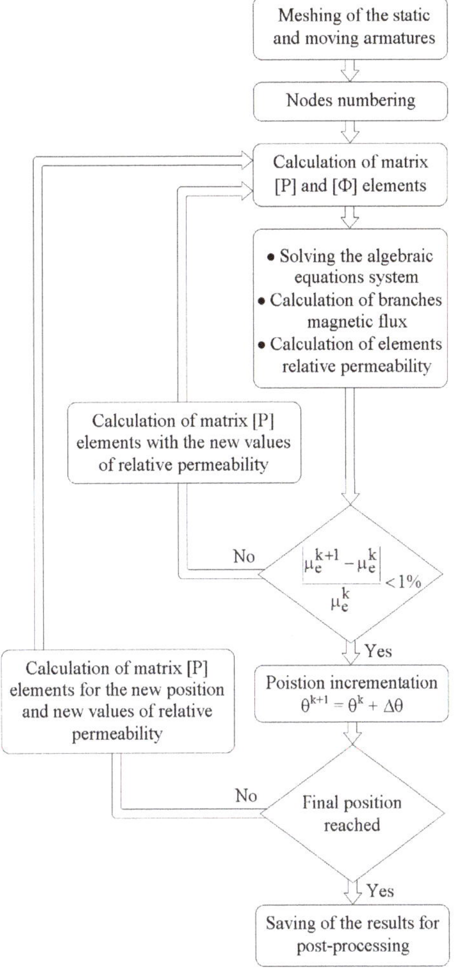

Figure 20. Mesh-based generated MEC method algorithm.

5.2.2. Modeling of Magnetic Field Sources

The modeling of permanent magnets and electric coils is discussed in this section. Permanent magnets can either be modeled by a flux source in parallel with a permeance or a magneto-motive force (MMF) in series with a permeance, as shown in Figure 21. Expressions of the different parameters of a permanent magnet region model depend on PM characteristics and the region geometry and dimensions. These expressions for a parallelogram PM region are given by Equation (5) where, B_r and μ_r are, respectively, the magnetic remanence and the relative permeability of the permanent magnet. l_{pm}, h_{pm} and w_{pm} are, respectively, PM length, height and width. Φ_{pm} and F_{pm} are PM flux source and PM MMF source.

$$\begin{cases} P_{pm} = \mu_0 \mu_r \frac{l_{pm} w_{pm}}{h_{pm}} \\ \Phi_{pm} = B_r l_{pm} w_{pm} \\ F_{pm} = \frac{B_r}{\mu_0 \mu_r} h_{pm} \end{cases} \quad (5)$$

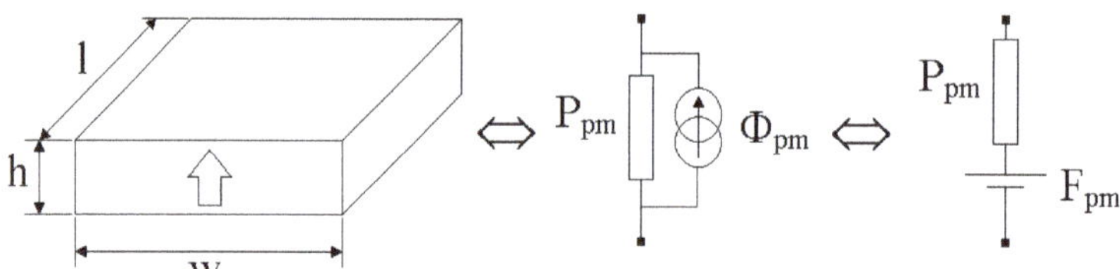

Figure 21. Permanent magnet region modeling.

There are two types of electric coils in double excitation machines: Armature windings and wound field excitation coils. Armature windings are often distributed in slots and can be divided into two parts: Conductors contained in the slot volumes and armature end-windings. Figure 22a shows the MMF variation with w coordinate for the coil and the maximum MMF value is equal to the product of the number of turns and current in one conductor ($F_{mMax} = N_t I$, where N_t is the number of turns and I the value of armature current in one conductor in the slot). Values of MMF sources for the different elements, es1, es2, es3 and ey1, depend on the geometric dimensions of these elements and the value of armature current in the slot as illustrated in Figure 22b,c. Expressions of MMF sources in these elements are given by Equation (6).

$$\begin{cases} Fm_{es1} = \frac{(w_1 - w_0)(v_1 - v_0)}{4(w_2 - w_0)(v_2 - v_0)} Fm_{Max} \\ Fm_{es2} = \frac{(w_1 - w_0)(v_2 - v_1)}{4(w_2 - w_0)(v_2 - v_0)} Fm_{Max} \\ Fm_{es3} = \frac{(w_2 + w_1 - 2w_0)}{4(w_2 - w_0)} Fm_{Max} \\ Fm_{ey1} = \frac{Fm_{Max}}{2} \end{cases} \tag{6}$$

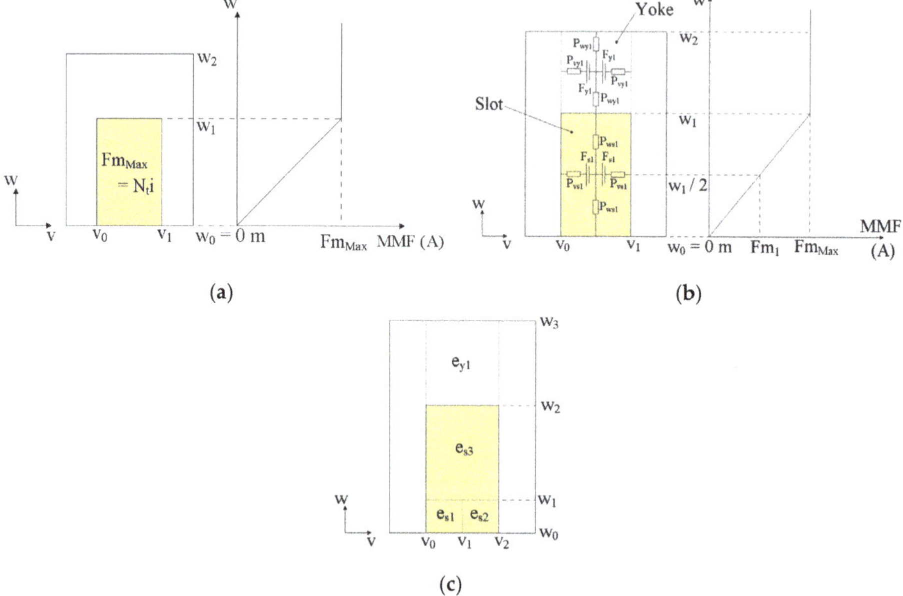

Figure 22. Magneto-motive force (MMF) coil variation.

The same technique is used for modeling the end-windings part of the armature coil and for both annular excitation coils (see Figure 23). MMF will vary accordingly with the variation of current at each motion step.

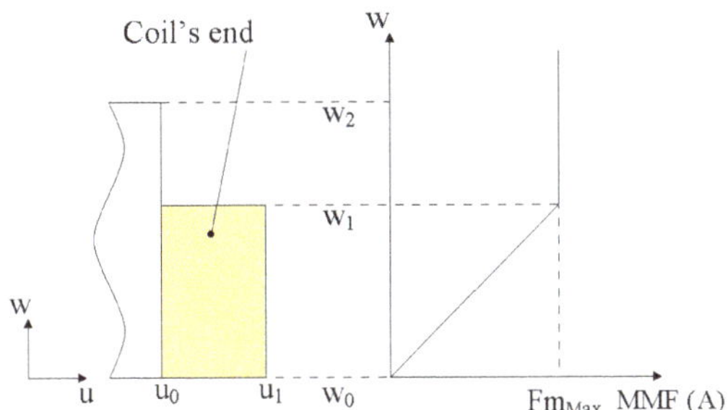

Figure 23. Armature end-windings modeling.

5.2.3. Air-Gap Modeling

In order to model the air-gap, two types of reluctance elements can be used: Unidirectional (radial direction for radial flux rotating machines) or multi-directional reluctance blocks (see Figures 9 and 17). This depends on the global quantity which is sought (flux or torque), the calculation time constraint and the sought precision [25,82]. For the study of flux control capability the use of unidirectional reluctances is largely enough. Furthermore, the use of unidirectional reluctances implies a reduced number of nodes, as compared to multi-directional reluctances, and as a consequence a reduced calculation time.

Figure 24 illustrates, on a simple 2D case, how the value of the reluctance between a static element and a moving element is calculated as a function of the moving armature relative position. For clarity reasons, only one static element e_1 from the stator and one moving element e_2 from the moving armature are represented (Figure 24a). The value of the unidirectional permeance between the two elements (see Figure 24b) is given by Equation (7) where, l_{ag} is the air-gap length, L_a is the elements axial length (it is supposed to be the same for both elements in 3D problems), and Δv is given by Equation (8). α and β values are given by Equation (9).

$$P_{e_1e_2} = \mu_0 \frac{L_a \Delta v}{l_{ag}} \tag{7}$$

$$\Delta v = \alpha(v_{e_1 2} - v_{e_2 1}) + \beta(v_{e_2 2} - v_{e_1 1}) - \alpha\beta(v_{e_1 2} - v_{e_1 1}) \tag{8}$$

$$\begin{cases} \alpha = \begin{cases} 1 & \text{if } v_{e_1 1} \leq v_{e_2 1} \leq v_{e_1 2} \\ 0 & \text{otherwise} \end{cases} \\ \beta = \begin{cases} 1 & \text{if } v_{e_1 1} \leq v_{e_2 2} \leq v_{e_1 2} \\ 0 & \text{otherwise} \end{cases} \end{cases} \tag{9}$$

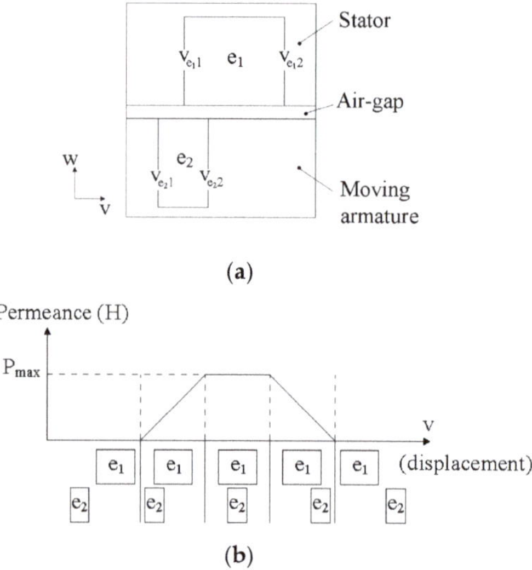

Figure 24. Air-gap permeance (reluctance) calculation.

5.2.4. Equations System Solution and Magnetic Saturation

Figure 25 illustrates how elements of the matrix [P] and [Φ] are determined from Kirchhoff's laws. According to Kirchhoff's laws, it can be established that the sum of fluxes going into each node is null and the magnetic potential difference of two nodes is equal to the flux of the branch linking both nodes divided by the permeance of the same branch as shown in Equations (10) and (11), respectively. Elements of matrix [P] and [Φ] can be directly determined from Equation (11). For the nodes that are not directly connected to the i_{th} node, the values of P_{ij}, Fms_{ij} and Φs_{ij} are null.

$$\begin{cases} \sum\limits_{\substack{j=1 \\ j\neq i}}^{n} \Phi_{ij} = 0\ \mathrm{Wb} \\ U_i - U_j = Fms_{ij} - \frac{(\Phi_{ij} - \Phi s_{ij})}{P_{ij}} \end{cases} \tag{10}$$

$$\left(\sum_{\substack{j=1 \\ j\neq i}}^{n} P_{ij} \right) U_i + \sum_{\substack{j=1 \\ j\neq i}}^{n} (-P_{ij}) U_j = \sum_{\substack{j=1 \\ j\neq i}}^{n} (\Phi s_{ij} + P_{ij} Fms_{ij}) \tag{11}$$

In order to take into account the magnetic saturation, the equations system is solved iteratively by adjusting the value of the permeance matrix [P] elements at each iteration. The convergence criterion is given by Equation (12) where $\mu_e{}^k$ and $\mu_e{}^{k+1}$ correspond to the value of relative permeability in the reluctance element e respectively at the kth and (kth + 1) iteration of the iterative process described earlier (see Figure 20).

$$\frac{\left| u_e^{k+1} - u_e^k \right|}{u_e^k} < 1\% \tag{12}$$

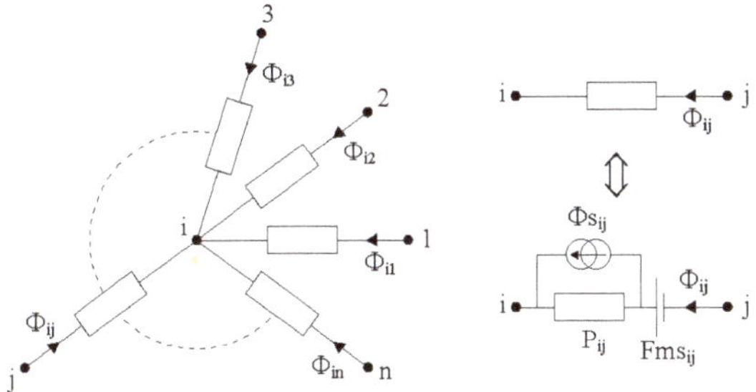

Figure 25. Mesh equation setting for the ith node.

The open-circuit flux in a phase is estimated, for each position, by averaging radial flux passing through the tooth region covered by the concentrated winding (Figure 26). The average flux passing through the concentrated winding is calculated using elements e_{t9} to e_{t14}, which are the elements belonging to the tooth region covered by the concentrated winding in Figure 26a. Previous to describing how the open-circuit flux linkage is calculated, the radial flux passing through an element should be defined (Figure 26b); it is given by Equation (13). The flux passing through the concentrated winding is then calculated as follows:

- The radial flux (w direction in 0) passing through the layer containing elements e_{t9} to e_{t12} is first calculated as shown by Equation (14);
- Then, the radial flux passing through the second layer containing elements e_{t13} and e_{t14} is calculated as shown by Equation (15);
- Finally, the average flux linkage, per turn, passing through the concentrated winding is given by Equation (16).

$$\Phi_e = \frac{\Phi_{w1} + \Phi_{w2}}{2} \tag{13}$$

$$\Phi_{l1} = \Phi_{e_{t9}} + \Phi_{e_{t10}} + \Phi_{e_{t11}} + \Phi_{e_{t12}} \tag{14}$$

$$\Phi_{l2} = \Phi_{e_{t13}} + \Phi_{e_{t14}} \tag{15}$$

$$\Phi_w = \frac{(w_2 - w_1)\Phi_{l1} + (w_3 - w_2)\Phi_{l2}}{(w_3 - w_1)} \tag{16}$$

$$\text{EMF} = N_t \frac{d\Phi_w}{d\theta} \Omega \tag{17}$$

$$T = N_t \sum_{i=1}^{3} \frac{d\Phi_{w\_i}}{d\theta} I_i \tag{18}$$

The electromotive force is calculated by the flux derivative as can be shown by Equation (17). The hybrid torque is calculated as can be shown by Equation (18). Torque estimation can also be based either on the variation of magnetic energy [98] or is evaluated via the maxwell stress tensor (MST) method. In order to calculate the torque based on the MST method [82,98,106], access is needed to both normal and tangential components of air-gap flux density. Since the air-gap is modeled using unidirectional reluctances, the use of MST is not possible. However, it is possible to estimate the hybrid component of the torque of the machine [107,108]. The product of current gives the hybrid torque estimation and EMF as shown by Equation (18) where T is the hybrid torque, N_t is the number of turns of armature windings, Ω is the rotational speed, and I is the phase current. The number of nodes for the 3D mesh-based generated MEC model, of the hybrid excitation machine, is equal to 8680. This is fifteen times lower than the number of nodes in the 3D finite element model. Since unidirectional reluctances are used for the modeling of the air-gap, the number of nodes is kept constant for all rotor/stator relative positions.

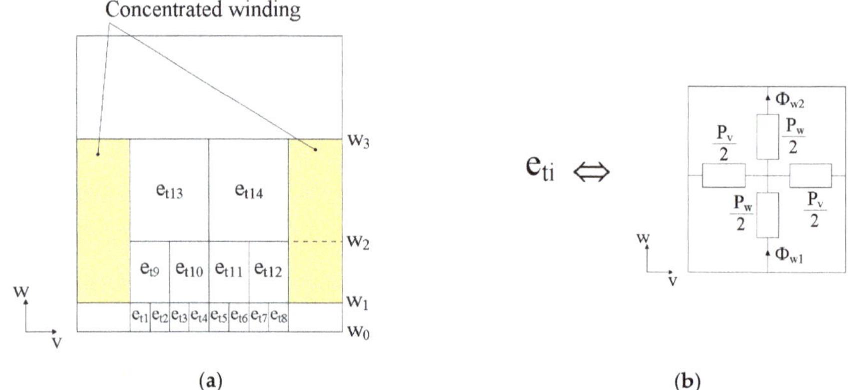

Figure 26. Open-circuit radial flux linkage calculation.

6. Experimental Validation, Comparison and Discussion

In this section, results from both modeling methods: The finite element model (FEM) and the magnetic equivalent circuit model (MEC), are compared to each other and to experimental measurements.

Figure 27 compares the measured open-circuit line-to-line electromotive force (EMF) per-turn waveform for a null value of excitation current (I_{exc} = 0 A) to corresponding waveforms obtained from the FEM and the MEC model. Figure 28a,b shows the same comparison, i.e., open-circuit line-to-line EMF per-turn for $I_{exc} = -4$ A and I_{exc} = 4 A, respectively. As can be seen, a fairly good agreement is achieved between measurements and both modeling methods. Measurements are done for a rotational speed of 170 rpm. It should be noticed that the computation time for the MEC and FEM methods are respectively 4.7 s and 1560 s for one position; computations being done with the same computer. Figure 29 compares the measured RMS value line-to-line flux linkage variations with excitation current to corresponding variations obtained from the 3D FEM and the MEC model. The measured RMS values of line-to-line flux linkage are obtained by first integrating line-to-line EMF waveforms and then calculating the RMS value.

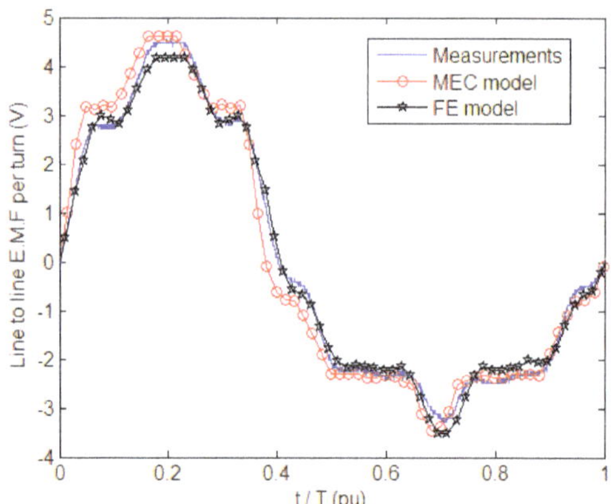

Figure 27. Open-circuit line-to-line EMF per-turn for I_{exc} = 0 A.

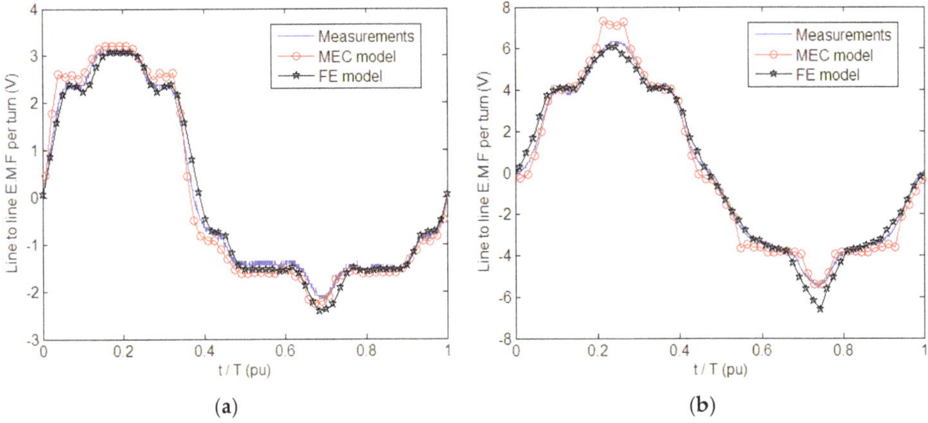

Figure 28. Open-circuit line-to-line EMF per-turn. (**a**) For I_{exc} = − 4 A; (**b**) for I_{exc} = + 4 A.

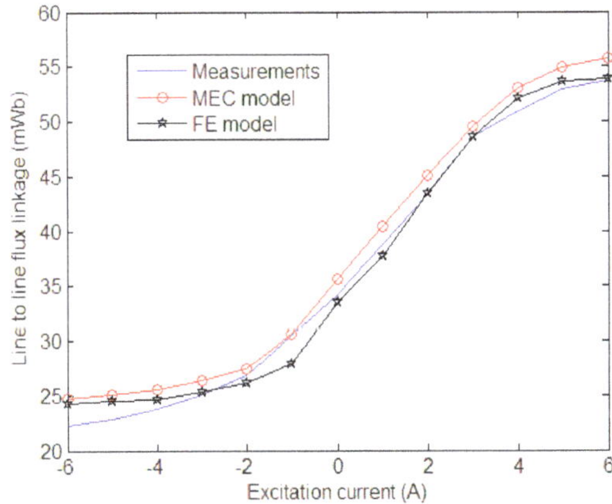

Figure 29. Open-circuit line-to-line maximum flux linkage variations with I_{exc}.

From Figure 29, it can be seen that a wide range of air-gap flux control can be achieved. The air-gap flux changes with a variation of +57% when air-gap flux is enhanced and −35% when it is weakened, with respect to the no-field excitation flux (I_{exc} = 0 A). Numerous reasons can explain the models discrepancies on global quantities such as flux and back-EMF. First, the mesh (spatial discretization) is not the same on both models. Another difference is that the 3D-FEA model is developed on a commercial FEA software and the implemented numerical methods (i.e., derivatives calculations, non-linear behaviors considerations) are not the same as those of the 3D-MEC model developed in MATLAB (see Figure 20).

Figure 30 compares the developed maximum hybrid torque evaluated by the FEA and MEC models with an armature current density of J_{max}= 10 A/mm^2 at a rotational speed of 170 rpm for I_{exc} = −4 A, I_{exc} = 0 A and I_{exc} = +4 A, respectively. It can be seen from Figure 30 that magnetic saturation effects induce an increased error between models. The best agreement between both modeling techniques is obtained at I_{exc} = −4 A when the air-gap flux is weakened and magnetic saturation is low. When air-gap flux is enhanced (I_{exc} = +4 A) the difference between the MEC and FEA models is the greatest.

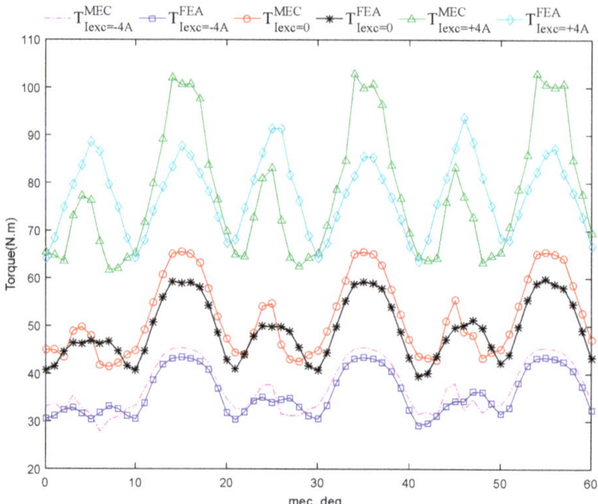

Figure 30. Torque vs. rotor position (MEC and FEA comparison) (170 rpm, J_{max} = 10 A/mm^2).

7. Conclusions

This paper allowed presenting a double excitation machine and studying its field weakening capability. A prototype has been built and delivered to a car manufacturer. The excitation flux control characteristic has been studied experimentally before delivering the prototype. This study has shown the relatively good flux control capability of the prototype. The machine has been modeled using two different modeling methods: The FEA method and mesh-based generated MEC method. The two modeling methods are complementary for a design optimization process. As it has been shown in Section 6 the mesh-based generated MEC model was fairly accurate when compared to the FEM method while necessitating less time. Computation time for the MEC and FEM methods were respectively 4.7 s and 1560 s for one position (computations done with the same computer were divided by $\approx$ 330). Its genericity and time saving, makes it well adapted for optimal design studies in the pre-design stage of electromagnetic devices. These features are even more noticeable in the case of complicated 3D structures as the hybrid excitation structure studied in this paper.

Author Contributions: The work presented here was a cooperative effort by all the authors.

Funding: This research received no external funding.

Acknowledgments: This work was supported in part by "Le Groupement de Recherche Systèmes d'Energie Electrique dans leur Dimension Sociétale" (GdR SEEDS). This scientific study is related to the project Groupe de Travail (GT) "Modèles analytiques". The authors would like to thank GdR SEEDS (France) for the help they provided in this work and on this project.

Conflicts of Interest: The authors declare no conflict of interest.

References

1. Amara, Y.; Vido, L.; Gabsi, M.; Hoang, E.; Ben Ahmed, A.H.; Lecrivain, M. Hybrid Excitation Synchronous Machines: Energy-Efficient Solution for Vehicle Propulsion. *IEEE Trans. Veh. Technol.* **2009**, *58*, 2137–2149. [CrossRef]
2. Finken, T.; Hameyer, K. Study of hybrid excited synchronous alternators for automotive applications using coupled FE and circuit simulations. *IEEE Trans. Magn.* **2008**, *44*, 1598–1601. [CrossRef]
3. Al-Adsani, A.S.; Schofield, N. Hybrid permanent magnet generators for electric vehicle applications. In Proceedings of the 2009 IEEE International Electric Machines and Drives Conference, Miami, FL, USA, 3–6 May 2009; pp. 1754–1761. [CrossRef]

4. Zhao, C.; Yan, Y. A review of development of hybrid excitation synchronous machine. In Proceedings of the IEEE International Symposium on Industrial Electronics 2005, Dubrovnik, Croatia, 20–23 June 2005. [CrossRef]
5. Nedjar, B.; Hlioui, S.; Amara, Y.; Vido, L.; Gabsi, M.; Lécrivain, M. A new parallel double excitation synchronous machine. *IEEE Trans. Magn.* **2011**, *47*, 2252–2260. [CrossRef]
6. Kamiev, K.; Nerg, J.; Pyrhönen, J.; Zaboin, V.; Hrabovcova, V.; Rafajdus, P. Hybrid excitation synchronous generators for island operation. *LET Electr. Power Appl.* **2012**, *6*, 1–11. [CrossRef]
7. Wang, Y.; Deng, Z. Hybrid excitation topologies and control strategies of stator permanent magnet machines for DC power system. *IEEE Trans. Ind. Electron.* **2012**, *59*, 4601–4616. [CrossRef]
8. Yang, H.; Zhu, Z.Q.; Chu, W. Flux adjustable permanent magnet machines: A technology status review. *Chin. J. Electr. Eng.* **2016**, *2*, 14–30. [CrossRef]
9. Wang, Q.; Niu, S. Overview of flux-controllable machines: Electrically excited machines, hybrid excited machines and memory machines. *Renew. Sustain. Energy Rev.* **2017**, *68*, 475–491. [CrossRef]
10. Zhang, Z.; Dai, J.; Dai, C.; Yan, Y. Design considerations of a hybrid excitation synchronous machine with magnetic shunt rotor. *IEEE Trans. Magn.* **2013**, *49*, 5566–5573. [CrossRef]
11. Chen, Z.; Wang, B.; Chen, Z.; Yan, Y. Comparison of Flux Regulation Ability of the Hybrid Excitation Doubly Salient Machines. *IEEE Trans. Ind. Electron.* **2014**, *61*, 3155–3166. [CrossRef]
12. Geng, W.; Zhang, Z.; Member, S.; Jiang, K.; Yan, Y. A New Parallel Hybrid Excitation Machine: Machine with Bidirectional Field-Regulating Capability. *IEEE Trans. Ind. Electron.* **2015**, *62*, 1372–1381. [CrossRef]
13. Wang, H.; Qu, Z.; Tang, S.; Pang, M.; Zhang, M. Analysis and optimization of hybrid excitation permanent magnet synchronous generator for stand-alone power system. *J. Magn. Magn. Mater.* **2017**, *436*, 117–125. [CrossRef]
14. Tapia, J.A.; Leonardi, F.; Lipo, T.A. Consequent-pole permanent-magnet machine with extended field-weakening capability. *IEEE Trans. Ind. Appl.* **2003**, *39*, 1704–1709. [CrossRef]
15. Kosaka, T.; Sridharbabu, M.; Yamamoto, M.; Matsui, N. Design studies on hybrid excitation motor for main spindle drive in machine tools. *IEEE Trans. Ind. Electron.* **2010**, *57*, 3807–3813. [CrossRef]
16. Chen, J.T.; Zhu, Z.Q.; Iwasaki, S.; Deodhar, R.P. A novel hybrid-excited switched-flux brushless AC machine for EV/HEV applications. *IEEE Trans. Veh. Technol.* **2011**, *60*, 1365–1373. [CrossRef]
17. Kamiev, K.; Pyrhonen, J.; Nerg, J.; Zaboin, V.; Tapia, J. Modeling and Testing of an Armature-Reaction-Compensated (PM) Synchronous Generator. *IEEE Trans. Energy Convers.* **2013**, *28*, 849–859. [CrossRef]
18. Giulii Capponi, F.; De Donato, G.; Borocci, G.; Caricchi, F. Axial-flux hybrid-excitation synchronous machine: Analysis, design, and experimental evaluation. *IEEE Trans. Ind. Appl.* **2014**, *50*, 3173–3184. [CrossRef]
19. Hua, H.; Zhu, Z.Q. Novel Parallel Hybrid Excited Machines with Separate Stators. *IEEE Trans. Energy Convers.* **2016**, *31*, 1212–1220. [CrossRef]
20. Bash, M.L.; Pekarek, S.D. Modeling of salient-pole wound-rotor synchronous machines for population-based design. *IEEE Trans. Energy Convers.* **2011**, *26*, 381–392. [CrossRef]
21. Bash, M.L.; Pekarek, S. Analysis and validation of a population-based design of a wound-rotor synchronous machine. *IEEE Trans. Energy Convers.* **2012**, *27*, 603–614. [CrossRef]
22. Sudhoff, S.D.; Shane, G.M.; Suryanarayana, H. Magnetic-equivalent-circuit-based scaling laws for low-frequency magnetic devices. *IEEE Trans. Energy Convers.* **2013**, *28*, 746–755. [CrossRef]
23. Rasmussen, C.B.; Ritchie, E. A magnetic equivalent circuit approach for predicting PM motor performance. In Proceedings of the IAS'97 Conference Record of the 1997 IEEE Industry Applications Conference Thirty-Second IAS Annual Meeting, New Orleans, LA, USA, 5–9 October 1997. [CrossRef]
24. Demenko, A.; Lech, N.; Szelag, W. Reluctance Network Formed by Means of Edge Element Method. *IEEE Trans. Magn.* **1998**, *34*, 2485–2488. [CrossRef]
25. Perho, J. *Reluctance Network for Analysing Induction Machines*; Helsinki University of Technology: Espoo, Finland, 2002.
26. Amrhein, M.; Krein, P.T. Induction machine modeling approach based on 3-D magnetic equivalent circuit framework. *IEEE Trans. Energy Convers.* **2010**, *25*, 339–347. [CrossRef]
27. Kou, B.; Jin, Y.; Zhang, H.; Zhang, L.; Zhang, H. Analysis and design of hybrid excitation linear eddy current brake. *IEEE Trans. Energy Convers.* **2014**, *29*, 496–506. [CrossRef]

28. Li, W.; Ching, T.W.; Chau, K.T. Design and analysis of a new parallel-hybrid-excited linear vernier machine for oceanic wave power generation. *Appl. Energy* **2017**, *208*, 878–888. [CrossRef]
29. Hsu, J.S. Direct control of air-gap flux in permanent-magnet machines. *IEEE Trans. Energy Convers.* **2000**, *15*, 361–365. [CrossRef]
30. Yildiriz, E.; Gulec, M.; Aydin, M. An Innovative Dual-Rotor Axial-Gap Flux-Switching Permanent-Magnet Machine Topology with Hybrid Excitation. *IEEE Trans. Magn.* **2018**. [CrossRef]
31. Nasr, A.; Hlioui, S.; Gabsi, M.; Mairie, M.; Lalevee, D. Design Optimization of a Hybrid-Excited Flux-Switching Machine for Aircraft safe DC Power Generation using a Diode Bridge Rectifier. *IEEE Trans. Ind. Electron.* **2017**, *64*, 9896–9904. [CrossRef]
32. Druant, J.; Vansompel, H.; De Belie, F.; Sergeant, P. Optimal Control for a Hybrid Excited Dual Mechanical Port Electric Machine. *IEEE Trans. Energy Convers.* **2017**, *32*, 599–607. [CrossRef]
33. Druant, J.; Vansompel, H.; De Belie, F.; Melkebeek, J.; Sergeant, P. Torque Analysis on a Double Rotor Electrical Variable Transmission with Hybrid Excitation. *IEEE Trans. Ind. Electron.* **2017**, *64*, 60–68. [CrossRef]
34. Liu, Y.; Zhang, Z.; Zhang, X. Design and Optimization of Hybrid Excitation Synchronous Machines with Magnetic Shunting Rotor for Electric Vehicle Traction Applications. *IEEE Trans. Ind. Appl.* **2017**. [CrossRef]
35. Berkoune, K.; Ben Sedrine, E.; Vido, L.; Le Ballois, S. Robust control of hybrid excitation synchronous generator for wind applications. *Math. Comput. Simul.* **2017**, *131*, 55–75. [CrossRef]
36. Wang, Y.; Deng, Z. A Controllable Power Distribution Strategy for Open Winding Hybrid Excitation Generator System. *IEEE Trans. Energy Convers.* **2017**, *32*, 122–136. [CrossRef]
37. Wang, Q.; Niu, S. A Novel Hybrid-Excited Dual-PM Machine with Bidirectional Flux Modulation. *IEEE Trans. Energy Convers.* **2017**, *32*, 424–435. [CrossRef]
38. Beik, O.; Schofield, N. High-Voltage Hybrid Generator and Conversion System for Wind Turbine Applications. *IEEE Trans. Ind. Electron.* **2018**, *65*, 3220–3229. [CrossRef]
39. Mseddi, A.; Le Ballois, S.; Aloui, H.; Vido, L. Robust control of a wind conversion system based on a hybrid excitation synchronous generator: A comparison between H∞ and CRONE controllers. *Math. Comput. Simul.* **2018**. [CrossRef]
40. Cheraghi, M.; Karimi, M. Optimal design of a Hybrid Excited Doubly Salient Permanent Magnet generator for wind turbine application. In Proceedings of the 2017 8th Power Electronics, Drive Systems & Technologies Conference (PEDSTC), Mashhad, Iran, 14–16 February 2017; pp. 19–24. [CrossRef]
41. Hoang, E.; Lecrivain, M.; Gabsi, M. A new structure of a switching flux synchronous polyphased machine with hybrid excitation. In Proceedings of the 2007 European Conference on Power Electronics and Applications, Aalborg, Denmark, 2–5 September 2007; pp. 1–8. [CrossRef]
42. Sulaiman, E.; Kosaka, T.; Matsui, N. Design and analysis of high-power/high-torque density dual excitation switched-flux machine for traction drive in HEVs. *Renew. Sustain. Energy Rev.* **2014**, *34*, 517–524.
43. Owen, R.L.; Zhu, Z.Q.; Jewell, G.W. Hybrid-excited flux-switching permanent-magnet machines with iron flux bridges. *IEEE Trans. Magn.* **2010**, *46*, 1726–1729. [CrossRef]
44. Hua, W.; Cheng, M.; Zhang, G. A novel hybrid excitation flux-switching motor for hybrid vehicles. *IEEE Trans. Magn.* **2009**, *45*, 4728–4731. [CrossRef]
45. Dupas, A.; Hlioui, S.; Hoang, E.; Gabsi, M.; Lecrivain, M. Investigation of a New Topology of Hybrid-Excited Flux-Switching Machine with Static Global Winding: Experiments and Modeling. *IEEE Trans. Ind. Appl.* **2016**, *52*, 1413–1421. [CrossRef]
46. Zhang, Z.; Tao, Y.; Yan, Y. Investigation of a New Topology of Hybrid Excitation Doubly Salient Brushless DC Generator. *IEEE Trans. Ind. Electron.* **2012**, *59*, 2550–2556. [CrossRef]
47. Du, Y.; Lu, W.; Zhu, X.; Quan, L. Optimal Design and Analysis of Partitioned Stator Hybrid Excitation Doubly Salient Machine. *IEEE Access* **2018**, *6*, 57700–57707. [CrossRef]
48. Syverson, D. Hybrid Alternator. U.S. Patent US5397975A, 14 March 1995.
49. Zhang, Z.; Yan, G.; Yang, S.; Bo, Z. Principle of operation and feature investigation of a new topology of hybrid excitation synchronous machine. *IEEE Trans. Magn.* **2008**, *44*, 2174–2180. [CrossRef]
50. Kosaka, T.; Matsui, N. Hybrid Excitation Machines with Powdered Iron Core for Electrical Traction Drive Applications. In Proceedings of the 2008 International Conference on Electrical Machines and Systems (ICEMS), Wuhan, China, 17–20 October 2008; pp. 2974–2979.
51. Wang, Y.; Deng, Z.-Q. A Position Sensorless Method for Direct Torque Control with Space Vector Modulation of Hybrid Excitation Flux-Switching Generator. *IEEE Trans. Energy Convers.* **2012**, *27*, 912–921. [CrossRef]

52. Fu, X.; Zou, J. Numerical analysis on the magnetic field of hybrid exciting synchronous generator. *IEEE Trans. Magn.* **2009**, *45*, 4590–4593. [CrossRef]
53. Wu, Y.; Sun, L.; Zhang, Z.; Miao, Z.; Liu, C. Analysis of Torque Characteristics of Parallel Hybrid Excitation Machine Drives with Sinusoidal and Rectangular Current Excitations. *IEEE Trans. Magn.* **2018**, *54*, 1–5. [CrossRef]
54. Shi, M.; Zhou, B.; Wei, J.; Zhang, Z.; Mao, Y.; Han, C. Design and practical implementation of a novel variable-speed generation system. *IEEE Trans. Ind. Electron.* **2011**, *58*, 5032–5040. [CrossRef]
55. Zhang, Z.; Ma, S.; Dai, J.; Yan, Y. Investigation of hybrid excitation synchronous machines with axial auxiliary air-gaps and non-uniform air-gaps. *IEEE Trans. Ind. Appl.* **2014**, *50*, 1729–1737. [CrossRef]
56. Greif, H.; Nguyen, N.-T.; Mueller, A.; Reutlinger, K. Electric Machine with a Rotor with Hybrid Excitation. International Patent WO2011036135A1, 31 March 2011.
57. Fang, Y.; Liu, Q.; Ma, J.; Meng, L.; Wang, L.; Xia, C.; Yao, Y. Hybrid Excitation Structure. Chinese Patent CN101814821 (A), 25 August 2010.
58. Xie, S.J.; Xu, Z. Multitooth Magnetic Bridge Type Hybrid Excitation Magnetic Flux Switching Motor. Chinese Patent CN101834474 (A), 15 September 2010.
59. Xia, G.; Zeng, Q.; Cai, Y. Tangential-Set Magnet Double Salient Hybrid Excitation Motor. Chinese Patent CN201403037 (Y), 10 February 2010.
60. Akemakou, D. Double-Excitation Rotating Electrical Machine for Adjustable Defluxing. U.S. Patent US20060119206A1, 8 June 2006.
61. Moynot, V.; Chabot, F.; Lecrivain, M.; Gabsi, M.; Hlioui, S. Rotating Electric Machine with Homopolar Double Excitation. International Patent WO2010052439A2, 14 May 2010.
62. Wang, G. A Hybrid Excitation Synchronous Generator. International Patent WO2009082875A1, 9 September 2009.
63. Reutlinger, K. Electric Machine Comprising a Rotor with Hybrid Excitation. International Patent WO2008148621A1, 19 August 2008.
64. Mizutani, R.; Tatematsu, K.; Yamada, E.; Matsui, N.; Kosaka, T. Rotating Electric Motor. International Patent WO2008093865A1, 7 August 2008.
65. Hoang, E.; Lecrivain, M.; Gabsi, M. Flux-Switching Dual Excitation Electrical Machine. U.S. Patent US7868506B2, 11 February 2011.
66. Aydin, M.; Lipo, T.A.; Huang, S. Field Controlled Axial Flux Permanent Magnet Electrical Machine. U.S. Patent US20070046124A1, 1 March 2007.
67. Yu, H.; Hu, M.; Shi, L.; Qin, F. Hybrid Excitation Linear Synchronous Motor Using Halbach Permanent Magnet. Chinese Patent CN101594040 (A), 2 December 2009.
68. Wang, H.; Zhao, C.; Guo, H. Hybrid Excitation Brushless Synchronous Motor. Chinese Patent CN101752969 (A), 23 June 2010.
69. Ahmad, M.Z.; Sulaiman, E.; Haron, Z.A.; Kosaka, T. Impact of Rotor Pole Number on the Characteristics of Outer-rotor Hybrid Excitation Flux Switching Motor for In-wheel Drive EV. *Procedia Technol.* **2013**, *11*, 593–601. [CrossRef]
70. Liu, J.; Yang, X.; Zhang, J.; Xue, J.; Gao, F.; Xiao, C.; Gao, Y.; Tan, F. Hybrid Excitation Motor for New Energy Automobile. Chinese Patent CN108429421 (A), 21 August 2018.
71. Zhao, J.; Jing, M.; Quan, X.; Sun, X. Direct Predictive Power Control Method for Hybrid Excitation Synchronous Motor. Chinese Patent CN108390602 (A), 10 August 2018.
72. Diao, T. Alternating-Current and Permanent Magnet Hybrid Excitation Doubly-Fed Wind Power Generator and Power Generation System. Chinese Patent CN108282064 (A), 13 July 2018.
73. Chao, Z.; Yuefei, Z.U.O.; Feng, L.I.; Zixuan, X. Hybrid Excitation Direct-Drive Motor. Chinese Patent CN108336837 (A), 27 July 2018.
74. Zhao, J.; Jing, M.; Quan, X.; Sun, X. Non-Salient Pole Type Hybrid Excitation Motor Constant-Power Loss Model Prediction Control Method. Chinese Patent CN108418485 (A), 17 August 2018.
75. Han, W. Control Method for Electric Vehicle Hybrid Excitation Type Internal Combustion Power Generation Range Extending System. Chinese Patent CN108407624 (A), 17 August 2018.
76. King, E.I. Equivalent Circuits for Two-Dimensional Magnetic Fields: II—The Sinusoidally Time-Varying Field. *IEEE Trans. Power Appar. Syst.* **1966**, *PAS-85*, 936–945. [CrossRef]
77. King, E.I. Equivalent Circuits for Two-Dimensional Magnetic Fields: I—The Static Field. *IEEE Trans. Power Appar. Syst.* **1966**, *PAS-85*, 927–935. [CrossRef]

78. Carpenter, C.J. Finite-element network models and their application to eddy-current problems. *Proc. Inst. Electr. Eng.* **1975**, *122*, 455–462. [CrossRef]
79. Liu, G.; Ding, L.; Zhao, W.; Chen, Q.; Jiang, S. Nonlinear Equivalent Magnetic Network of a Linear Permanent Magnet Vernier Machine With End Effect Consideration. *IEEE Trans. Magn.* **2018**, *54*, 1–9. [CrossRef]
80. Tang, Y.; Paulides, J.J.H.; Lomonova, E.A. Analytical Modeling of Flux-Switching In-Wheel Motor Using Variable Magnetic Equivalent Circuits. *ISRN Automot. Eng.* **2014**, *2014*, 530260. [CrossRef]
81. Benlamine, R.; Hamiti, T.; Vangraefschèpe, F.; Dubas, F.; Lhotellier, D. Modeling of a coaxial magnetic gear equipped with surface mounted PMs using nonlinear adaptive magnetic equivalent circuits. In Proceedings of the 2016 XXII International Conference on Electrical Machines (ICEM), Lausanne, Switzerland, 4–7 September 2016; pp. 1888–1894. [CrossRef]
82. Asfirane, S.; Hlioui, S.; Amara, Y.; Barriere, O.D.L.; Barakat, G.; Gabsi, M. Global Quantities Computation Using Mesh-Based Generated Reluctance Networks. *IEEE Trans. Magn.* **2018**, *54*, 1–4. [CrossRef]
83. Amrhein, M.; Krein, P.T. 3-D magnetic equivalent circuit framework for modeling electromechanical devices. *IEEE Trans. Energy Convers.* **2009**, *24*, 397–405.
84. Benmessaoud, Y.; Dubas, F.; Hilairet, M.; Beniamine, R. Three-dimensional automatic generation magnetic equivalent circuit using mesh-based formulation. In Proceedings of the 2017 20th International Conference on Electrical Machines and Systems (ICEMS), Sydney, Australia, 11–14 August 2017; pp. 1–6. [CrossRef]
85. Laoubi, Y.; Dhifli, M.; Barakat, G.; Amara, Y. Hybrid analytical modeling of a flux switching permanent magnet machines. In Proceedings of the 2014 International Conference on Electrical Machines (ICEM), Berlin, Germany, 2–5 September 2014; pp. 1018–1023. [CrossRef]
86. Ilhan, E.E.; Gysen, B.L.J.; Paulides, J.J.H.; Lomonova, E.A. Analytical Hybrid Model for Flux Switching Permanent Magnet Machines. *IEEE Trans. Magn.* **2010**, *46*, 1762–1765. [CrossRef]
87. Pluk, J.W.K.; Jansen, J.W.; Lomonova, E.A. 3-D Hybrid Analytical Modeling: 3-D Fourier Modeling Combined With Mesh-Based 3-D Magnetic Equivalent Circuits. *IEEE Trans. Magn.* **2015**, *51*, 1–14. [CrossRef]
88. Martins, D.; Araujo, C.J.-L.; Delinchant, B.; Chadebec, O. A Hybrid Method BEM-NRM for Magnetostatics Problems. *J. Microw. Optoelectron. Electromagn.* **2013**, *12*, 555–568.
89. Philips, D.A. Coupling Finite Elements and Magnetic Networks in Magnetostatics. *Int. J. Numer. Methods Eng.* **1992**, *35*, 1991–2002.
90. Petrichenko, D. Calculation and Simulation of Turbogenerators Using Permeance Network. Optimization Application. Ph.D. Thesis, École Centrale de Lille, Villeneuve-d'Ascq, France, 2007.
91. ANSYS®. Electromagnetics Suite, Release 16.2. Available online: https://www.ansys.com/ (accessed on 27 March 2019).
92. Miller, T.J.E.; McGilp, M.; Wearing, A. Motor Design Optimisation Using SPEED CAD Software. In Proceedings of the IEE Seminar Practical Electromagnetic Design Synthesis, London, UK, 11 February 1999.
93. Du Peloux, B.; Gerbaud, L.; Wurtz, F.; Leconte, V.; Dorschner, F. Automatic generation of sizing static models based on reluctance networks for the optimization of electromagnetic devices. *IEEE Trans. Magn.* **2006**, *42*, 715–718.
94. De Saint Romain, B.D.; Gerbaud, L.; Wurtz, F.; Morin, E. A method and a tool for fast transient simulation of electromechanical devices: application to linear actuators. In Proceedings of the 14th Brazilian Microwave and Optoelectronics Symposium and the 9th Brazilian Conference on Electromagnetics (MOMAG 2010), Vila Velha, Brazil, 29 August–1 September 2010.
95. Demenko, A.; Sykulski, J.K.; Wojciechowski, R. On the Equivalence of Finite Element and Finite Integration Formulations. *IEEE Trans. Magn.* **2010**, *46*, 3169–3172. [CrossRef]
96. Amrhein, M.; Krein, P.T. Magnetic Equivalent Circuit Modeling of Induction Machines Design-Oriented Approach with Extension to 3-D. In Proceedings of the 2007 IEEE International Electric Machines & Drives Conference, Antalya, Turkey, 3–5 May 2007; pp. 1557–1563. [CrossRef]
97. Benhamida, M.A.; Ennassiri, H.; Amara, Y.; Barakat, G.; Debbah, N. Study of switching flux permanent magnet machines using interpolation based reluctance network model. In Proceedings of the 2016 International Conference on Electrical Sciences and Technologies in Maghreb (CISTEM), Antalya, Turkey, 3–5 May 2007. [CrossRef]
98. Nedjar, B.; Hlioui, S.; Vido, L.; Amara, Y.; Gabsi, M. Hybrid Excitation Synchronous Machine modeling using magnetic equivalent circuits. In Proceedings of the 2011 International Conference on Electrical Machines and Systems, Beijing, China, 20–23 August 2011; pp. 1–6. [CrossRef]

99. Bekhaled, C.; Hlioui, S.; Vido, L.; Gabsi, M.; Lecrivain, M.; Amara, Y. 3D magnetic equivalent circuit model for homopolar hybrid excitation synchronous machines. In Proceedings of the 2007 International Aegean Conference on Electrical Machines and Power Electronics, Bodrum, Turkey, 10–12 September 2007; pp. 575–580. [CrossRef]
100. Tolyat, H.; Hong, J.P.; Hur, J. Dynamic Analysis of Linear Induction Motors Using 3-D Equivalent Magnetic Circuit Network (EMCN) Method. *Electr. Power Compon. Syst.* **2001**, *29*, 531–541.
101. Belalahy, C.; Rasoanarivo, I.; Sargos, F.M. Using 3D reluctance network for design a three phase synchronous homopolar machine. In Proceedings of the 2008 34th Annual Conference of IEEE Industrial Electronics, Orlando, FL, USA, 10–13 November 2008; pp. 2067–2072. [CrossRef]
102. Ostović, V. *Dynamics of Saturated Electric Machines*, 1st ed.; Springer: New York, NY, USA, 1989.
103. Nedjar, B. Modélisation Basée sur la Méthode des Réseaux de Perméances en vue de L'optimisation de Machines Synchrones à Simple et à double Excitation. Ph.D. Thesis, École Normale Supérieure de Cachan—ENS Cachan, Cachan, France, 2011.
104. Kosaka, T.; Pollock, C.; Matsui, N. 3 Dimensional finite element analysis of hybrid stepping motors taking inter-lamination gap into account. In Proceedings of the Second International Conference on Power Electronics, Machines and Drives (PEMD 2004), Edinburgh, UK, 31 March–2 April 2004; pp. 534–539. [CrossRef]
105. Benlamine, R.; Dubas, F.; Randi, S.; Lhotellier, D.; Espanet, C. 3-D Numerical Hybrid Method for PM Eddy-Current Losses Calculation: Application to Axial-Flux PMSMs. *IEEE Trans. Magn.* **2015**, *51*, 1–10. [CrossRef]
106. Amrhein, M.; Krein, P.T. Force Calculation in 3-D Magnetic Equivalent Circuit Networks with a Maxwell Stress Tensor. *IEEE Trans. Energy Convers.* **2009**, *24*, 587–593. [CrossRef]
107. Le Huy, H.; Perret, R.; Feuillet, R. Minimization of Torque Ripple in Brushless DC Motor Drives. *IEEE Trans. Ind. Appl.* **1986**, *IA-22*, 748–755. [CrossRef]
108. Clenet, S.; Lefevre, Y.; Sadowski, N.; Astier, S.; Lajoie-Mazenc, M. Compensation of permanent magnet motors torque ripple by means of current supply waveshapes control determined by finite element method. *IEEE Trans. Magn.* **1993**, *29*, 2019–2023. [CrossRef]

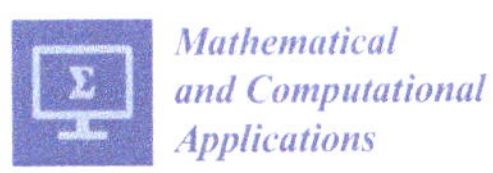
Mathematical and Computational Applications

Article

Steady State and 2D Thermal Equivalence Circuit for Winding Heads—A New Modelling Approach

Julien Petitgirard [1,2], Tony Piguet [1], Philippe Baucour [1,*], Didier Chamagne [1], Eric Fouillien [2,*] and Jean-Christophe Delmare [2]

1 ENERGIE Department, FEMTO-ST Institute, CNRS, Univ. Bourgogne Franche-Comte, 2 av. Jean Moulin, 90000 Belfort, France; Julien.petitgirard@femto-st.fr (J.P.); tony.piguet@femto-st.fr (T.P.); didier.chamagne@univ-fcomte.fr (D.C.)

2 Groupe PSA, Technical Center of Velizy, 78140 Velizy-Villacoublay, France; jeanchristophe.delmare@mpsa.com

* Correspondence: philippe.baucour@univ-fcomte.fr (P.B.); eric.fouillien@mpsa.com (E.F.)

Received: 22 September 2020; Accepted: 14 October 2020; Published: 18 October 2020

Abstract: The study concerns the winding head thermal design of electrical machines in difficult thermal environments. The new approach is adapted for all basic shapes and solves the thermal behaviour of a random wire layout. The model uses the nodal method but does not use the common homogenization method for the winding slot. The layout impact can be precisely studied to find different hotspots. To achieve this a Delaunay triangulation provides the thermal links between adjoining wires in the slot. Voronoï tessellation gives a cutting to estimate thermal conductance between adjoining wires. This thermal behaviour is simulated in cell cutting and it is simplified with the thermal bridge notion to obtain a simple solving of these thermal conductances. The boundaries are imposed on the slot borders with Dirichlet condition. Then solving with many Dirichlet conditions is described. Some results show different possible applications with rectangular and round shapes, one ore many boundaries, different limit condition values and different layouts. The model can be integrated into a larger model that represents the stator to have best results.

Keywords: thermal equivalence circuit; Voronoï tessellation; winding heads; nodal method; thermal resistances

---

1. Introduction

The study of increasingly compacted electrical machines in severe thermal environments is today an important tendency in electrical engineering [1,2]. The electrical machines with concentrated windings exhibit many advantages like high slot-filling factor, short end-winding, high fault tolerance capability, and automated winding process. Those advantages allow the high power density applications like electrical vehicles, electric aircraft, and wind turbines [3]. For such applications, accurate thermal models are necessary to describe the system behaviour. One of the main problems in the thermal study of electrical machines concerns their winding, where the temperature rises to its maximum value [2,4]. Moreover the study of thermal field becomes more and more important because the electrical designs are more compacted with more electrical density. The use of numerical tools, like the finite element method (FEM), to estimate thermal field and find hot spots in coils leads to excessive simulation time. Thus, some methods such as the nodal method like lumped thermal model have been introduced to solve quickly a thermal field in end-windings [3–8]. The objective of the present study is then to create an adaptable winding model, that reproduces a similar thermal behaviour. This model can solve all simple slot shapes with random wire layouts. Moreover, this study does not use the homogenisation technique commonly used in winding thermal calculation methods. This technique provides a homogeneous distribution of temperature while a random layout distorts this distribution and can create other hot spots. To do this, a specific tessellation obtained via the Voronoï diagram is used. This tessellation allows evaluating thermal conductance between each adjoining wire and between a wire and its adjoining boundary. The solving is given and different applications show the results for different shapes, different layouts and different boundary conditions. The advantages of this method are to keep the fast solving from a nodal method with the exact layout of wires in winding in all simple shapes.

First, we describe four problems to be solved, the geometric choices and simplifications as well as the different simplifications applied to the model. The graph of the nodal network and the boundary conditions are given as showed with green flow-chart Figure 1. Then, in the second step, we provide two different way to estimate the thermal conductance that will be applied in the network. The first way describes a simple equation based on the shortest distance between adjoining wires. The second way gives a numerical integration which is more adapted as showed with blue flow-chart Figure 1. In a third step, the network solving is described thanks to adapted matrices. The temperature of each wire and the heat flux between each adjoining wire are solved as showed with red flow-chart Figure 1. Finally, the last step solves the thermal nodal networks of the four examples. The tool process used is described. Moreover, a comparison with Finite Volume Method (FVM) is provided to evaluate the nodal model.

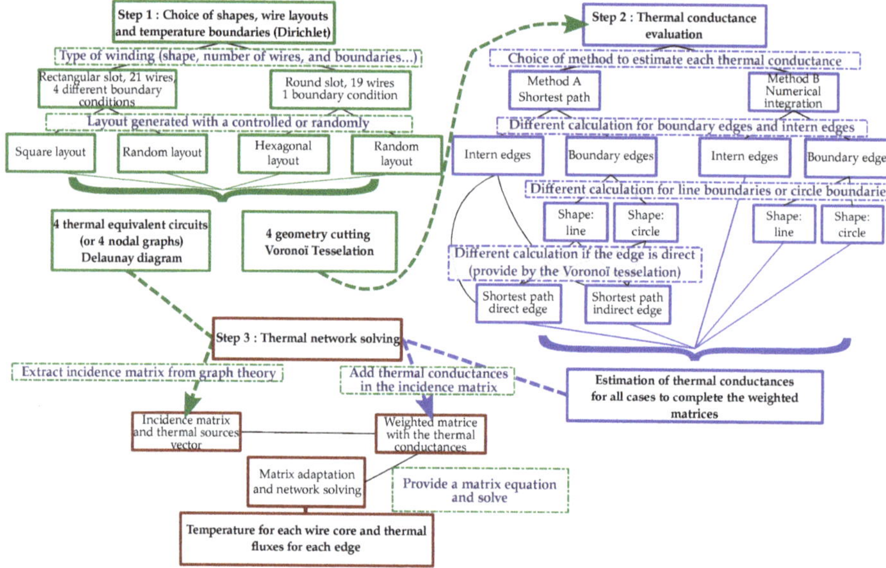

Figure 1. Flow-chart to define different steps of the thermal model with its different possibilities.

2. Thermal Equivalent Circuit

2.1. Application to a Lot of Slot Shapes with a Random Layout of Round Wires

To the best of our knowledge, the analytic solutions for the thermal modelling of an electrical winding slot in 2D use homogenisation techniques [4–6]. These techniques use the homogeneous conductivities [9–11] for the material slot and a homogeneously distributed thermal power. A numerical solution such as the finite element method, boundary element method, finite volume method or finite difference method should be implemented. These methods refine the results and obtain a more realistic thermal distribution and also to answer the random character of wire layout in the slots. Although these solutions are accurate, they cannot meet the industrial request in terms of speed, flexibility and design purpose. For this situation, the best solving method consists of a nodal network method. This approach is known to be faster than the others [12]. The heat sources are injected in nodes like current source in an electrical field. The difficult parts are to cut the slot into regions (stated as cells) with a homogeneous temperature and to evaluate correctly thermal resistances [13]. The thermal resistances depend on geometrical dimensions and thermal material properties. Between two nodes i and j, in steady-state, the thermal Ohm's law is applied with ΔT_{ij} as temperature gradient, Rth_{ij} as thermal resistance and q_{ij} as heat flow:

$$\Delta T_{ij} = Rth_{ij} \times q_{ij}. \tag{1}$$

Many geometrical and physical parameters characterise a coil. Our study concerns a section in a slot. The wires have the same geometrical and thermal characteristics. The wires are composed of insulation materials and core materials. The thin protection paper used around wires in the slot is neglected and the materials around wires can be air or resin.

A winding is composed of wires tightened to each other. However, their arrangement is not perfect. The part of wire section ($\sum_{i=1}^{I} s_{wire,i}$) compared to the part of the slot section (S_{slot}) is obtained with the ratio in Equation (2).

$$\tau_{slot} = \frac{\sum_{i=1}^{I} s_{wire,i}}{S_{slot}}. \tag{2}$$

The ratio between slot the surface and the wire cross-section surfaces is between 0.5 and 0.8 depending on the manufacturing process [14]. The improvement of electrical machines' slot filling factors is still studied today [15]. The wire layout is hardly controlled on the machine-made end-windings. So each coil is different for the same product and the ratio is not optimal. More compact layouts are possible for the hand-made end-windings. For specific electrical machines, a specific tool can be used to obtain a flat wire layout [16]. However, this study considers a constant layout along the wire axis. Heat transfer appears only in wire layout sections. Thus the problem is reduced to a 2D study in a cross-section

If the winding is not in resin, the air is trapped between the wires. We consider during all this study that low values of $1 - \tau_{slot}$ create only small air cavities. If the buoyancy forces created by the heat flow through these air cavities cannot overcome the viscous forces [17], then the air trapped in the winding is supposed to be motionless. Convection can be neglected which is ensured by a low number of Rayleigh that is, $Ra < 1708$. So, our model considers only conductive thermal transfers.

The big advantage of this model is the possibility of solving any wire layout for any slot shape, some examples are presented in Figure 2. This flexibility makes it possible to test a large number of random cases in order to detect the worst and best cases in terms of thermal heating. This gives the designer a possibility of decision for the design choices of these electric winding.

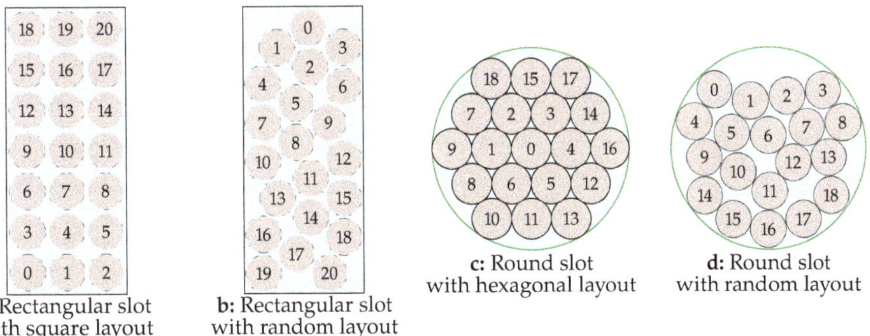

Figure 2. Example of different possibilities of shapes and layouts.

To create an efficient thermal nodal network, some regions are identified as uniform in temperature. Computable thermal resistances must be found between regions. Each core, made of copper, has its thermal source and its thermal conductivity k_{core} and each core is surrounded by insulation. With $k_{core} \gg k_{ins}$, we will suppose that cores have a homogeneous temperature and that they will be each one a node. The temperature at the slot border in real conditions is dependant of the type of electrical machine. It should be noted that the method has been successfully modded to take into account Neumann or Robin boundary conditions. The goal of the present study is to describe the core of the method. Therefore a Dirichlet boundary condition is taken here and each boundary temperature is labelled as $T_{bnd,i}$. It should be noticed that the temperature imposed may vary spatially.

The temperature T_i rises due to copper losses (Joule effect) inside each core. The heat sources Q_i from Joule losses is calculated with Equation (3) where the electrical resistivity ρ is supposed to be constant. I_i is the nominal current of each wire in steady-state and $s_{core,\,i}$ the core section of each wire.

$$Q_i = \frac{\rho \cdot l}{s_{core,\,i}} \times I_i^2. \tag{3}$$

2.2. Creation of the Internal Thermal Circuit

The heat fluxes along the wire axis are neglected. The resolution only considers the heat flow in a 2D section. This network is created thanks to Delaunay triangulation [18].

In the Delaunay diagram, each node represents a wire. All of the nodes are positioned thanks to the wire layouts. It should be noted that the determination of the layout for an industrial case is not easy and should be made via a circle packing algorithm [19]. Delaunay triangulation is applied to find all the links between the adjoining wires (blue lines in Figure 3). In the network, these links are labelled edges and they are connected to the nodes corresponding to each wire.

The nodes are the temperature potential in the corresponding wire cores and the edges could be seen as the heat fluxes between two nodes. The resolution of this thermal circuit is analogous to the resolution of an electrical circuit. So, each edge is arbitrarily oriented (blue arrow cf. Figure 3) and weighted with thermal conductances. Joules losses are added on each node like a thermal source. This thermal network is simple and its solving is very fast.

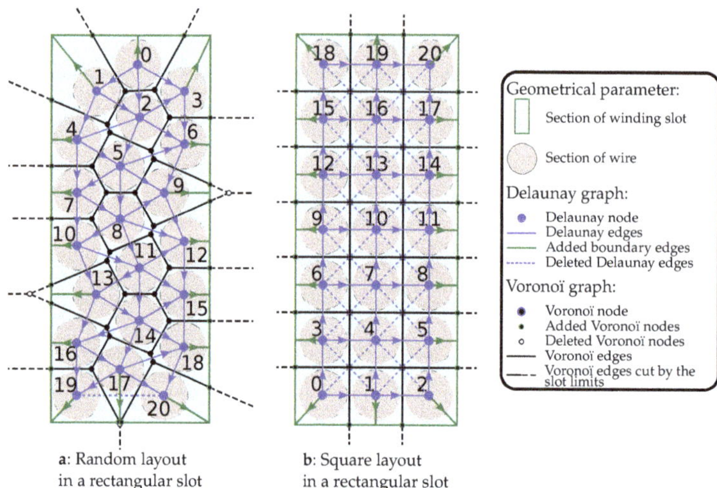

Figure 3. The equivalent thermal circuit for a 2D view of a rectangular slot with a random layout (**a**) and a square layout (**b**).

These thermal conductances are dependent on the materials between each temperature gradient and the thermal characteristics of these materials. The thermal conductances solving is presented in Section 3.

Some modifications of the Delaunay diagram will allow simplifying a special case and adding appropriate boundary conditions. To do this, the dual graph of Delaunay triangulation, Voronoï tessellation, is used [20] (black lines Figure 3). The Voronoï cells give several details. First of all, each node which is out of the end-winding slot is deleted and its links to each other node are cut at the slot border. Three examples are given in Figure 3a for triangles (node numbers: 10, 13, 16), (6, 9, 12) and (17, 19, 20). The semi-infinite edges of Voronoï (dotted black line) can enable to identify all the Delaunay nodes which are located at the periphery of Delaunay graph. Each peripheral Delaunay node is then connected with a new edge (green edge) to one or several boundaries. These new edges are weighted with thermal conductances adapted to the shape of the slot. In Figure 3 the green line represents the border.

Another simplifying can be done when a local square layout of Delaunay node exits, as on Figure 3b. It is considered no thermal flux between diagonal wires. So, each crossed Delaunay edge is just deleted. For example, all deleted edges are symbolized with blue dotted lines in Figure 3b.

2.3. Selection of Boundary Conditions to the Thermal Circuit

As part of this work, the boundary conditions are added directly to the inner edge of the slot. However, this nodal network can easily be added to another thermal network which solves thermal field in an electric machine. In this study, the boundary conditions are imposed at the inner edges of the slot. The thermal phenomenon on geometry borders in a thermal circuit is translated by these boundary conditions: Dirichlet condition which represents a known temperature or a Robin condition which represents a convection phenomenon between temperature fluid and walls surface. In this study, only resolutions based on Dirichlet conditions are presented. However, Robin's condition can also be used following Saulnier's recommendations [13] and an iterative resolution will have to be implemented.

A single boundary condition as presented in Figure 3 requires that all the walls of the slot are at the same known temperature. To be the most representative of the thermal environment, it is possible to add several boundary conditions. For example Figure 4 shows a possibility to apply different boundary conditions. It is possible to cut the border in another way and add as many boundary conditions as desired. It is important to check if the two boundary conditions are separated by a node of the Voronoï graph to have distinct Delaunay edges for the different boundary conditions. If not, like on each slot angle in Figure 4, the corresponding Delaunay node must be linked to the two boundary conditions in a distinct Delaunay edge. Two edges instead of one are used. A node corresponding to the boundary separation is added to the Voronoï graph. It will be used to determine the corresponding thermal conductances in Section 3.

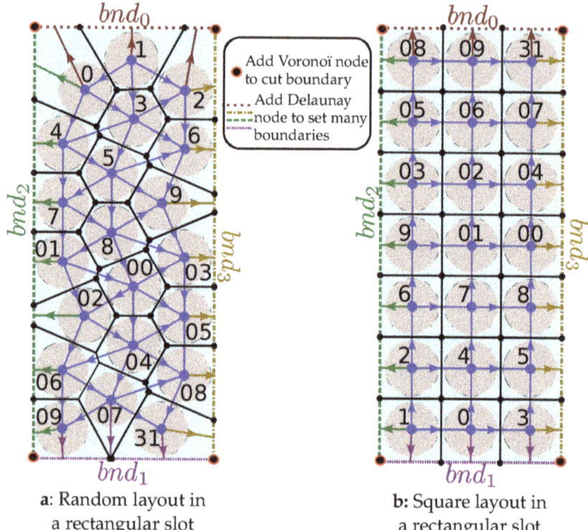

Figure 4. Several boundary conditions for a rectangular slot with a random layout (**a**) and a square layout (**b**).

3. Thermal Conductance Determination

In a thermal network, the real difficulty is to find the thermal conductance between each node to estimate the heat transfer between each adjoining wire and calculate the temperature rise created by the Joule losses. This study proposes an estimation of thermal resistances which represent the inverse of thermal conductances ($Rth = 1/G$). It is based on the principle of the thermal bridge and provides a direct calculation of resistances.

The Voronoï tessellation is used to find the thermal conductances corresponding to each Delaunay edge (cf. Section 2). Each node is now included in a cell and each Delaunay edge can intersect with a Voronoï's cell edge. We consider 2 cases:

- A direct case when the 2 edges intersect.
- An indirect case when the 2 edges do not intersect. This case appears when the shortest distance between 2 adjacent nodes does not coincide with the Delaunay edge.

Figure 5a presents the first case and Figures 5b and 6 presents the second case. Also, when a cell is connected to the boundary, the Delaunay node is directly linked to the boundary and the Voronoï cell is truncated (see Figure 7). The Voronoï edge is not necessary a segment and can respect a circular slot border as shown on Figure 7c,d. In these figures, the proportion of the insulation is increased for a better understanding. It is assumed that the heat flow between two wires is only exchanged through their shared Voronoï edge.

The Delaunay nodes give uniform core temperatures of wires. The edges represent the heat flux across insulation and around media (trapped air or resin). Each material gives a thermal resistance.

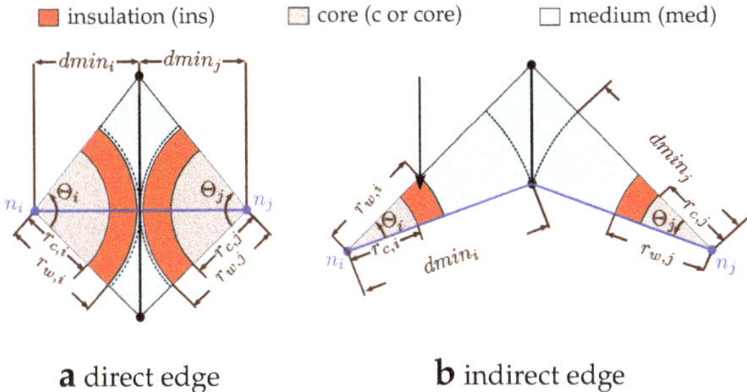

Figure 5. Cutting an internal cell with the 2 possible cases: direct edge (**a**) and indirect edge (**b**).

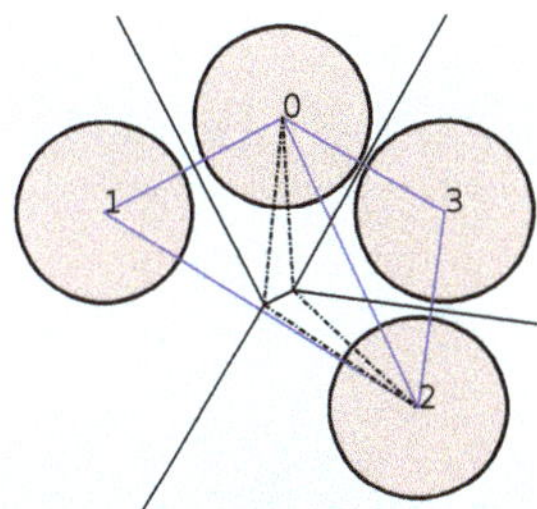

Figure 6. Layout where wire 0 and 2 have an indirect edge.

For an internal edge (cf. Figure 5), the sum of these resistances in series provides their thermal resistances as follows:

$$Rth_{eij} = Rth_{ins,ij} + Rth_{med_ij} + Rth_{med_ji} + Rth_{ins,ji}. \tag{4}$$

It is assumed that the heat flow from the core to the Voronoï edge is only radial. The wires are all identical and the cells on both side of a Voronoï edge are symmetrical. So, on the same edge, the two thermal resistances of insulation are equal (i.e., $Rth_{ins,ij} = Rth_{ins,ji}$). The insulation resistance is given with the cylindrical known resistances [17] as follow:

$$Rth_{ins,ij} = \frac{1}{\Theta_i k_{ins}} \cdot \ln\left(\frac{r_{w,i}}{r_{c,i}}\right). \tag{5}$$

To determine the medium thermal resistance in a cell, it is assumed that the Voronoï edge is at a uniform temperature. With the symmetry, it is determined thermal resistances between the insulation and the Voronoï edge (Rth_{med_ij} and Rth_{med_ji}). Two geometrical cases are possible like direct internal edge (Figure 5a) or indirect internal edges (Figure 5b). Two methods to estimate medium thermal resistances are implemented. The first method (Method A) is based on a very simple assumption which is based on the shortest path. The second method (Method B) consider that there is an infinite sum of elementary resistance in parallel between the Voronoï edge and the insulation edge.

3.1. Method A: Shortest Path

The thermal behaviour shows that the heat flow favours the easiest path. So in a uniform domain, it is the shortest path. In the *a* case, the shortest path is the distance between the two wires. So, in this segment, it is defined two identical lengths $dmin_i$ and $dmin_j$. In the *b* case, the segment between two nodes does not cut the Voronoï edge, then the lengths $dmin_i$ and $dmin_j$ are defined by the shortest segment between the Delaunay nodes and the Voronoï edges. It corresponds to the shortest cell border. Then it is assumed to simplify the domain with the previous cylindrical resistances which symbolize the thermal resistance between the dotted arc and the insulation arc as follow:

$$Rth_{med,ij} = Rth_{med,ji} = \frac{1}{\Theta_i k_{med}} \cdot \ln\left(\frac{dmin_i}{r_{w,i}}\right). \quad (6)$$

So with symmetrical context and Equations (4)–(6), the internal thermal resistance of edge is:

$$Rth_{eij} = 2 \times Rth_{ins,ij} + 2 \times Rth_{med,ij} = \frac{2}{\Theta_i k_{ins}} \cdot \ln\left(\frac{r_{w,i}}{r_{c,i}}\right) + \frac{2}{\Theta_i k_{med}} \cdot \ln\left(\frac{dmin_i}{r_{w,i}}\right). \quad (7)$$

For boundary edges, the thermal resistances follow the same principle with some adaptations. The cell studied is defined by the Delaunay node at the wire centre and the two Voronoï nodes added at the slot limit. Moreover, the Voronoï edge can be a segment or an arc according to the slot shape. Then we have to check which is the shortest path to best determine the $dmin_i$ as shown in the Figure 7.

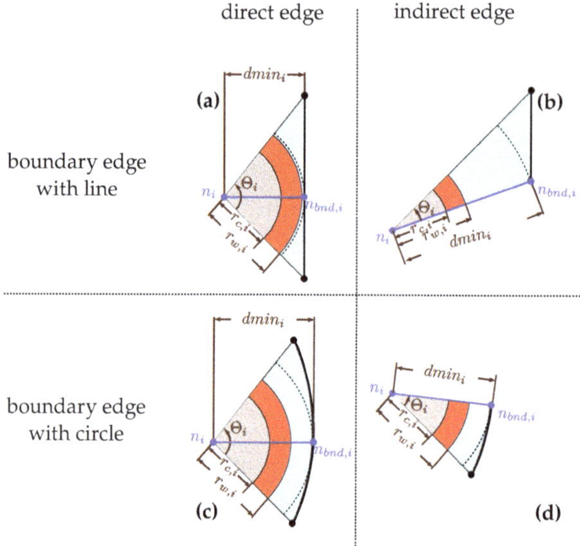

Figure 7. Cutting a boundary cell with the 4 possible cases: direct edge with a line border (**a**) or an arc border (**c**) and indirect edge with a line border (**b**) or an arc border (**d**).

When the $dmin_i$ is found, the thermal resistances could be evaluated as follows:

$$Rth_{e(bnd-i)} = Rth_{ins,bnd-i} + Rth_{med,bnd-i} = \frac{1}{\Theta_i k_{ins}} \cdot \ln\left(\frac{r_{w,i}}{r_{c,i}}\right) + \frac{1}{\Theta_i k_{med}} \cdot \ln\left(\frac{dmin_i}{r_{w,i}}\right). \quad (8)$$

3.2. Method B: Numerical Integration

This first method gives the smallest conceivable resistance but the real value is mandatory higher. A second method is proposed to find a more precise resistance. These resistances are not directly soluble. They need a numerical integration.

The assumption made is that the flow is radial from the insulation to Voronï line. The principle is using an infinity of parallel resistances to represent the flow as shown in Figure 8. With this supposition, the effect of minimal distance is represented. Indeed, the resistance discretized for the minimum distance has the smallest value than the other radial discretized resistances. The solving of parallel discretized resistance show a value slightly higher than the smallest value.

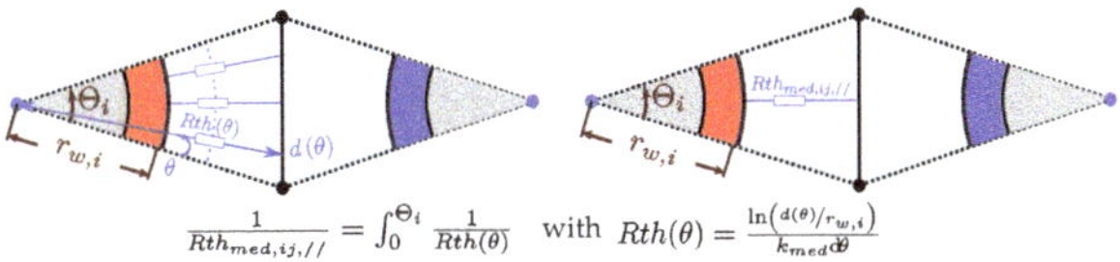

Figure 8. Determination of parallel thermal resistance in a Voronoï cell.

The integral which represents this parallel resistance is written in function of angle θ, the origin is imposed at the centre of the study wire and the x-axis at the bottom line of the cells, as follows:

$$\frac{1}{Rth_{med,ij}} = \int_0^{\Theta_i} \frac{1}{Rth\,(\theta)} \tag{9}$$

with

$$Rth(\theta) = \frac{\ln\left(^{d(\theta)}/_{r_{w,i}}\right)}{k_{med}\,\mathrm{d}\theta} \tag{10}$$

The distance $d\,(\theta)$ represents the distance between centre of studied wire and Voronï line. For internal edges or boundary edges composed by a line, this distance is written as follow:

$$d\,(\theta) = \frac{p}{\sin\theta - m \times \cos\theta} \tag{11}$$

m and p represent the line coefficients like $y = m.x + p$. If the Voronoï nodes in the local coordinate system is at this position: $N_{vor,1}\,(x_{vor,1}, y_{vor,1})$ and $N_{vor,2}\,(x_{vor,2}, y_{vor,2})$ so $m = {}^{y_{vor,2}-y_{vor,1}}/_{x_{vor,2}-x_{vor,1}}$ and $p = y_{vor,1} - m.x_{vor,1}$.

For boundaries where the slot shape is circular, the distance $d\,(\theta)$ is written as follow:

$$d\,(\theta) = \max\left(r_{cnt}.\cos\,(\theta - \varphi) \pm 0.5\sqrt{4.R_{bnd} - 4.r_{cnt}{}^2.\sin^2\,(\theta - \varphi)}\right). \tag{12}$$

As previously, the origin of the coordinate system is at the studied wire centre and the x-axis at the bottom line of the cells. R_{bnd} is the shape radius, r_{cnt} is the distance between the origin and the centre shape position, φ is the cylindrical angle of shape centre from the coordinate system.

In this second method, the resistance of internal edges is deduced from the numerical estimation of Equation (9) with adapt application of $d\,(\theta)$ and the Equation (7) as follow:

$$Rth_{eij} = 2 \times Rth_{ins,i} + 2 \times Rth_{med,ij} = \frac{2}{\Theta_i k_{ins}} \cdot \ln\left(\frac{r_{w,i}}{r_{c,i}}\right) + 2 \times \left(\int_0^{\Theta_i} \frac{1}{Rth\,(\theta)}\right)^{-1}. \tag{13}$$

For the resistance of boundary edges with the same deduction from Equation (9) and Equation (8) as follow:

$$Rth_{e(bnd-i)} = Rth_{ins,i} + Rth_{med,ij} = \frac{1}{\Theta_i k_{ins}} \cdot \ln\left(\frac{r_{w,i}}{r_{c,i}}\right) + \left(\int_0^{\Theta_i} \frac{1}{Rth\,(\theta)}\right)^{-1}. \tag{14}$$

4. Nodal Network Solving

The heat sources in each electrical wire are generated by Joule losses and are noted Q_i. This heat source Q_i is incoming on the node n_i representing a wire on a Delaunay graph. The temperatures corresponding to Dirichlet conditions $T_{bnd,i}$ will be imposed at the boundary nodes. The temperatures at each internal node are not known and will be noted T_i. The heat flows through edges will be noted

q_{ij}. These flows q_{ij} are algebraic terms and can be negative or positive. In this paper we -arbitrarily- impose the orientation of the flow from node i to node j with $j > i$. In this section, each linear system and matrix respects the network of Figure 4a. The nodes are listed starting with the internal nodes first then the boundary nodes like $[n_0, n_1, \cdots, n_{10}, \cdots, n_{20}, n_{bnd0}, \cdots n_{bnd3}]$. For the edges we use a double increasing system starting with the internal edges and then the edges cutting the boundaries like $[e_{0,1}, e_{0,2}, \cdots, e_{1,2}, \cdots, e_{19,20}, e_{0,bnd0}, e_{1,bnd0}, \cdots e_{20,bnd3}]$.

To find the temperatures, we make an energy balance on all the nodes. The linear system obtained is composed of as many equations as there are nodes.

$$Q_i = \sum q_{eij,out} - \sum q_{eij,in}. \tag{15}$$

The Equation (15) applied for random layout in a rectangular slot (Figure 4a) gives a linear system as follows:

$$\begin{array}{l} n_0 \\ n_1 \\ \vdots \\ n_{20} \\ n_{bnd0} \\ \vdots \\ n_{bnd3} \end{array} \left\{ \begin{aligned} Q_0 &= q_{0,1} + q_{0,2} + q_{0,3} + q_{0,bnd0} \\ Q_1 &= -q_{0,1} + q_{1,2} + q_{1,4} + q_{1,bnd0} + q_{1,bnd2} \\ &\vdots \\ Q_{20} &= -q_{17,20} - q_{18,20} + q_{20,bnd1} + q_{20,bnd3} \\ -Q_{bnd0} &= -q_{0,bnd0} - q_{1,bnd0} - q_{3,bnd0} \\ &\vdots \\ -Q_{bnd3} &= -q_{3,bnd3} - q_{6,bnd3} - q_{9,bnd3} - q_{12,bnd3} - q_{15,bnd3} - q_{18,bnd3} - q_{20,bnd3} \end{aligned} \right. \tag{16}$$

A specificity in thermal balance is written in boundary equation. The heat sources Q_{bndi} correspond to the heat flow which leaves the system. Q_{bndi} is not know but the energy conservation provide this equation: $\sum Q_i = \sum Q_{bnd}$.

If the thermal conductance is defined as $G_{ij} = Rth_{ij}^{-1}$, the heat flux is developed as follows:

$$q_{i,j} = G_{i,j} \times (T_j - T_i) = G_{i,j}.T_j - G_{i,j}.T_i. \tag{17}$$

The linear system (16) combined with Equation (17) applied to all nodes gives a linear system where the unknowns are the temperatures T_i. To simplify the solving, the system is written with matrix thanks to the graph theory. First, the incidence matrix ($[Inc]$ in Equation (18)) that connects edges and nodes with a sign convention. $Inc_{i,j} = 1$ if the heat flux leaves the node i and respectively $Inc_{i,j} = -1$ if it enters into node i. It should be noted that the transposed incidence matrix $[Inc]^T$ gives the two nodes connected by a specific edge. And the weighted incidence matrix $[G]$ (Equation (19)) gives the thermal conductance oriented and connected at each node according to edges.

$$[Inc] = \begin{array}{c} \begin{array}{cccccccc} e_{0,1} & e_{0,2} & \cdots & e_{18,20} & e_{bnd0,0} & \cdots & e_{bnd3,20} & \end{array} \\ \left[\begin{array}{cccc|ccc} 1 & 1 & \cdots & 0 & 1 & \cdots & 0 \\ -1 & 0 & \cdots & 0 & 1 & \cdots & 0 \\ \vdots & \vdots & \ddots & \vdots & \vdots & \ddots & \vdots \\ 0 & 0 & \cdots & -1 & 0 & \cdots & 1 \\ \hline 0 & 0 & \cdots & 0 & -1 & \cdots & 0 \\ \vdots & \vdots & \ddots & \vdots & \vdots & \ddots & \vdots \\ 0 & 0 & \cdots & 0 & 0 & \cdots & -1 \end{array}\right] \begin{array}{l} n_0 \\ n_1 \\ \vdots \\ n_{20} \\ n_{bnd0} \\ \vdots \\ n_{bnd3} \end{array} \end{array} \tag{18}$$

$$[G] = \begin{array}{c} \\ \left[\begin{array}{cccc|ccc} G_{0,1} & G_{0,2} & \cdots & 0 & G_{bnd0,0} & \cdots & 0 \\ -G_{0,1} & 0 & \cdots & 0 & G_{bnd0,0} & \cdots & 0 \\ \vdots & \vdots & \ddots & \vdots & \vdots & \ddots & \vdots \\ 0 & 0 & \cdots & -G_{bnd18,20} & 0 & \cdots & G_{bnd3,20} \\ \hline 0 & 0 & \cdots & 0 & -G_{bnd0,0} & \cdots & 0 \\ \vdots & \vdots & \ddots & \vdots & \vdots & \ddots & \vdots \\ 0 & 0 & \cdots & 0 & 0 & \cdots & -G_{bnd3,20} \end{array}\right] \end{array} \tag{19}$$

(columns: $e_{0,1}$, $e_{0,2}$, …, $e_{18,20}$, $e_{bnd0,0}$, …, $e_{bnd3,20}$; rows: n_0, n_1, …, n_{20}, n_{bnd0}, …, n_{bnd3})

With these different matrices which come from the graph theory, the vector of edges temperature potential is given with Equation (20). This vector coupled with matrix $[G]$ lets us write the previous linear system to matrix system as Equation (21) with an efficient and computable tool:

$$\left([Inc]^T \cdot [T]\right)^T = \begin{bmatrix} T_0 - T_1 & T_0 - T_2 & \cdots & T_0 - T_{bnd0} & \cdots & T_{20} - T_{bnd3} \end{bmatrix} \tag{20}$$

(columns: $e_{0,1}$, $e_{0,2}$, …, $e_{bnd0,0}$, …, $e_{bnd3,20}$)

$$\underbrace{[G] \cdot [Inc]^T}_{[M]} \cdot [T] = [Q]\,. \tag{21}$$

This matrix system is detailed in Equation (22). It can not be solved directly. In the vector $[Q]$, the Joule losses are known but the heat flows out of the system on boundary nodes are unknown ($Qbnd_i$). However, the number of unknowns (temperature of the internal nodes) corresponds to the number of internal nodes. It is not mandatory to keep the corresponding equations at the boundary nodes to solve. So, the first simplification, we remove equations from boundary nodes in $[M]$ and $[Q]$ (i.e., equations from n_{bnd0} to n_{bnd3} in grey part). Then the system is horizontally split into 2 matrices labelled M^{L*} and M^{R*}. This allows to separate the unknown internal temperatures (red part) from the unknown boundary condition temperatures (green part).

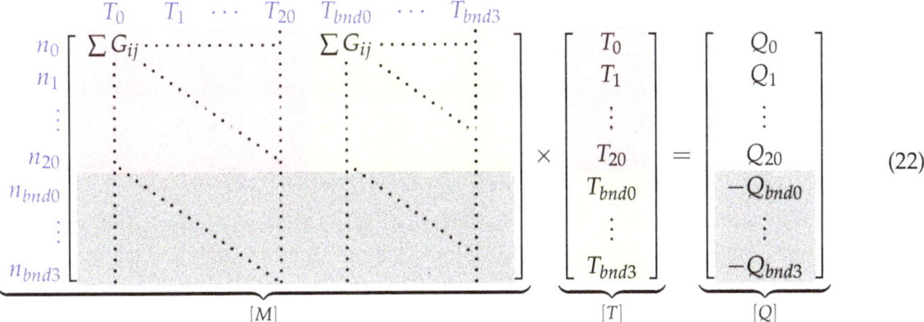

(22)

Equation (22) could be rewritten as:

$$M^{L*} \times T^L + M^{R*} \times T^R = Q^*. \tag{23}$$

Equation (23) could be easily transform a now solvable Equation (24) with $\left[M^{L*}\right]$, $[Q^*]$, $\left[M^{R*}\right]$, $\left[T^R\right]$ known and $\left[T^L\right]$ unknown.

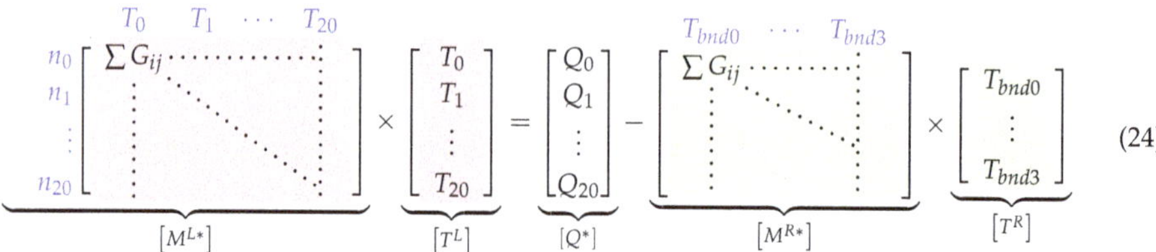

(24)

To determine all the heat fluxes in the system $\left[q_{edge}\right]$ we could rely on the vector $[T]$ previously determined. We just create the conductance vector $\left[G_{edge}\right]$

$$\left[G_{edge}\right]^T = [G_{0,1}, G_{0,2}, \cdots, G_{1,2}, \cdots, G_{18-20}, G_{bnd0,0}, \cdots, G_{bnd3,20}] \quad (25)$$

with the Equations (17), (20) and the known vector $[T]$, we could determine $\left[q_{edge}\right]$ thanks to the following equation:

$$\left[q_{edge}\right] = [Inc]^T \cdot [T] \cdot \left[G_{edge}\right] \quad (26)$$

5. Application and Validation

Four applications corresponding to Figure 2 with the two thermal conductance determinations (cf. Section 3) are given in this section. The thermal properties, slot dimensions and wire positions are mentioned in Appendix A. These results show different temperature fields with different coil implementations. These cases are not representative of existing electric coils in an electrical machine in terms of dimensions but they tend to prove the ability of the method to be applied to several applications.

Figure 9 shows a clear understanding of tool process presented in this paragraph. A comparison with a Finite Volume Method (FVM) used in ANSYS FLUENT is done with the same boundaries and assumptions. All materials are solid domains, so only the thermal equation is solved. The domain discretization uses triangle elements and it is applied with GMSH meshing [21]. The common interface between the different materials is meshed with conforming mesh. The mesh is refined at core and insulation boundaries until the results are independent of the mesh. For the nodal model, to apply the mathematical process, the PYTHON code is used with some libraries. First, the random layouts are generated from a 2-dimensional real-time rigid body physics engine: PYMUNK library as described in a previous work [22]. The Delaunay and Voronoï networks are created thanks to Nocaj study [20] with different library tools in SCIPY library like Convexhull. The NETWORKX library [23] is used to save and transform Delaunay and Voronoï network. Each node and each edge is provided with full data like position, weight, dual edges id ... These tools provide the incidence matrix and the weighted incidence matrix. The Quadpack routine [24] via the SCIPY library is used to solve numerical integrals in the determination of thermal resistances. Finally, Equation (24) is solved with a standard linear algebra routine from Scipy based on the GESV Lapack routine [25].

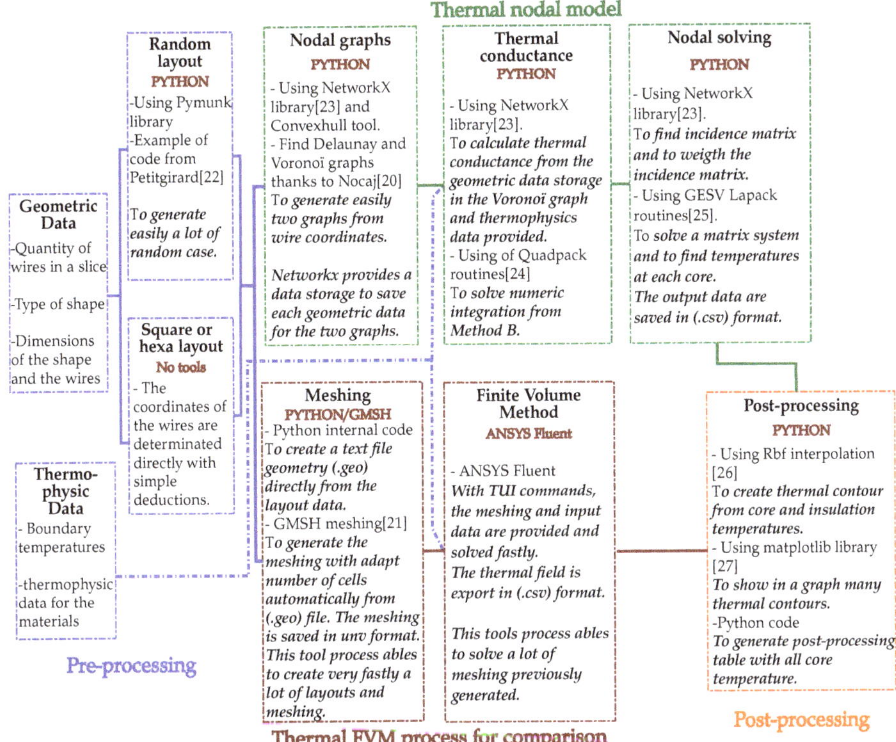

Figure 9. Software, programs and codes used to implement models, methods and comparison.

All the contours of the nodal model results are obtained with a radial basis function (Rbf) interpolation [26] on a sufficiently refined grid. The Rbf interpolation is directly available in SCIPY library and could be easily plotted via the MATPLOTLIB library [27].

To guide the interpolation, all the node's temperatures are known as the boundary condition ones. Also, we could determine several temperatures on the insulation of each node. As an example to determine the temperature of the insulation of the node i along the Delaunay edge between node i and j ($T_{ins,i,j}$), we could use the heat flux q_{ij} and $R_{ins,i}$.

$$T_{ins,i,j} = T_i + R_{ins,i} \times q_{ij}. \tag{27}$$

These additional temperatures are used in the interpolation process to obtain a more detailed contour, as shown on Figure 10 and in Appendix B.

5.1. Comparison between Two Conductance Methods and a Commercial Software

To identify the different results the indices A and B correspond to the model result with respectively the first method and the second method cited in the section of conductance determination. The acronym FVM corresponds to commercial software results. In all this section, each heat-up and each relative difference are based on the temperature: T_{bnd0}.

Table 1 shows that the model A with thermal conductances based on the minimum length underestimates the temperature compared to the results of the FVM. This solution is not protective and the gaps are large. The relative gaps are around 30.8% on the random layout and 18.3% on the square layout. These gaps are up to 30 °C on a heating estimated by the FVM of 137 °C (wire 10 of the square layout). Despite these gaps, the qualitative distribution of the temperature is respected (see contour Figure A1 and Figure A4).

Table 1. Results for square and random layout in rectangular slot with the model A, B and Finite Volume Method (FVM) compared. Corresponding thermal contour in Appendix B respectively in Figure A1 and Figure A2.

Square Layout in Rectangular Slot cf. Figure 2a						**Random Layout in Rectangular Slot cf. Figure 2b**					
T [°C]	*Model A*	*Model B*	*FVM*	$\frac{\text{Model A}-\text{FVM}}{\text{FVM}-T_{bnd0}}$	$\frac{\text{Model B}-\text{FVM}}{\text{FVM}-T_{bnd0}}$	T [°C]	*Model A*	*Model B*	*FVM*	$\frac{\text{Model A}-\text{FVM}}{\text{FVM}-T_{bnd0}}$	$\frac{\text{Model B}-\text{FVM}}{\text{FVM}-T_{bnd0}}$
T_0	110.9	127.7	121.5	−14.9%	8.6%	T_0	109.0	146.3	135.5	−31.0%	12.7%
T_1	125.2	150.5	140.6	−16.9%	10.9%	T_1	117.9	162.8	149.7	−31.9%	13.2%
T_2	109.7	126.5	119.7	−14.3%	9.8%	T_2	126.2	173.1	159.2	−30.2%	12.7%
T_3	125.8	153.2	140.8	−16.5%	13.6%	T_3	102.8	136.4	126.2	−30.7%	13.5%
T_4	147.0	189.4	168.9	−18.5%	17.2%	T_4	102.4	137.7	128.4	−33.2%	11.9%
T_5	124.4	151.6	139.9	−17.2%	13.1%	T_5	129.4	180.1	166.1	−31.6%	12.0%
T_6	131.7	162.9	148.6	−17.2%	14.5%	T_6	109.7	146.3	135.2	−29.9%	13.0%
T_7	155.5	204.6	183.9	−21.2%	15.5%	T_7	103.5	139.6	130.4	−33.5%	11.4%
T_8	130.1	161.2	148.1	−18.3%	13.4%	T_8	131.6	183.2	168.6	−31.2%	12.4%
T_9	132.8	165,0	151.4	−18.3%	13.5%	T_9	129.7	182.1	166.6	−31.6%	13.3%
T_{10}	157.1	207.9	187,0	−21.8%	15.2%	T_{10}	103.9	140.4	131.4	−33.8%	11.1%
T_{11}	131.2	163.3	149.2	−18.1%	14.3%	T_{11}	133.8	185.4	171.1	−30.8%	11.8%
T_{12}	129.9	161.2	147.9	−18.4%	13.6%	T_{12}	112.9	151.2	140.4	−30.4%	12.0%
T_{13}	152.0	202.0	181.7	−21.8%	15.4%	T_{13}	124.0	174.3	160.4	−33.0%	12.6%
T_{14}	128.3	159.5	146.4	−18.7%	13.6%	T_{14}	129.4	178.5	164.8	−30.8%	12.0%
T_{15}	120.9	148.3	137.1	−18.6%	12.9%	T_{15}	111.1	147.8	137.9	−30.5%	11.2%
T_{16}	140.1	182.3	165.9	−22.3%	14.2%	T_{16}	99.4	130.1	122.9	−32.2%	9.9%
T_{17}	119.4	146.7	135,0	−18.4%	13.7%	T_{17}	116.1	157.2	144.7	−30.2%	13.2%
T_{18}	98.6	114.9	108.4	−16.8%	11.2%	T_{18}	105.5	138.2	129.9	−30.5%	10.5%
T_{19}	109.7	133.9	124.8	−20.2%	12.2%	T_{19}	91.7	110.5	105.7	−25.1%	8.5%
T_{20}	97.4	113.7	106.9	−16.7%	11.9%	T_{20}	104.8	133.1	123.7	−25.6%	12.7%
T_{bnd0}		50.0				T_{bnd0}		50.0			
T_{bnd1}		80.0				T_{bnd1}		80.0			
T_{bnd2}		65				T_{bnd2}		65			
T_{bnd3}		62				T_{bnd3}		62			
I_i [A]		7.5				I_i [A]		7.5			
Q_i [W]		13.15				Q_i [W]		13.15			

Model B always overestimates the temperature. For the square layout the deviation is approximately 13.3% and for the random layout is around 12.0%. In the random case this gaps ranging from 4.8 °C (wire 19) to 15.6 °C (wire 9) for heat-ups ranging from 55.7 °C (wire 19) to 121.1 °C (wire 11).

The observations for the round slots are similar (Table 2). With method A, the temperatures are underestimated. The relative gaps for the hexagonal layout are around 63.5% and the relative gaps for the random layout are around 42.5%. In the random case, the heat-up differences between the method A and FVM range between 26.4 °C (wire 17) and 61.5 °C (wire 6). The FVM corresponding heat-ups range between 64.7 °C (wire 8) and 145.9 °C (wire 6).

The B model is still protective for this slot shape. For the hexagonal layout, the relative deviation is around 6.5%. The heat-up differences range between 3 and 4.5 °C for heat-up between 43.3 and 79.7 °C. For the random layout, the relative gap is around 12.1% with heat-up differences of 8.0 °C (wire 8) to 16.4 °C (wire 6) for FVM heat-up between 64.7 °C (wire 8) and 145.9 °C (wire 6).

Method A has the advantage of being very fast with a direct calculation of thermal conductances. It qualitatively represents the thermal behaviour but greatly underestimates the temperatures. Method B is always protective, more precise and the temperature distribution is also preserved (Appendix B). Its disadvantage is the estimation of the integral when calculating the thermal conductance. This solution requires a bit more IT resources but remains faster than the solving of finite element methods.

Table 2. Results for hexagonal and random layout in round slot with the model A, B and FVM compared. Corresponding thermal contour in Appendix B respectively in Figure A3 and Figure A4.

Hexa. Layout in Round Slot cf. Figure 2c						Random Layout in Round Slot cf. Figure 2d					
T [°C]	*Model A*	*Model B*	*FVM*	$\frac{\text{Model A}-\text{FVM}}{\text{FVM}-T_{bnd1}}$	$\frac{\text{Model B}-\text{FVM}}{\text{FVM}-T_{bnd1}}$	T [°C]	*Model A*	*Model B*	*FVM*	$\frac{\text{Model A}-\text{FVM}}{\text{FVM}-T_{bnd0}}$	$\frac{\text{Model B}-\text{FVM}}{\text{FVM}-T_{bnd0}}$
T_0	81.9	134.2	129.7	−60.0%	5.6%	T_0	90.3	132.0	122.6	−44.4%	13.0%
T_1	78.5	127.1	122.7	−60.8%	6.1%	T_1	124.5	198.7	181.7	−43.5%	12.9%
T_2	78.5	127.2	122.9	−60.8%	5.9%	T_2	117.4	187.2	171.7	−44.7%	12.7%
T_3	78.5	127.2	122.9	−60.8%	5.9%	T_3	90.5	133.6	123.8	−45.1%	13.2%
T_4	78.5	127.1	122.9	−60.8%	5.8%	T_4	90.8	131.7	122.2	−43.5%	13.2%
T_5	78.5	127.2	122.9	−60.8%	5.8%	T_5	127.4	198.6	183.1	−41.9%	11.6%
T_6	78.5	127.2	122.8	−60.8%	6.0%	T_6	134.4	212.3	195.9	−42.2%	11.2%
T_7	73.7	117.7	113.4	−62.5%	6.9%	T_7	117.7	184.4	170.6	−43.94%	11.4%
T_8	73.7	117.7	113.2	−62.5%	7.1%	T_8	81.5	122.7	114.7	−51.4%	12.4%
T_9	64.2	96.2	93.3	−67.2%	6.7%	T_9	109.1	169.2	156.1	−44.3%	12.4%
T_{10}	64.5	96.7	93.7	−66.8%	7.0%	T_{10}	128.1	195.2	181.0	−40.4%	10.8%
T_{11}	74.0	118.1	113.7	−62.3%	7.0%	T_{11}	125.0	188.5	174.9	−40.0%	10.9%
T_{12}	73.7	117.7	113.3	−62.5%	6.9%	T_{12}	127.2	199.4	184.2	−42.5%	11.3%
T_{13}	64.5	96.7	93.7	−66.7%	7.0%	T_{13}	110.1	171.4	158.0	−44.4%	12.4%
T_{14}	73.7	117.7	113.5	−62.6%	6.7%	T_{14}	87.0	124.6	116.4	−44.3%	12.4%
T_{15}	74.0	118.1	113.7	−62.3%	6.9%	T_{15}	93.7	129.4	121.0	−38.4%	11.9%
T_{16}	64.2	96.2	93.3	−67.1%	6.9%	T_{16}	96.1	132.6	123.7	−37.4%	12.2%
T_{17}	64.5	96.7	93.7	−66.8%	7.0%	T_{17}	94.8	129.9	121.3	−37.1%	12.1%
T_{18}	64.5	96.7	93.7	−66.8%	6.9%	T_{18}	93.4	130.0	121.2	−39.0%	12.4%
T_{bnd0}		50.0				T_{bnd0}		50.0			
I_i [A]		7.5				I_i [A]		7.5			
Q_i [W]		13.15				Q_i [W]		13.15			

5.2. Comparison between the Different Shapes and the Wire Layouts

For the next results, only the method B is presented and discussed. For the Figure 10a,b four distinct boundary conditions are applied such as $T_{bnd,0} = 50$ °C, $T_{bnd,1} = 80$ °C, $T_{bnd,2} = 65$ °C and $T_{bnd,3} = 62$ °C. For the Figure 10c,d, only one boundary temperature is imposed at $T_{bnd,0} = 50$ °C. For all cases in this figure, the electrical current in each wire is 7.5 A which corresponds to a heat flow of 13.15 W to Joule losses.

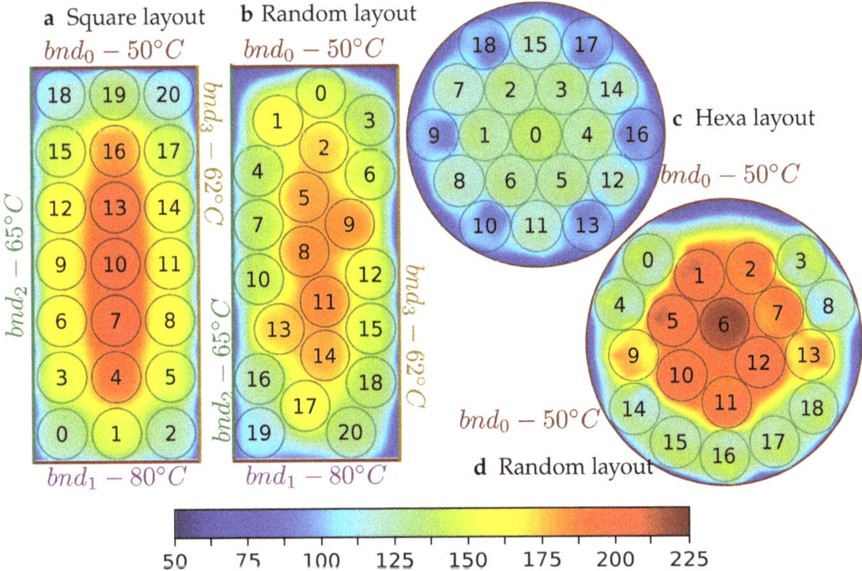

Figure 10. Temperatures contour for 2 shapes and 2 layouts. Results data from Tables 1 and 2.

The results prove that the modelling technique can provide individual wire temperatures. With a central area hotter than the periphery, the thermal field between (a) and (b) are very similar: the hotspot is in the centre of rectangle, that is, on wires 7, 10, 13 for the square layout and on wires 5, 8, 9, 11, 13 and 14 for the random layout. However, in detail, the number of wires in the hot spot is 3 for a square layout against 6 for the random layout. So, in the random layout, the temperature is lower (180 to 185 °C against 202 to 207 °C) but the hot spot is spread over a larger area. The boundary conditions greatly influence the temperature field. Consequently, in a random geometry, a bigger distance to the border decreases the influence of the boundary temperatures. This influence is visible between wires 18, 19, 20 of Figure (a) and wires 0, 1, 3 of Figure (b). Indeed in Figure b wire 0 is very close and has a minimal temperature influenced by the boundary conditions while wires 1 and 3 are warmer. The same phenomenon is visible in Figures (c) and d when we consider wires 15, 17, 18 to (c) and 0, 1, 2, 3 to (d).

The heat-up between the hexagonal (c) and random (d) layout for the round shapes is not of the same order of magnitude. The heat-up in the hexagonal layout is very lower than the random layout. Indeed the hexagonal layout optimises a small and constant space between each wire. The insulation created by the air is minimised. Many wires are close to boundary and they help the heat flow to escape.

The comparison between the shapes is interesting despite the lower number of wires in the round geometry (19 wires) than in the rectangular geometry (21 wires). The comparison shows that the round shape has the lowest (c) and the biggest (d) global temperature between all the cases. This shows the importance of the wire layout and the choice of shapes to design the coil.

6. Conclusions

This study provides a methodology for analysing the heat-up of any set of wires in a end-winding or any other device. Based on geometrical assumptions (Delaunay triangulation and Voronoï tessellation), the model creates a thermal network that could solve easily the temperature field. The evaluation of the thermal transfer thanks to thermal resistances and conductances is also provided. A second estimation more precise of thermal resistances and conductances is provided. Network set-up, adaptation, matrix writing and resolution are detailed. The model has been compared with FVM and several experiments are planned. To refine the model several issues should be tackled:

- Add convection by using the Robin type conditions or control the outgoing heat flow at the limit with Neumann conditions.
- Refine the determination of the thermal resistance between each wire (i.e., Rth)
- Integrate this end-winding slot model into a larger model which includes the stator.
- Thermophysical data can be made temperature dependent with an iterative convergence process in which the matrices containing the resistances and thermal conductances are updated synchronously.
- Finally, the model could be transformed into a transient model by the addition of thermal capacitors at each node.

Author Contributions: Conceptualization, E.F. and P.B.; methodology, J.P. and T.P.; software, J.P., T.P. and P.B.; validation, D.C.; investigation, J.P. and T.P.; data curation, T.P.; writing—original draft preparation, J.P.; writing—review and editing, P.B., D.C. and E.F.; visualization, D.C.; supervision, D.C. and E.F.; project administration, J.-C.D. and D.C.; funding acquisition, J.-C.D. All authors have read and agreed to the published version of the manuscript.

Funding: This research received no external funding.

Acknowledgments: The authors express their gratitude to the Groupe PSA, to the EIPHI University Research School (contract "ANR-17-EURE-0002") and to the National Association for Research and Technology (ANRT convention "2017/1091") for their support in carrying out this work.

Conflicts of Interest: The authors declare no conflict of interest.

Appendix A. Application: Materials Properties, Dimensions and Wire Positions

Appendix A.1. Materials Properties

- Media thermal conductivity (trapped air): 0.028 W/mK
- Insulation thermal conductivity: 0.2 W/mK
- Copper electrical resistivity: 18.7×10^{-9} Ωm

Appendix A.2. Slot Dimension, Slot Properties and Wire Positions

The Table A1 gives all geometry data of wires and slots corresponding at rectangular and round slot.

Table A1. Type and dimension of slot application.

	Rectangular slot		Round slot
Figure	2a,b	Figure	2c,d
wire number	21	wire number	19
wire section $[mm^2]$	0.785	wire section $[mm^2]$	0.785
insulation thickness $[\mu m]$	1,25	insulation thickness $[\mu m]$	1,25
Surface ratio	0.62	Surface ratio	0.73 Fig. c 0.62 Fig. d
slot height $[mm]$	7.88	slot diameter $[mm]$	5.13 Fig. c 5.55 Fig. d
slot width $[mm]$	3.38		

The Table A2 gives all wire positions for the 4 slot applications.

Table A2. Wire positions in 4 slot applications cf. Figures 2 and 10, the origin point is at the slot gravity center.

	Rectangular Slot		Round Slot	
	position (x, y) $[mm]$			
Wire	*Square layout*	*Random layout*	*Hexa layout*	*Random layout*
0	(−1.1, −3.3375)	(0.14, 3.35)	(0.0, 0.0)	(−1.53, 1.61)
1	(0.0, −3.3375)	(−0.74, 2.8)	(−1.025, 0.0)	(−0.5, 1.3)
2	(1.1, −3.3375)	(0.19, 2.27)	(−0.512, 0.888)	(0.54, 1.43)
3	(−1.1, −2.225)	(1.13, 2.8)	(0.512, 0.888)	(1.57, 1.58)
4	(0.0, −2.225)	(−1.15, 1.81)	(1.025, 0.0)	(−2.11, 0.74)
5	(1.1, −2.225)	(−2.2, 1.28)	(0.512, −0.888)	(−1.04, 0.42)
6	(−1.1, −1.1125)	(1.13, 1.73)	(−0.512, −0.888)	(0, 0.35)
7	(0.0, −1.1125)	(−1.15, 0.74)	(−1.538, 0.888)	(1.1, 0.57)
8	(1.1, −1.1125)	(−0.22, 0.2)	(−1.538, −0.888)	(2.14, 0.7)
9	(−1.1, 0.0)	(0.71, 0.74)	(−2.05, 0.0)	(−1.83, −0.25)
10	(0.0, 0.0)	(−1.15, −0.34)	(−1.025, −1.775)	(−0.86, −0.65)
11	(1.1, 0.0)	(0.19, −0.79)	(0.0, −1.775)	(0.04, −1.18)
12	(−1.1, 1.1125)	(1.13, −0.25)	(1.538, -0.888)	(0.71, −0.4)
13	(0.0, 1.1125)	(−0.74, −1.33)	(1.025, −1.775)	(1.73, −0.26)
14	(1.1, 1.1125)	(0.19, −1.86)	(1.538, 0.888)	(−1.82, −1.3)
15	(−1.1, 2.225)	(1.13, −1.33)	(0.0, 1.775)	(−1, −1.95)
16	(0.0, 2.225)	(−1.15, −2.32)	(2.05, −0.0)	(−0.01, −2.22)
17	(1.1, 2.225)	(−0.22, −2.86)	(1.025, 1.775)	(1,−1.95)
18	(−1.1, 3.3375)	(1.13, −2.4)	(−1.025, 1.775)	(1.8,−1.3)
19	(0.0, 3.3375)	(−1.15, −3.39)		
20	(1.1, 3.3375)	(0.71, −3.39)		

Appendix B. Results Data

The Figures A1 and A2 give respectively a temperature comparison between the two methods and FVM for the rectangular shape with a square layout and the rectangular shape with a random layout. The Figures A3 and A4 give respectively a temperature comparison between the two methods and FVM for the round shape with a hexagonal layout and the round shape with a random layout.

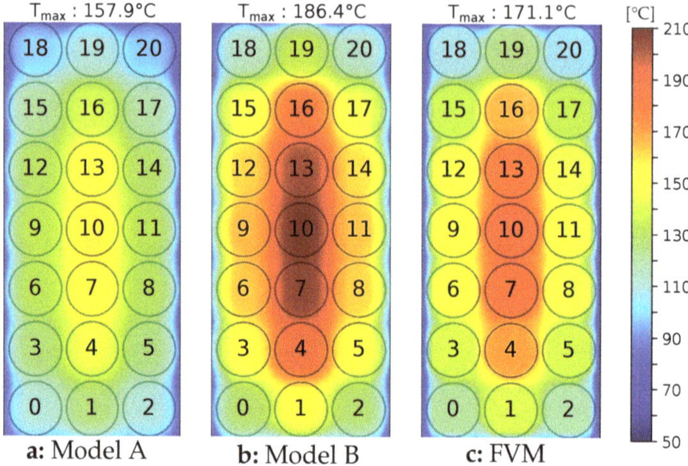

Figure A1. Comparison of thermal contours for rectangular shape with a square layout between method A (**a**), method B (**b**) and FVM (**c**). Results data from Table 1.

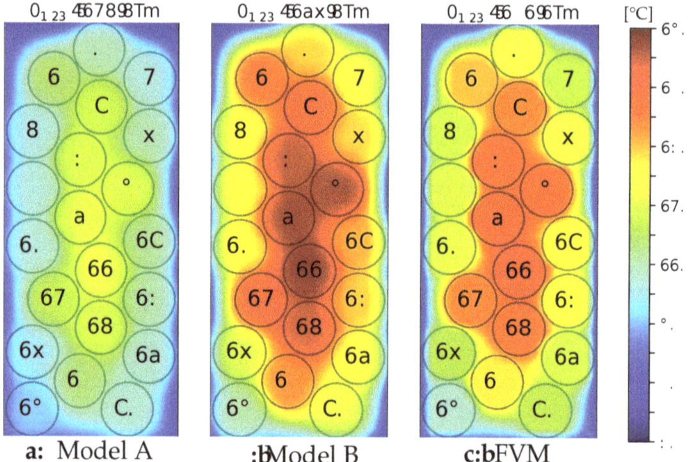

Figure A2. Comparison of thermal contours for rectangular shape with a random layout between method A (**a**), method B (**b**) and FVM (**c**). Results data from Table 1.

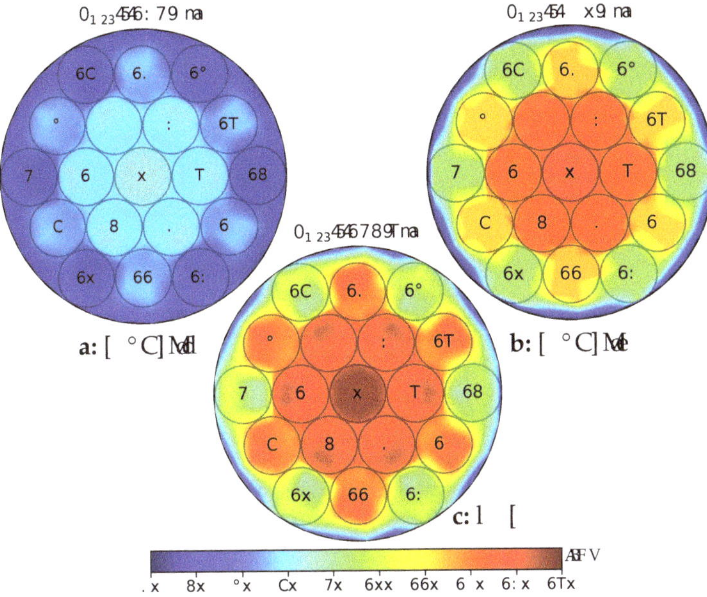

Figure A3. Comparison of thermal contours for round shape with a hexagonal layout between method A (**a**), method B (**b**) and FVM (**c**). Results data from Table 2.

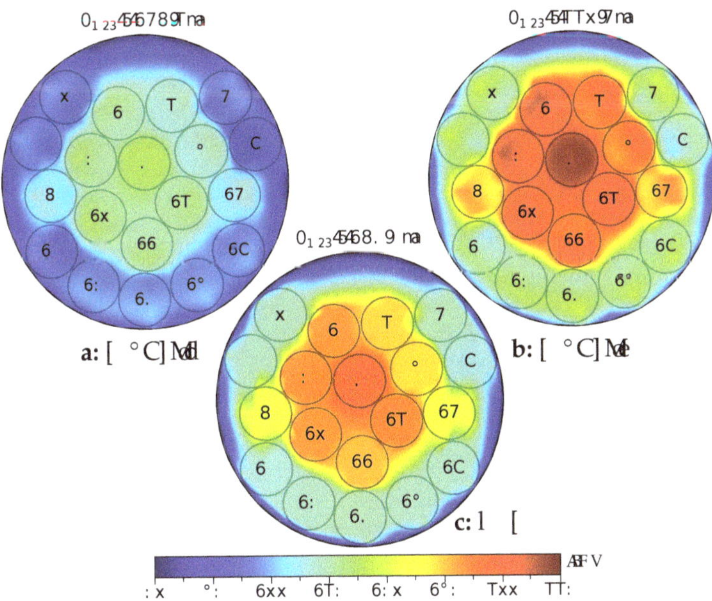

Figure A4. Comparison of thermal contours for round shape with a random layout between method A (**a**), method B (**b**) and FVM (**c**). Results data from Table 2.

References

1. Mellor, P.H.; Roberts, D.; Turner, D.R. Lumped parameter thermal model for electrical machines of TEFC design. *IEE Proc. B Electr. Pow. Appl.* **1991**, *138*, 205–218. [CrossRef]
2. Trigeol, J.F.; Bertin, Y.; Lagonotte, P. Thermal modeling of an induction machine through the association of two numerical approaches. *IEEE Trans. Energy Convers.* **2006**, *21*, 314–323. [CrossRef]
3. Fan, X.; Li, D.; Qu, R.; Wang, C. A dynamic multilayer winding thermal model for electrical machines with concentrated windings. *IEEE Trans. Ind. Electron.* **2019**, *66*, 6189–6199. [CrossRef]
4. Idoughi, L.; Mininger, X.; Bouillault, F.; Bernard, L.; Hoang, E. Thermal model with winding homogenization and FIT discretization for stator slot. *IEEE Trans. Magn.* **2011**, *47*, 4822–4826. [CrossRef]
5. Rasid, M.A.H.; Ospina, A.; El Kadri Benkara, K.; Lanfranchi, V. Thermal Model of Stator Slot for Small Synchronous Reluctance Machine. In Proceedings of the International Conference on Electrical Machines (ICEM), Berlin, Germany, 2–5 September 2014.
6. Hoang, E.; Lécrivain, M.; Hlioui, S.; Ahmed, H.B.; Multon, B. Element of slot thermal modelling. *SPEEDAM* **2010**, 303–305. [CrossRef]
7. Dannier, A.; Pizzo, A.D.; Di Noia, L.P.; Rizzo, R. Equivalent Thermal Model of Stator Slots using FEA. In Proceedings of the 2020 IEEE Texas Power and Energy Conference (TPEC), Texas, TX, USA, 6–7 February 2020.
8. Sciascera, C.; Giangrande, P.; Papini, L.; Gerada, C.; Galea, M. Analytical thermal model for fast stator winding temperature prediction. *IEEE Trans. Ind. Electr.* **2017**, *64*, 6116–6126. [CrossRef]
9. Milton, G.W. Bounds on the transport and optical properties of a two-component composite material. *J. Appl. Phys.* **1981**, *52*, 5294–5304. [CrossRef]
10. Hashin, Z.; Shtrikman, S. A variational approach to the theory of the elastic behaviour of multiphase materials. *J. Mech. Phys. Solids* **1963**, *11*, 127–140. [CrossRef]
11. Perrins, W.T.; McKenzie, D.R.; McPhedran, R.C. Transport properties of regular arrays of cylinders. *Proc. R. Soc. Lond. A Math. Phys. Eng. Sci.* **1979**, *369*, 207–225.
12. Stafford, J.; Grimes, R.; Newport, D. Development of compact thermal–fluid models at the electronic equipment level. *J. Therm. Sci. Eng. Appl.* **2012**, *4*, 031007. [CrossRef]
13. Saulnier, J.B.; Alexandre, A. La modélisation thermique par la méthode nodale. *Rev. Gén. Therm.* **1985**, *3*, 363–371.
14. Stenzel, P.; Dollinger, P.; Richnow, J.; Franke, J. Innovative Needle Winding Method using Curved Wire Guide in Order to Significantly Increase the Copper Fill Factor. In Proceedings of the 17th International Conference on Electrical Machines and Systems (ICEMS), Hangzhou, China, 24–25 October 2014.
15. Dietz, A.; Tommaso, A.O.D.; Marignetti, F.; Miceli, R.; Nevoloso, C. Enhanced flexible algorithm for the optimization of slot filling factors in electrical machines. *Energies* **2020**, *13*, 1041. [CrossRef]
16. Charih, F.; Dubas, F.; Espanet, C.; Bernard, R. Étude de la réduction des pertes électromagnétiques d'un MSAP à fort couple et à encoches ouvertes. In Proceedings of the Electrotechnique du Futur, Belfort, France, 14–15 December 2011.
17. Incropera, F.P.; Bergman, T.L.; DeWitt, D.P.; Lavine, A.S. Free Convection. In *Fundamentals of Heat and Mass Transfer*; Wiley: Hoboken, NJ, USA, 2013.
18. Delaunay, B. Sur la sphère vide. À la mémoire de Georges Voronoï. In *Bulletin de l'Académie des Sciences de l'URSS. Classe des Sciences Mathématiques et Naturelles*. Available online: http://mi.mathnet.ru/eng/izv4937 (accessed on 16 October 2020).
19. Szabó, P.G.; Markót, M.C.; Csendes, T.; Specht, E.; Casado, L.G. In *New Approaches to Circle Packing in a Square*; Springer-Verlag GmbH: Wien, Austria, 2007.
20. Nocaj, A.; Brandes, U. Computing Voronoi Treemaps: Faster, Simpler, and Resolution-independent. *Comput. Gr. Forum* **2012**, *31*, 855–864. [CrossRef]
21. Geuzaine, C.; Remacle, J.F. Gmsh: A 3-D finite element mesh generator with built-in pre- and post-processing facilities. *Int. J. Numer. Method. Eng.* **2009**, *79*, 1309–1331. [CrossRef]
22. Petitgirard, J.; Baucour, P.; Chamagne, D.; Fouillien, E. Étude 2D de l'échauffement d'un faisceau électrique pour une multitude de dispositions aléatoires de fils. Available online: https://www.sft.asso.fr/actes.html (accessed on 16 October 2020).

23. Hagberg, A.A.; Schult, D.A.; Swart, P.J. Exploring Network Structure, Dynamics, and Function Using NetworkX. In Proceedings of the 7th Python in Science Conference, Pasadena, CA, USA, 19–24 August 2008.
24. Piessens, R.; de Doncker-Kapenga, E.; Überhuber, C.W.; Kahaner, D.K. *Quadpack*; Springer: Heidelberg, Germany, 1983.
25. Anderson, E.; Bai, Z.; Bischof, C.; Blackford, S.; Demmel, J.; Dongarra, J.; Du Croz, J.; Greenbaum, A.; Hammarling, S.; McKenney, A.; et al. *LAPACK Users' Guide*, 3 ed.; Society for Industrial and Applied Mathematics: Philadelphia, PA, USA, 1999.
26. Buhmann, M.D. *Radial Basis Functions: Theory and Implementations*; Cambridge University Press: London, UK, 2003.
27. Hunter, J.D. Matplotlib: A 2D graphics environment. *Comput. Sci. Eng.* **2007**, *9*, 90–95. [CrossRef]

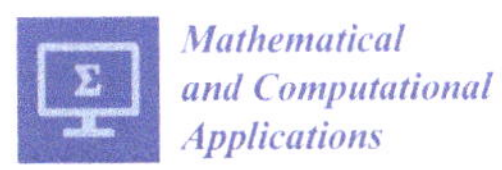

Article

Reduced-Order Model of Rotor Cage in Multiphase Induction Machines: Application on the Prediction of Torque Pulsations

Abdelhak Mekahlia [1,2,*], Eric Semail [1], Franck Scuiller [3] and Hussein Zahr [2]

[1] Univ. Lille, Arts et Metiers Institute of Technology, Centrale Lille, HEI, L2EP—Laboratoire d'Electrotechnique et d'Electronique de Puissance-ULR2697, HESAM Université, F-59000 Lille, France; Eric.SEMAIL@ENSAM.EU
[2] Institut Vedecom, 23 bis Allée des Marronniers, 78000 Versailles, France; hussein.zahr@vedecom.fr
[3] Naval Academy Research Institute (IRENav), Ecole Navale, F-29240 Brest, France; franck.scuiller@ecole-navale.fr
* Correspondence: abdelhak.mekahlia@vedecom.fr or mekahlia.abdelhak@gmail.com; Tel.: +33-650-27-1449

Received: 31 December 2019; Accepted: 27 January 2020; Published: 29 January 2020

Abstract: For three-phase induction machines supplied by sinusoidal current, it is usual to model the n-bar squirrel-cage by an equivalent two-phase circuit. For a multiphase induction machine which can be supplied with different harmonics of current, the reduced-order model of the rotor must be more carefully chosen in order to predict the pulsations of torque. The proposed analysis allows to avoid a wrong design with non-sinusoidal magnetomotive forces. An analytical approach is proposed and confirmed by Finite-Element modelling at first for a three-phase induction machine and secondly for a five-phase induction machine.

Keywords: multiphase induction machine; reduced order; rotor cage; torque pulsations

---

1. Introduction

In electromechanical energy conversion, the three-phase induction machines are the most used. The rules and methods of the design of this kind of machines have been widely investigated in [1] and [2].

The mathematical modeling of induction machines is characterized by the complexity of the squirrel cage rotor. Modeling and parameters determination of this kind of rotors is investigated in [3].

The three-phase induction machines with classical design and sinusoidal currents are limited regarding torque density and speed range [4,5]. With multiphase induction machines, non-sinusoidal stator current harmonics allow to improve the torque density [6–9]. The speed range can be extended thanks to the sequential injection of current harmonics [10,11]. Several works have also investigated the speed range extension in multiphase induction machines with two polarities by pole-phase modulation [12–14].

Thanks to the aforementioned advantages, multiphase induction machines are becoming attractive in several sectors especially aeronautics [15], naval [16], and automotive industries [17]. With these increasing demands on multiphase induction machines, the design rules and theories, classically developed for the case of three phases, must be reconsidered.

In fact, the induction machine is characterized by complex harmonic interactions due to the squirrel cage rotor structure, if this one is not well-designed dangerous phenomena could occur, especially torque ripple and mechanical vibrations. For the three-phase machines several design rules have been developed [2,18]. It must be mentioned that in the case of a three-phase machine, only one stator current harmonic is used to produce torque, which is not the case for a multiphase machine.

With the possibility of injecting more than one harmonic, more degrees of freedom regarding the machine supply and control are appearing [19,20]. However, on the other hand, most of the classical design approaches, adapted for three-phase machines, cannot be directly used for multiphase machines due to the impact of injecting several stator harmonics on the design parameters [21]. Furthermore, it becomes more difficult to design a squirrel cage rotor without the risk of parasitic phenomena occurring during all the supplying modes (with different sequences). As an example, a previous work has shown that the rotor bars number has to be carefully chosen to avoid important torque ripple under different supplying harmonics (sequences) [22]. Hence, it is necessary to understand deeply the interactions between time and space harmonics and their impact on torque production in multiphase induction machine [23].

Generalized mathematical modeling, based on voltage and flux equations, allows to interpret these harmonic interactions. Thanks to the symmetrical component transformations, the study of space and time harmonics is easier [24]. This mathematical approach requires to identify the machine parameters, inductances and resistances, several works were done in this topic [25,26].

In three-phase induction machines, the stator winding, whose dimension is $N_{ph} = 3$, can be modeled by one two-phase equivalent circuit corresponding to one α-β plane and a zero-sequence (not excited for balanced stator supply) thanks to α-β-0 Concordia Transformation. This mathematical transformation is also applied on the squirrel cage rotor, whose dimension is N_{bar}, which gives:

- $\frac{N_{bar}-1}{2}$ α-β rotor planes, excited by independent harmonic sets, and **one** zero-sequence if N_{bar} is odd
- $\frac{N_{bar}}{2} - 1$ α-β rotor planes and **two** zero-sequences if N_{bar} is even.

However, in the classical simplified modeling approaches, the rotor cage is modeled by only one equivalent two-phase circuit, which corresponds to the α-β plane excited by the fundamental sequence (only the first space harmonic considered). This simplified approach supposes a sinusoidal distribution of the winding, so other space harmonics can be neglected comparing to the fundamental. In fact, this simplification is not always available, especially for the windings with a non-sinusoidal distribution, as the tooth concentrated windings, where all space harmonics exist and with significant amplitudes, which induce in the rotor bars more than one current harmonic. Hence different α-β rotor planes are excited.

Furthermore, in the case of multiphase machines (more than 3 phases), the stator winding is modeled by:

- $\frac{N_{ph}-1}{2}$ α-β stator planes, and **one** zero-sequence if N_{ph} is odd;
- $\frac{N_{ph}}{2} - 1$ α-β stator planes, and **two** zero-sequences if N_{ph} is even.

Every α-β stator plane can be excited by a stator sequence, usually several current harmonics are injected to improve torque production in multiphase machines, these injected harmonics belong to independent sets, so they excite different α-β stator planes. All these excited α-β planes must be considered in modeling, so the stator winding is represented by more than one equivalent two-phase circuit, each one interacts with one α-β rotor plane at least, depending on the induced rotor current harmonics and the importance of their amplitudes.

Considering all the α-β rotor planes, excited by important space harmonics, is important to predict torque pulsations due to harmonic interactions. Recent work has investigated the analytical estimation of torque ripples, due to interactions between stator and rotor space harmonics, for three-phase induction machine [23].

In this paper, a mathematical approach allowing to find the excited α-β rotor planes which must be considered in the modelling of multiphase induction machine is proposed. This approach allows to justify the reduction of the rotor dimension, which is "N_{bar}" in the natural base model (Non-transformed electrical equations). Firstly, the generalized mathematical model of induction machine is presented. Then, the α-β transformation is applied on the stator and rotor equations.

Thanks to the transformed stator and rotor current vectors and the mutual inductance matrix, a new arithmetic approach of prediction of torque pulsating components frequencies is proposed.

This arithmetic tool is used to predict pulsating components frequencies for firstly a three-phase machine, and secondly a five-phase machine supplied by two different sequences. The results are validated by F-E simulations.

2. Induction Machine Model

Based on inductance and resistance matrices in stator and rotor, voltage equations can be written as follows [27]:

$$\underline{V}_s = \underline{R_s}.\underline{I}_s + \frac{d\underline{\phi_s}}{dt}, \tag{1}$$

$$\underline{V}_r = 0 = \underline{R_r}.\underline{I}_r + \frac{d\underline{\phi_r}}{dt}, \tag{2}$$

$$\underline{\phi}_s = \underline{L_{ss}}.\underline{I}_s + \underline{L_{sr}}(\theta).\underline{I_r}, \tag{3}$$

$$\underline{\phi}_r = \underline{L_{rr}}.\underline{I}_r + \underline{L_{rs}}(\theta).\underline{I_s}, \tag{4}$$

The stator matrix can be written as follows:

$$\underline{\mathrm{R_s}} = \begin{bmatrix} \mathrm{R_s} & 0 & \dots & 0 \\ 0 & \mathrm{R_s} & 0 & \vdots \\ \vdots & \ddots & \ddots & 0 \\ 0 & \dots & 0 & \mathrm{R_s} \end{bmatrix}, \tag{5}$$

$$\underline{\mathrm{L_{ss}}} = \begin{bmatrix} \mathrm{L_{s11}} + \mathrm{L_{ls}} & \mathrm{L_{s12}} & \dots & \dots & \mathrm{L_{s1N_{ph}}} \\ \mathrm{L_{s1N_{ph}}} & \mathrm{L_{s11}} + \mathrm{L_{ls}} & \mathrm{L_{s12}} & \dots & \mathrm{L_{s1(N_{ph}-1)}} \\ \vdots & \ddots & \ddots & \ddots & \vdots \\ \mathrm{L_{s12}} & \dots & \dots & \mathrm{L_{s1N_{ph}}} & \mathrm{L_{s11}} + \mathrm{L_{ls}} \end{bmatrix}, \tag{6}$$

The rotor cage is represented as shown in Figure 1. Each rotor loop (composed by two adjacent half-bars and two ring portions) is considered like a phase [28].

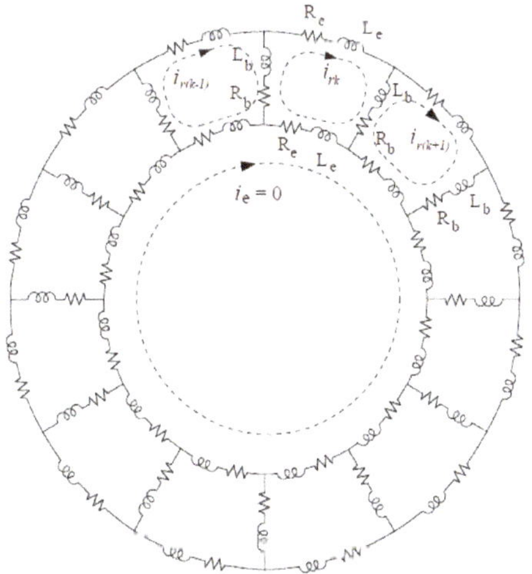

Figure 1. Squirrel cage rotor representation, figure taken from [29] (p. 60).

According to the rotor representation approach shown in Figure 1, rotor resistance and inductance matrices can be written as follows:

$$\underline{R_r} = \begin{bmatrix} 2(R_b+R_e) & -R_b & 0 & \cdots & -R_b \\ -R_b & 2(R_b+R_e) & -R_b & \ddots & 0 \\ \vdots & \ddots & \ddots & \ddots & \vdots \\ \vdots & \ddots & \ddots & \ddots & \vdots \\ -R_b & 0 & \cdots & -R_b & 2(R_b+R_e) \end{bmatrix}, \quad (7)$$

$$\underline{L_{rr}} = \begin{bmatrix} L_{r11}+2(L_b+L_e) & L_{r12}-L_b & L_{r13} & \cdots & L_{r1(N_{bar}-1)} & L_{r1N_{bar}}-L_b \\ L_{r21}-L_b & L_{r11}+2(L_b+L_e) & L_{r23}-L_b & L_{r24} & \cdots & L_{r2N_{bar}} \\ \vdots & \ddots & \ddots & \ddots & \ddots & \vdots \\ L_{rN_{bar}1}-N_b & L_{rN_{bar}2} & \cdots & L_{rN_{bar}(N_{bar}-2)} & L_{rN_{bar}(N_{bar}-1)}-N_b & L_{r11}+2(L_b+L_e) \end{bmatrix}, \quad (8)$$

Mutual inductance between stator and rotor can be written as follows:

$$\underline{L_{sr}}(\theta) = \sum_k \text{real}\left((\hat{M_{sr}})_k \begin{bmatrix} e^{-j.kp\theta} & \cdots & e^{j.k.(-p(N_{bar}-1)\frac{2\pi}{N_{bar}}-p\theta)} \\ \vdots & \ddots & \vdots \\ e^{j.k.((N_{ph}-1)\frac{2\pi}{N_{ph}}-p\theta)} & \cdots & e^{j.k.((N_{ph}-1)\frac{2\pi}{N_{ph}}-p(N_{bar}-1)\frac{2\pi}{N_{bar}}-p\theta)} \end{bmatrix}\right) \quad (9)$$

Mutual inductance between a stator phase and a rotor phase (rotating) is a periodic function depending on the electrical angle (mechanical angle multiplied by the number of pole-pairs) and whose period is an electrical turn. This periodic spatial function contains harmonics, called space harmonics. Hence, Fourier transformation of the mutual inductance spatial function allows to find the complex coefficients $(\hat{M_{sr}})_k$, where "k.p" is the space harmonic range (k is positive integer).

As we can see in Equation (9), to simplify the expression of the mutual inductance matrix, the complex coefficients $(\hat{M_{sr}})_k$ are used. For example, the first element (1,1) of the matrix $L_{sr}(\theta)$ from Equation (9) is: $real\left((\hat{M_{sr}})_k.e^{-j.kp\theta}\right)$, which is equal to: $M_{srk}.cos\left(-kp\theta+\varphi_{sr_k}\right)$, where M_{srk} is the amplitude of the complex coefficient $(\hat{M_{sr}})_k$, and φ_{sr_k} is its angle.

3. α-β-0 Transformed Model

α-β-0 generalized transformation matrix [24], for the **odd** dimension "n", is defined as follows:

$$\underline{A_n} = \sqrt{\frac{2}{n}} \begin{bmatrix} \frac{1}{\sqrt{2}} & \frac{1}{\sqrt{2}} & \cdots & \frac{1}{\sqrt{2}} \\ 1 & \cos\left(\frac{2\pi}{n}\right) & \cdots & \cos\left(\frac{2(n-1)\pi}{n}\right) \\ 0 & \sin\left(\frac{2\pi}{n}\right) & \cdots & \sin\left(\frac{2(n-1)\pi}{n}\right) \\ \vdots & \ddots & \cdots & \vdots \\ 1 & \cos\left(\frac{(n-1)\pi}{n}\right) & \cdots & \cos\left(\frac{(n-1)^2\pi}{n}\right) \\ 0 & \sin\left(\frac{(n-1)\pi}{n}\right) & \cdots & \sin\left(\frac{(n-1)^2\pi}{n}\right) \end{bmatrix}, \quad (10)$$

If the dimension "n" is **even**, a new line is added at the end of the matrix $\underline{A_n}$ (last line). This line has this form: $\frac{2}{\sqrt{n}}\begin{bmatrix} \frac{1}{\sqrt{2}} & -\frac{1}{\sqrt{2}} & \cdots & \frac{1}{\sqrt{2}} & -\frac{1}{\sqrt{2}} \end{bmatrix}$.

Equations (1)–(4) are transformed as follows:

$$\underline{V_{s_{\alpha\beta}}} = \underline{A_s}.\underline{R_s}.\underline{A_s}^{-1}.\underline{I_{s_{\alpha\beta}}} + \frac{d\underline{\phi_{s\alpha\beta}}}{dt}, \quad (11)$$

$$\underline{V}_{r_{\alpha\beta}} = 0 = \underline{A_r}.\underline{R_r}.\underline{A_r}^{-1}.\underline{I_{r_{\alpha\beta}}} + \frac{d\underline{\phi_{r_{\alpha\beta}}}}{dt}, \tag{12}$$

$$\underline{\phi}_{s_{\alpha\beta}} = \underline{A_s}.\underline{L_{ss}}.\underline{A_s}^{-1}.\underline{I_{s_{\alpha\beta}}} + \underline{A_s}.\underline{L_{sr}}(\theta).\underline{A_r}^{-1}.\underline{I_{r_{\alpha\beta}}}, \tag{13}$$

$$\underline{\phi}_{r_{\alpha\beta}} = \underline{A_r}.\underline{L_{rr}}.\underline{A_r}^{-1}.\underline{I_{r_{\alpha\beta}}} + \underline{A_r}.\underline{L_{rs}}(\theta).\underline{A_s}^{-1}.\underline{I_{s_{\alpha\beta}}}, \tag{14}$$

where $\underline{A_s}$ is α-β-0 transformation matrix with the stator dimension "N_{ph}", and $\underline{A_r}$ is related to the rotor dimension "N_{bar}".

To investigate the rotor dimension reduction, this transformation is applied on the stator-rotor mutual inductance matrix $\underline{L_{sr}}(\theta)$. The use of the transformed matrix, and the transformed stator and rotor currents vectors, lead to the prediction of torque components frequencies.

3.1. *α-β-0 Transformation of the Stator Current Vector*

The stator winding can be supplied by different sequences "u", defined as follows:

$$u \in \left[0,\ 1 \cdots N_{ph} - 1\right] \tag{15}$$

For example, in the case of classical three-phase machines, the sequence "u = 0" corresponds to the zero-sequence, the sequence "u = 1" corresponds to the direct order of supplying the phases and the sequence "u = 2" corresponds to the inverse order.

If we consider that the winding is supplied by one sequence "u", the current of the stator phase "i" in the natural base, can be expressed as follows:

$$I_{s_i} = I_{s_u}.\sin\left(2\pi.f_{s_u}.t - (i-1).u.\frac{2\pi}{N_{ph}} + \varphi_{s_u}\right), \tag{16}$$

The stator current vector in the natural base can be expressed as follows:

$$\underline{I_s} = \begin{pmatrix} I_{s_0} \\ I_{s_1} \\ \vdots \\ I_{s_{N_{ph}-1}} \end{pmatrix}, \tag{17}$$

If only one sequence "u" is supplying the stator, Concordia transformation reduces the dimension of the current vector from N_{ph} to 2 (corresponding to one α-β stator plane). The different cases of the transformed stator current vector $\underline{I_{s_{\alpha\beta}}}$, which depend on the imposed stator sequence, are shown in Table 1.

This table shows the projection of four different cases of sequences (u_0, u_1, u_2, u_3), where:

- $u_0 = N_{ph}$ (Zero-sequence) → Example of 6 phases: $u_0 = 5$;
- $u_1 < \text{floor}\left(\frac{N_{ph}}{2}\right)$ (≤ if N_{ph} is odd) → Example of 6 phases: $u_1 = 1$ or 2;
- $u_2 > \text{floor}\left(\frac{N_{ph}}{2}\right)$ → Example of 6 phases: $u_2 = 4$ or 5;
- $u'_0 = \frac{N_{ph}}{2}$ (considered as a "second zero-sequence", exists only when N_{ph} is even) → Example of 6 phases: $u'_0 = 3$.

Table 1. Transformed stator current vector $\underline{I_{s_{\alpha\beta}}}$.

Index	Transformed Current vector $\underline{I_{s_{\alpha\beta}}}$
1	$\sqrt{N_{ph}}.I_{s_{u_0}}.\sin\left(2\pi f_{s_{u_0}}.t+\varphi_{s_{u_0}}\right)$
$\vdots$	$\vdots$
$2.u_1$	$\sqrt{\frac{N_{ph}}{2}}.I_{s_{u_1}}.\sin\left(2\pi f_{s_{u_1}}.t+\varphi_{s_{u_1}}\right)$
$2.u_1+1$	$-\sqrt{\frac{N_{ph}}{2}}.I_{s_{u_1}}.\cos\left(2\pi f_{s_{u_1}}.t+\varphi_{s_{u_1}}\right)$
$\vdots$	$\vdots$
$2.\left(N_{ph}-u_2\right)$	$\sqrt{\frac{N_{ph}}{2}}.I_{s_{u_2}}.\sin\left(2\pi f_{s_{u_2}}.t+\varphi_{s_{u_2}}\right)$
$2.\left(N_{ph}-u_2\right)+1$	$\sqrt{\frac{N_{ph}}{2}}.I_{s_{u_2}}.\cos\left(2\pi f_{s_{u_2}}.t+\varphi_{s_{u_2}}\right)$
$\vdots$	$\vdots$
N_{ph}	$\sqrt{N_{ph}}.I_{s_{u'_0}}.\sin\left(2\pi f_{s_{u'_0}}.t+\varphi_{s_{u'_0}}\right)$

3.2. α-β-0 Transformation of Mutual Inductance Matrix

Stator to rotor mutual inductance matrix $\underline{L_{sr}}(\theta)$ is characterized by the presence of several space harmonics, α-β-0 transformation allows to separate these space harmonics into different planes.

The transformed matrix $\underline{L_{sr_{\alpha\beta}}}(\theta)$ has a dimension of $N_{ph}\times N_{bar}$, its general form is presented in Figure 2.

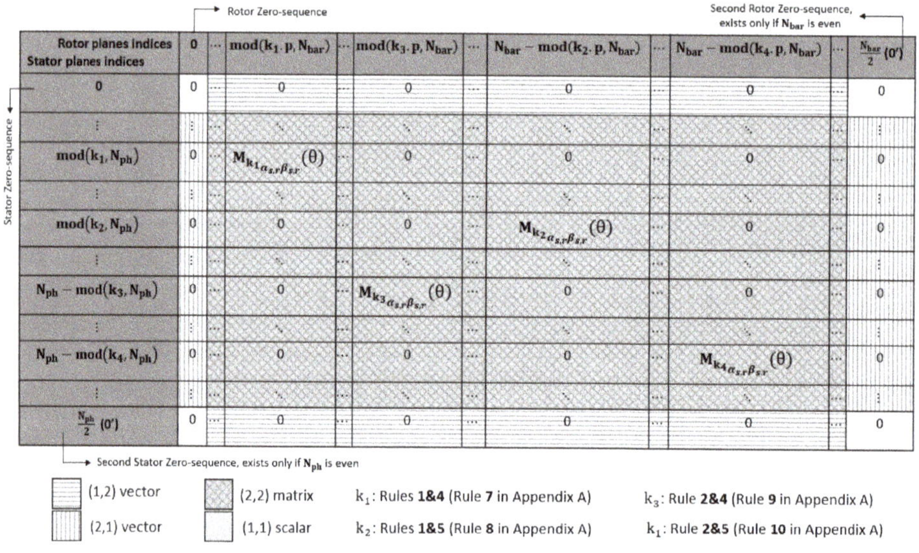

Rotor Zero-sequence

Second Rotor Zero-sequence, exists only if N_{bar} is even

Rotor planes indices / Stator planes indices	0	···	$mod(k_1.p, N_{bar})$	···	$mod(k_3.p, N_{bar})$	···	$N_{bar}-mod(k_2.p, N_{bar})$	···	$N_{bar}-mod(k_4.p, N_{bar})$	···	$\frac{N_{bar}}{2}$ (0')
0	0	···	0	···	0	···	0	···	0	···	0
⋮	⋮	···	⋱	···	⋱	···	⋱	···	⋱	···	⋮
$mod(k_1, N_{ph})$	0	···	$M_{k1_{\alpha_{s,r}\beta_{s,r}}}(\theta)$	···	0	···	0	···	0	···	0
⋮	⋮	···	⋱	···	⋱	···	⋱	···	⋱	···	⋮
$mod(k_2, N_{ph})$	0	···	0	···	0	···	$M_{k2_{\alpha_{s,r}\beta_{s,r}}}(\theta)$	···	0	···	0
⋮	⋮	···	⋱	···	⋱	···	⋱	···	⋱	···	⋮
$N_{ph}-mod(k_3, N_{ph})$	0	···	0	···	$M_{k3_{\alpha_{s,r}\beta_{s,r}}}(\theta)$	···	0	···	0	···	0
⋮	⋮	···	⋱	···	⋱	···	⋱	···	⋱	···	⋮
$N_{ph}-mod(k_4, N_{ph})$	0	···	0	···	0	···	0	···	$M_{k4_{\alpha_{s,r}\beta_{s,r}}}(\theta)$	···	0
⋮	⋮	···	⋱	···	⋱	···	⋱	···	⋱	···	⋮
$\frac{N_{ph}}{2}$ (0')	0	···	0	···	0	···	0	···	0	···	0

Stator Zero-sequence

Second Stator Zero-sequence, exists only if N_{ph} is even

(1,2) vector; (2,1) vector; (2,2) matrix; (1,1) scalar

k_1: Rules **1&4** (Rule **7** in Appendix A)

k_2: Rules **1&5** (Rule **8** in Appendix A)

k_3: Rule **2&4** (Rule **9** in Appendix A)

k_1: Rule **2&5** (Rule **10** in Appendix A)

Figure 2. Transformed Stator–Rotor mutual inductance matrix $\underline{L_{sr_{\alpha\beta}}}(\theta)$.

As we can see in Figure 2, each space harmonic appears only in one cell. The different cells in Figure correspond to two-by-two matrices, two-element vectors or scalars (as described in the legend in the bottom of Figure). A cell corresponding to two-by-two matrix represents the intersection between an α-β stator plane and an α-β rotor plane. The general form of these matrices can be written as follows:

$$\underline{M_{k\,\alpha_{s,r}\beta_{s,r}}}(\theta) = \frac{\sqrt{N_{ph}.N_{bar}}}{2}.M_{sr_k}\begin{pmatrix} \delta_{11}\cos\left(-kp\theta + \varphi_{sr_k}\right) & \delta_{12}\sin\left(-kp\theta + \varphi_{sr_k}\right) \\ \delta_{21}\sin\left(-kp\theta + \varphi_{sr_k}\right) & \delta_{22}\cos\left(-kp\theta + \varphi_{sr_k}\right) \end{pmatrix}, \tag{18}$$

where $\delta_{ij} = \pm 1$ depending on the category of the harmonic (see the rules in Appendix A).

The following rules (from 1 to 6) define the distribution of a space harmonic "k" in $\underline{L_{sr_{\alpha\beta}}}(\theta)$:

- Rule 1: If $\mathrm{mod}\left(k, N_{ph}\right) \leq \frac{N_{ph}-1}{2}$, the harmonic "k" is projected on the stator plane number $\mathrm{mod}\left(k, N_{ph}\right)$, so the two lines whose indices are $2.\mathrm{mod}\left(k, N_{ph}\right)$ and $2.\mathrm{mod}\left(k, N_{ph}\right) + 1$ in the matrix $\underline{L_{sr_{\alpha\beta}}}(\theta)$.
- Rule 2: If $\mathrm{mod}\left(k, N_{ph}\right) > \frac{N_{ph}-1}{2}$, the harmonic "k" is projected on the stator plane number $N_{ph} - \mathrm{mod}\left(k, N_{ph}\right)$, so the two lines whose indices are $2.\left(N_{ph} - \mathrm{mod}\left(k, N_{ph}\right)\right)$ and $2.\left(N_{ph} - \mathrm{mod}\left(k, N_{ph}\right)\right) + 1$ in the matrix $\underline{L_{sr_{\alpha\beta}}}(\theta)$.
- Rule 3: If $\mathrm{mod}\left(k, N_{ph}\right) = 0$, the harmonic "k" is projected on the first stator zero-sequence (0), so the first line in the matrix $\underline{L_{sr_{\alpha\beta}}}(\theta)$.
- Rule 3-bis (Only when N_{ph} is even): If $\mathrm{mod}\left(k, N_{ph}\right) = \frac{N_{ph}}{2}$, the harmonic "k" is projected on the second stator zero-sequence, so the last line in the matrix $\underline{L_{sr_{\alpha\beta}}}(\theta)$.
- Rule 4: If $\mathrm{mod}(k.p, N_{bar}) \leq \frac{N_{bar}-1}{2}$, the harmonic "k" is projected on the rotor plane number $\mathrm{mod}(k.p, N_{bar})$, so the two lines whose indices are $2.\mathrm{mod}(k.p, N_{bar})$ and $2.\mathrm{mod}(k.p, N_{bar}) + 1$ in the matrix $\underline{L_{sr_{\alpha\beta}}}(\theta)$.
- Rule 5: If $\mathrm{mod}(k.p, N_{bar}) > \frac{N_{bar}-1}{2}$, the harmonic "k" is projected on the rotor plane number $N_{bar} - \mathrm{mod}(k.p, N_{bar})$, so the two lines whose indices are $2.(N_{bar} - \mathrm{mod}(k.p, N_{bar}))$ and $2.\left(N_{ph} - \mathrm{mod}\left(k, N_{ph}\right)\right) + 1$ in the matrix $\underline{L_{sr_{\alpha\beta}}}(\theta)$.
- Rule 6: If $\mathrm{mod}(k.p, N_{bar}) = 0$, the harmonic "k" is projected on the first rotor zero-sequence (0), so the column whose index is 1, in the matrix $\underline{L_{sr_{\alpha\beta}}}(\theta)$.
- Rule 6-bis (Only when N_{bar} is even): If $\mathrm{mod}(k.p, N_{bar}) = \frac{N_{bar}}{2}$, the harmonic "k" is projected on the second rotor zero-sequence (0'), so the last column in the matrix $\underline{L_{sr_{\alpha\beta}}}(\theta)$.

When several harmonics are projected on the same stator and rotor planes, their matrices (or vectors or scalars, see Appendix A) $\underline{M_{k\,\alpha_{s,r}\beta_{s,r}}}(\theta)$ are superposed. For example, we consider the harmonics "k_1" and "k_2", if: $\mathrm{mod}\left(k_1, N_{ph}\right) = \mathrm{mod}\left(k_2, N_{ph}\right)$ and $\mathrm{mod}(k_1.p, N_{bar}) = \mathrm{mod}(k_2.p, N_{bar})$, then $\underline{M_{k_1\,\alpha_{s,r}\beta_{s,r}}}(\theta)$ and $\underline{M_{k_1\,\alpha_{s,r}\beta_{s,r}}}(\theta)$ are superposed in $\underline{L_{sr_{\alpha\beta}}}(\theta)$.

3.3. α-β-0 Transformation of Rotor Current Vector

For the windings with an integral number of slots per phase and per pole-pair, **2.spp = integral** (including the special fractional like spp = 0.5, 1.5 …), the sequence "u" generates a set of space harmonics listed in F_u, which is defined as follows [30]:

$$F_u = \left\{Z.N_{ph} + u\right\}p\ Z = 0,\ 1,\ -1,\ 2,\ -2,\ 3,\ -3\ldots \tag{19}$$

It should be mentioned that all the space harmonics are multiples of "p".

As examples:

- For $N_{ph} = 3$ and "u = 1", Fu = p * {1, −2, 4, −5, 7, −8, 10, −11, 13 … };

- For $N_{ph} = 5$: for "u = 1", Fu = p * {1, −4, 6, -9, 11, −14, 16, −19, 21 ... }; for u = 2, Fu = p * {2, −3, 7, −8, 12, −13, 17 ... }; for u = 3, Fu = p * {−2, 3, −7, 8, −12, 13, −17 ... }. As we can see, the sequences "u = 2" and "u = 3" generate the same set of space harmonics, but with opposite signs.

If the stator sequence "u" has a frequency "f_s", each space harmonic "v.p" ("v" is integer, can be positive or negative) belonging to the set "F_u" induces in the rotor bar currents a frequency, called "f_{r_v}", which can be determined as follows (taking into account the mechanical speed of the rotor):

$$f_{r_v} = f_s - v.p.\frac{\Omega_{mec}}{2\pi} \tag{20}$$

So, the current in a rotor bar "i", can be expressed as following:

$$I_{r_i} = \sum_{v}\left[I_{r_v}.\sin\left(2\pi.f_{r_v}.t - (i-1).v.p.\frac{2\pi}{N_{bar}} + \varphi'_{r_v}\right)\right] \tag{21}$$

where I_{r_v} is the amplitude of the harmonic range "v" of rotor current, and φ'_{r_v} is its angle.

The transformed rotor current vector contains several harmonics (according to Equation (19)), separated into different α-β rotor planes as shown in the Table 2. Each α-β rotor plane can be excited by more than one harmonic, but to simplify the presentation only one harmonic per plane is considered in the table.

Table 2. Transformed rotor current vector $\underline{I_{r_{\alpha\beta}}}$.

Index	Transformed Current vector $\underline{I_{r_{\alpha\beta}}}$
1	$\sqrt{N_{bar}}.I_{r_{v_0}}.\sin\left(2\pi\left\lvert f_{r_{v_0}}\right\rvert.t + \varphi_{r_{v_0}}\right)$
⋮	⋮
$2.\mathrm{mod}\left(\lvert v_1\rvert.p, N_{bar}\right)$	$\sqrt{\frac{N_{bar}}{2}}.I_{r_{v_1}}.\sin\left(2\pi\left\lvert f_{r_{v_1}}\right\rvert.t + \varphi_{r_{v_1}}\right)$
$2.\mathrm{mod}(\lvert v_1\rvert.p, N_{bar}) + 1$	$-\sqrt{\frac{N_{bar}}{2}}.I_{r_{v_1}}.\cos\left(2\pi\left\lvert f_{r_{v_1}}\right\rvert.t + \varphi_{r_{v_1}}\right)$
⋮	⋮
$2.\mathrm{mod}\left(\lvert v_2\rvert.p, N_{bar}\right)$	$\sqrt{\frac{N_{bar}}{2}}.I_{r_{v_2}}.\sin\left(2\pi\left\lvert f_{r_{v_2}}\right\rvert.t + \varphi_{r_{v_2}}\right)$
$2.\mathrm{mod}(\lvert v_2\rvert.p, N_{bar}) + 1$	$\sqrt{\frac{N_{bar}}{2}}.I_{r_{v_2}}.\cos\left(2\pi\left\lvert f_{r_{v_2}}\right\rvert.t + \varphi_{r_{v_2}}\right)$
⋮	⋮
$2.(N_{bar} - \mathrm{mod}(\lvert v_3\rvert.p, N_{bar}))$	$\sqrt{\frac{N_{bar}}{2}}.I_{r_{v_3}}.\sin\left(2\pi\left\lvert f_{r_{v_3}}\right\rvert.t + \varphi_{r_{v_3}}\right)$
$2.\left(N_{bar} - \mathrm{mod}\left(\lvert v_3\rvert.p, N_{bar}\right)\right)+1$	$\sqrt{\frac{N_{bar}}{2}}.I_{r_{v_3}}.\cos\left(2\pi\left\lvert f_{r_{v_3}}\right\rvert.t + \varphi_{r_{v_3}}\right)$
⋮	⋮
$2.(N_{bar} - \mathrm{mod}(\lvert v_4\rvert.p, N_{bar}))$	$\sqrt{\frac{N_{bar}}{2}}.I_{r_{v_4}}.\sin\left(2\pi\left\lvert f_{r_{v_4}}\right\rvert.t + \varphi_{r_{v_4}}\right)$
$2.\left(N_{bar} - \mathrm{mod}\left(\lvert v_4\rvert.p, N_{bar}\right)\right)+1$	$-\sqrt{\frac{N_{bar}}{2}}.I_{r_{v_4}}.\cos\left(2\pi\left\lvert f_{r_{v_4}}\right\rvert.t + \varphi_{r_{v_4}}\right)$
⋮	⋮
N_{bar}	$\sqrt{N_{bar}}.I_{r_{v_5}}.\sin\left(2\pi\left\lvert f_{r_{v_5}}\right\rvert.t + \varphi_{r_{v_5}}\right)$

This table shows the projection of four different cases of rotor harmonics (v_0, v_1, v_2, v_3, v_4, v_5), where:

- $\mathrm{mod}(\lvert v_0\rvert.p, N_{bar}) = N_{bar}$ (Rotor zero-sequence);
- $\mathrm{mod}(\lvert v_1\rvert.p, N_{bar}) < \mathrm{floor}\left(\frac{N_{bar}}{2}\right)$ ($\leq$ if N_{bar} is odd), and $v_1 > 0$;
- $\mathrm{mod}(\lvert v_2\rvert.p, N_{bar}) < \mathrm{floor}\left(\frac{N_{bar}}{2}\right)$ ($\leq$ if N_{bar} is odd), and $v_2 < 0$;
- $\mathrm{mod}(\lvert v_3\rvert.p, N_{bar}) > \mathrm{floor}\left(\frac{N_{bar}}{2}\right)$, and $v_3 > 0$;

- $\mathrm{mod}(|v_4|.p, N_{bar}) > \mathrm{floor}\left(\frac{N_{bar}}{2}\right)$, and $v_4 < 0$;
- $\mathrm{mod}\left(|v_5|.p, N_{bar}\right) = \frac{N_{bar}}{2}$ (only when N_{bar} is even).

When different harmonics are projected on the same α-β rotor plane, their α and β components are superposed. This property will be at the origin of torque ripples.

As we can see in the Table 2, the harmonic ranges "v" and their frequencies "f_{r_v}", are written in the absolute value. In fact, according to the Equation (19), "v" can be positive or negative. Therefore, the frequencies "f_{r_v}", determined by the Equation (20), can also have positive or negative values.

In the case when "f_{r_v}" is positive, the angle "φ_{r_v}" in Concordia base is equal to "φ'_{r_v}" in the natural base. When "f_{r_v}" is negative, $\varphi_{r_v} = \pi - \varphi'_{r_v}$.

4. Torque Calculation

The torque can be calculated in the natural base by the following expression [27]:

$$T = \underline{I'_s} \cdot \frac{dL_{sr}(\theta)}{d\theta} \cdot \underline{I_r} \tag{22}$$

In the case when the stator is supplied by a sequence "u", the torque can be expressed in α-β-0 base as following:

$$T = \begin{bmatrix} I_{s_\alpha} & I_{s_\beta} \end{bmatrix} \cdot \sum_{v} \frac{\underline{dM_{\alpha\beta_v}}}{d\theta} \cdot \underline{I_{r_{\alpha\beta_v}}} \tag{23}$$

Using the transformed stator and rotor current vectors (Tables 1 and 2), and the transformed mutual inductance matrix (Figure 2), the expressions of the torque developed from the interaction between space and time harmonics can be determined.

As described before, under stator sequence "u", a set of space harmonics "F_u" are excited (defined in Equation (19)). These space harmonics induce in the rotor bars currents time harmonics. Each rotor time harmonic is characterized by its amplitude "I_{r_v}", its frequency "f_{r_v}", its polarity "v.p" and its phase "φ_{r_v}" (defined in Concordia base).

A space harmonic "v.p" (represented by its projection on Concordia base $\underline{M_{v\,\alpha_{s,r}\beta_{s,r}}}(\theta)$) interacts with the rotor time harmonic of the same range to produce a constant torque (as shown in Equations (24), (25) and (26)).

The torque developed from the interaction the space and the rotor time harmonic "v = u" is expressed as follows:

$$T_{u-u} = \frac{N_{ph}.N_{bar}}{4}.p.u.I_{s_u}.I_{r_{v=u}}.(M_{sr})_{v=u}.\sin(\varphi_{s_u} - \varphi_{r_{v=u}} + \varphi_{sr_{v=u}}) \tag{24}$$

For the other induced rotor harmonics, the resultant harmonic constant torque depends on the sign of "v":

- if v < 0:

$$T_{v^- - v^-} = -\frac{N_{ph}.N_{bar}}{4}.p.|v|.I_{s_u}.I_{r_v}.(M_{sr})_v.\sin(\varphi_{s_u} - \varphi_{r_v} - \varphi_{sr_v}) \tag{25}$$

- if v > 0:

$$T_{v^+ - v^+} = -\frac{N_{ph}.N_{bar}}{4}.p.|v|.I_{s_u}.I_{r_v}.(M_{sr})_v.\sin(\varphi_{s_u} + \varphi_{r_v} + \varphi_{sr_v}) \tag{26}$$

On the other hand, when many space and rotor time harmonics are superposed on the same α-β rotor plane, they all interact with each other. For example, if two harmonic ranges "v_1" and "v_2" are in the same rotor Concordia plane, the space harmonic "$v_1.p$" interacts with the time harmonics "$v_2.p$"

and produces pulsating torque component. The expressions of pulsating torque components depend on the harmonic ranges, and are detailed in Appendix B.

5. Application

To validate the mathematical approach of prediction of pulsating torque components proposed in this paper, several multiphase induction machine topologies were inspected. Transient F-E simulations (Maxwell 2D software) are done to determine the steady-state torque temporal variation. The simulations are done by imposing sinusoidal stator currents.

5.1. Three-Phase Induction Machine

The machine parameters are shown in Appendix C.

Table 3 shows the imposed parameters in this simulation.

Table 3. Simulation parameters for the three-phase induction machine (IM).

Parameter	Value
Injected sequence	1
Stator current frequency	50 Hz
Peak current	80 A
Mechanical speed	1470 rpm

The simulated machine has a 36 stator slots, two investigated numbers of rotor bars 48 and 49, and two pole-pairs. The torque variation is extracted from the simulation results (for the machine with 48 bars), and a harmonic analysis is done and shown in Figure 3.

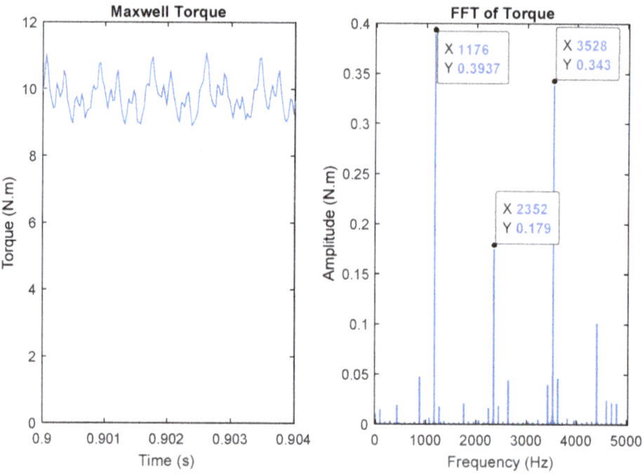

Figure 3. FFT of Finite-Element torque for the three-phase IM with 48 bars.

As we can see in Figure 3, there are three predominant pulsating frequencies (1176 Hz, 2352 Hz and 3528 Hz). To understand the origin of these frequencies, it is important to determine the interactions between space and time harmonics. The distribution of space harmonics (up to v = 50) in the matrix $\underline{L_{sr_{\alpha\beta}}}(\theta)$, according to Figure 2, is shown in Figure 4. To simplify the presentation, each space harmonic "v.p" is represented by the number "v" in Figure (v = 1, … , 50).

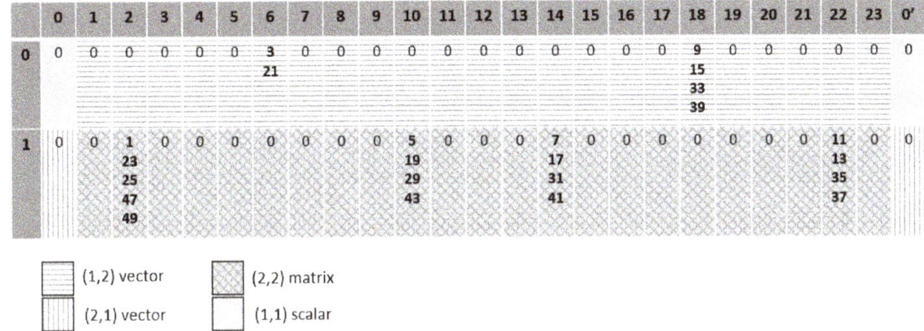

Figure 4. Space harmonics distribution in $L_{sr_{\alpha\beta}}(\theta)$ for the three-phase IM with 48 bars (in line 1, 4 planes among 23 are excited, each one by several harmonics which lead to important interactions).

All the even order harmonics are null, according to the winding distribution, so only odd order harmonics appear in the matrix.

The first line (0) in the table corresponds to the stator zero-sequence. For this zero-sequence, the harmonics 3 and 21 can excite the plane n°6 of the rotor and the harmonics 9, 15, 33 and 39 can excite the plane n°18 of the rotor. With a wye-connected three-phase machine, the zero-sequence is not excited. As consequence the plane n°6 and n°18 are not excited.

The second line (1) corresponds to the only possible stator plane in the case of this three-phase machine, which can be excited be sequences "u = 1" or "u = 2". For this stator plane, **four** different rotor planes are excited (n°2, n°10, n°14, n°22), each one of them contains several superposed harmonics. As example the rotor plane n°2 is excited by the space harmonics n°1, 23, 25, 47 and 49.

In Figure 4, the number of bars is even, so a second rotor zero-sequence appears in **column 0′**.

According to the Equations (20), (27) and Table A1 in Appendix B, the frequencies of pulsating torque components can be predicted (shown in Table 4).

Table 4. Predicted torque pulsating components frequencies for the three-phase IM with 48 bars (in Hz).

Rotor Plane		2						10			
		1p	23p	25p	47p	49p		5p	19p	29p	43p
	1p	0	1176	1176	2352	2352	5p	0	1176	1176	2352
	23p	1176	0	2352	3528	3528	19p	1176	0	2352	1176
	25p	1176	2352	0	3528	1176	29p	1176	2352	0	3528
	47p	2352	3528	3528	0	4704	43p	2352	1176	3528	0
	49p	2352	3528	1176	4704	0					
Rotor Plane		14						22			
		7p	17p	31p	41p			11p	13p	35p	37p
	7p	0	1176	1176	2352		11p	0	1176	1176	2352
	17p	1176	0	2352	1176		13p	1176	0	2352	1176
	31p	1176	2352	0	3528		35p	1176	2352	0	3528
	41p	2352	1176	3528	0		37p	2352	1176	3528	0

In this table, we find the same three predominant frequencies in Figure 3, which are 1176, 2352 and 3528 Hz. Each frequency results from many interactions between space and time harmonics. For example, the frequency 1176 Hz (the highest torque pulsation) results from the interactions: 1p-23p, 1p-25p, 25p-49p, 5p-19p, 5p-29p, 19p-43p, 7p-17p, 7p-31p, 17p-41p, 11p-13p, 11p-35p and 13p-37p.

According to the Table A3 in Appendix C, the 10 most important space harmonics (excited by the sequence "u = 1") regarding their amplitudes in the winding function are: v = 1, 5, 7, 11, 13, 17, 19, 23, 35, 37. An interaction between two of these harmonics is considered important regarding the amplitude of torque pulsation. These significant interactions are colored in gray in the Table 4.

We observe that most of these colored interactions generate a pulsation at 1176 Hz, which is consistent with F-E results (Figure 3), where the highest amplitude of torque ripple is at 1176 Hz.

Basing on Figure 3, this machine with 48 bars, presents many superpositions between space harmonics (in the same rotor planes). All these harmonics interactions can be avoided by taking 49 bars instead of 48. Figure 5 shows how the choice of 49 bars improves the separation between harmonics, all the harmonics below v = 50 are perfectly separated.

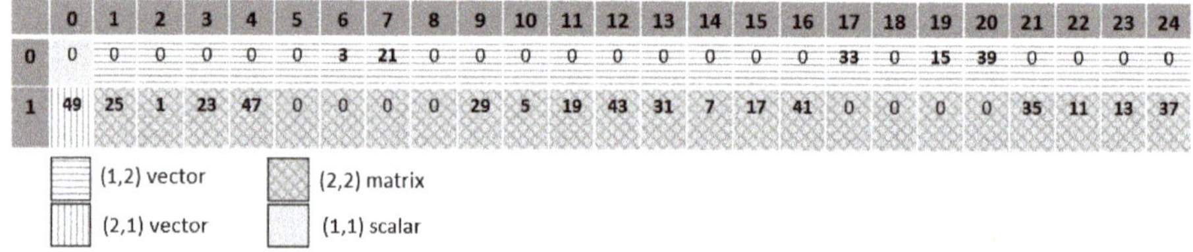

	0	1	2	3	4	5	6	7	8	9	10	11	12	13	14	15	16	17	18	19	20	21	22	23	24
0	0	0	0	0	0	0	3	21	0	0	0	0	0	0	0	0	0	33	0	15	39	0	0	0	0
1	49	25	1	23	47	0	0	0	0	29	5	19	43	31	7	17	41	0	0	0	0	35	11	13	37

(1,2) vector
(2,2) matrix
(2,1) vector
(1,1) scalar

Figure 5. Space harmonics distribution in $\underline{L_{sr_{\alpha\beta}}}(\theta)$ for the three-phase IM with 49 bars (in line 1, 17 planes among 24 are excited, each one by one harmonic, so without interactions).

The Figure 6 shows the comparison between the three-phase IM with 48 and 49 bars regarding the torque (F-E results). It proves that the torque is smoother with 49 bars thanks to the good separation between space harmonics into several α-β rotor planes (as shown in Figure 5).

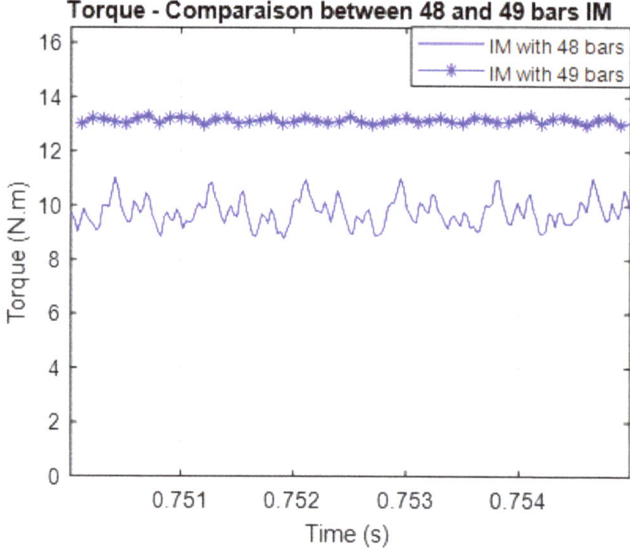

Figure 6. Comparison of the developed torque with 48 and 49 bars, three-phase IM.

This result is consistent with the rules related to the choice of rotor bars number described in the reference [2] (pp. 340–341). According to this reference, for this machine topology (three phases, 36 stator slots and two pole-pairs), the number of bars 48 presents harmful parasitic torque, which is explained thanks to this new approach of prediction of torque pulsations.

After this first example, we conclude that even for three-phase induction machines, the classical approach of rotor cage modeling (simply by one α-β rotor plane) can only be used if the combination of phase number and bar number does not lead to harmful interactions between space and time harmonics (like the combination $N_{ph} = 3$, $N_{bar} = 49$ and $p_1 = 2$). However, when the combination causes harmful interactions (like the one with 48 bars in this example), it is important to consider several α-β rotor planes in modeling (4 planes in this example, for 48 bars) to predict the parasitic torque pulsations.

5.2. Five-Phase Induction Machine with Double-Layer Tooth Concentrated Winding

The machine parameters are shown in Appendix D. The Table 5 shows simulation parameters.

Table 5. Simulation parameters for the five-phase IM with tooth concentrated winding.

Parameter	Value
Injected sequences	1, 3
Stator current frequencies	50, 150 Hz
Peak current	50 A_{peak}
Mechanical speed	744, 740.6 rpm

Two different stator sequences are imposed in separate simulations (u = 1 and u = 3). According to the Equation (19), the sequence "u = 1" induces in the rotor bars the set of harmonics "v = 1, −4, 6, −9, 11, −14, 16, −19 ... ". The sequence "u = 3" induces the set of rotor current harmonics "v = −2, 3, −7, 8, −12, 13, −17, 18 ... ".

This machine was studied in a previous paper [22]. In this paper it has been observed that the number of bars 64 presents important torque pulsations, especially referring to the number 65 characterized by a very low torque harmonics content. Nevertheless, according to the rules in the reference [2] (pp. 340–341), for this machine topology (5 phases, 20 stator slots and 4 pole-pairs) the number of bars 64 is not forbidden. In fact, the reported rules for selecting the rotor bars number are defined for three-phase machines and are not directly applicable to multiphase machines.

To explain the origin of torque pulsations for this machine with 64 bars, the developed torque under both sequences is extracted from the simulation results, and a harmonic analysis is done and shown in Figures 7 and 9.

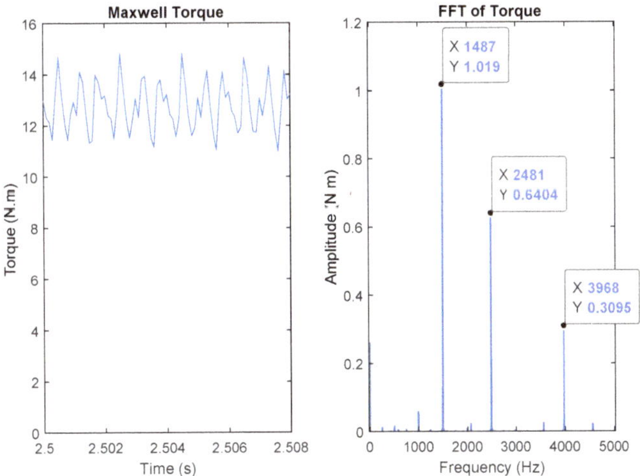

Figure 7. FFT of Finite-Element torque for the five-phase IM with 64 bars. Under sequence "u = 1" (the planes of line (1), Figure 8 are excited).

	0	1	2	3	4	5	6	7	8	9	10	11	12	13	14	15	16	17	18	19	20	21	22	23	24	25	26	27	28	29	30	31	0′
0	0	0	0	0	0	0	0	0	0	0	0	0	0	0	0	0	0	0	0	0	0	0	0	0	0	0	0	0	0	0	0	0	0
1	16	0	0	0	1 31 49	0	0	0	14 34 46	0	0	0	19 29	0	0	0	4 36 44	0	0	0	11 21	0	0	0	6 26	0	0	0	9 39 41	0	0	0	24
2	32 48	0	0	0	17 33 47	0	0	0	2 18	0	0	0	3 13	0	0	0	12 28	0	0	0	27 37 43	0	0	0	22 38 42	0	0	0	7 23	0	0	0	8

(1,2) vector (2,2) matrix
(2,1) vector (1,1) scalar

Figure 8. Space harmonics distribution in $\underline{L_{sr_{\alpha\beta}}}(\theta)$ for the five-phase IM with 64 bars (in each line, 7 rotor planes, among 31, and both zero-sequences are excited, which lead to harmful interactions).

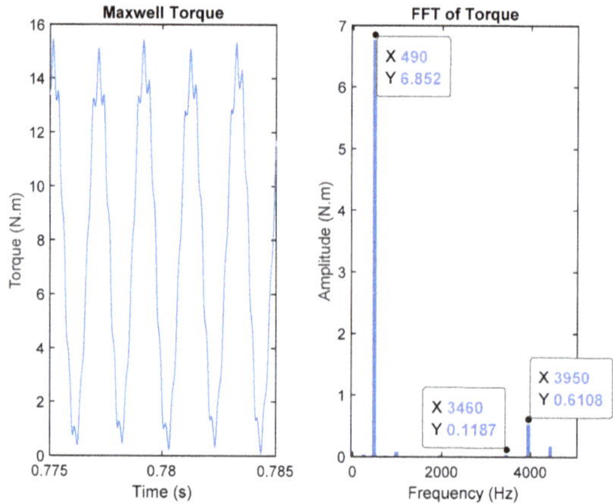

Figure 9. FFT of Finite-Element torque for the five-phase IM with 64 bars. Under sequence "u = 3" (the planes of line (2), Figure 8 are excited).

With this number of bars, the developed torque under both sequences contains important pulsating components. To understand the origin of these components, the distribution of space harmonics (up to v = 50) in the matrix $\underline{L_{sr_{\alpha\beta}}}(\theta)$, according to the Figure 2, is shown in the Figure 8.

This figure shows the interactions between space and time harmonics.

The space harmonics generated by the sequence "u = 1" are distributed in the second line of Figure, and the sequence "u = 3" corresponds to the third line. Under both stator sequences, **seven** rotor planes, the **first rotor zero-sequence** (column 0) and the **second rotor zero-sequence** (column 0′) are excited, each one of them contains several superposed harmonics.

According to the Appendix B, the frequencies of pulsating torque components can be predicted for the sequence "u = 1" in Table 6 and for the sequence "u = 3" in Table 7. According to the Table 6, the frequencies of torque pulsations for the sequence "u = 1" (space harmonics considered up to "v = 50") are: 1487, 2481 and 3968 Hz. These frequencies are the same as the predominant frequencies in F-E results (Figure 7). The interactions in the Table 7 produce three main frequencies: 490, 3460 and 3950 Hz, the same as F-E results in Figure 9.

Table 6. Predicted torque pulsating components frequencies for the five-phase IM with 64 bars (in Hz). Sequence "u = 1".

Rotor Plane	0 (First Rotor Zero-Sequence)				4				8			
		16p				1p	31p	49p		14p	34p	46p
	16p	0 & 1487			1p	0	1487	2481	14p	0	2481	1487
					31p	1487	0	3968	34p	2481	0	3968
					49p	2481	3968	0	46p	1487	3968	0
Rotor Plane	**12**				**16**				**20**			
		19p	29p			4p	36p	44p		11p	21p	
	19p	0	2481		4p	0	1487	2481	11p	0	1487	
	29p	2481	0		36p	1487	0	3968	21p	1487	0	
					44p	2481	3968	0				
Rotor Plane	**24**				**28**				**0′ (Second Rotor Zero-Sequence)**			
		6p	26p			9p	39p	41p		24p		
6p 26p	6p	0	1487		9p	0	2481	1487	24p	0 & 2481		
	26p	1487	0		39p	2481	0	3968				
					41p	1487	3968	0				

Table 7. Predicted torque pulsating components frequencies for the five-phase IM with 64 bars (in Hz). Sequence "u = 3".

Rotor Plane	0 (First Rotor Zero-Sequence)				4				8			
		32p	48p			17p	33p	47p		2p	18p	
	32p	0 & 3460	3950 & 490		17p	0	490	3460	2p	0	490	
					33p	490	0	3950	18p	490	0	
	48p	3950 & 490	0 & 4440		47p	3460	3950	0				
Rotor Plane	**12**				**16**				**20**			
		3p	13p			12p	28p			27p	37p	43p
	3p	0	490		12p	0	490		27p	0	3460	490
	13p	490	0		28p	490	0		37p	3460	0	3950
									43p	490	3950	0
Rotor Plane	**24**				**28**				**0′ (Second Rotor Zero-Sequence)**			
		22p	38p	42p		7p	23p			8p		
6p 26p	22p	0	490	3460	7p	0	490		8p	0 & 490		
	38p	490	0	3950	23p	490	0					
	42p	3460	3950	0								

In Appendix D, the 20 most important space harmonics regarding amplitude are indicated in the Table A4. To have an idea about the importance of a torque pulsation, every interaction between two harmonics among the 20 most important ones (Table A4) is colored in gray.

Under the sequence "u = 1", the only interaction between a space and a time harmonic whose amplitudes are among the 20 most important, is in the cell corresponding to the rotor zero-sequence containing the harmonic "16.p", whose interaction generates two components (as explained in Appendix B): a constant torque at 0 Hz, and a pulsating component of a frequency of 1487 Hz (which has the highest pulsation amplitude in Figure 7).

Under the sequence "u = 3", several interactions are colored in gray (interaction between two harmonics among the 20 most important in Table A4). All these marked interactions generate a frequency of 490 Hz, which has the highest pulsation amplitude in Figure 9.

The Figure 10 shows the space harmonics distribution in $\underline{L_{sr_{\alpha\beta}}}(\theta)$ for the machine with 65 bars. We observe that this number of bars allows to separate the harmonics better than the case of 64 bars. In fact, the harmonics with the most important amplitudes (Table A4) do not interact between each other. Furthermore, with this number of bars, the rotor zero-sequence is not excited under both sequences.

Figure 10. Space harmonics distribution in $L_{sr_{\alpha\beta}}(\theta)$ for the five-phase IM with 65 bars (in line 1, 13 planes among 32 are excited. In line 2, 12 planes among 32 are excited. No zero-sequence is excited. This leads to less interactions than 64 bars).

The Figure 11 shows the comparison of the developed torque between the machines with 64 and 65 bars. With 65 bars the torque is smoother under both sequences thanks to the separation of harmonics (shown in Figure 10). FFT analysis of torque for the new number of bars "65", is done and shown in Figure 12 for both sequences.

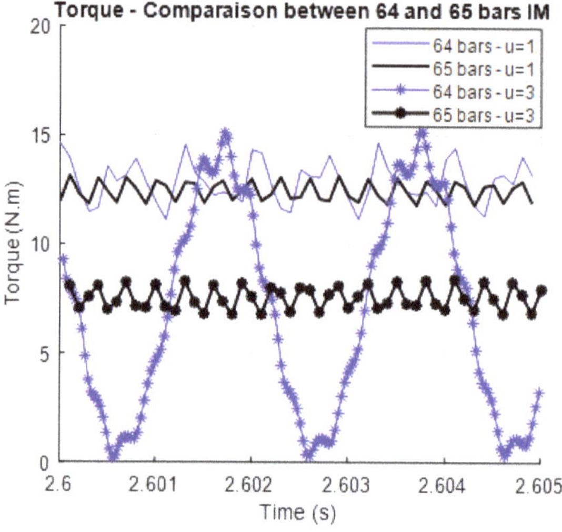

Figure 11. Comparison of the developed torque with 64 and 65 bars, five-phase IM.

Under the sequence "u = 1" (corresponds to the black curve in Figure 11), the predominant pulsating component is at the frequency of 3224 Hz. After the application of equations in Appendix B, it appears that this frequency of 3224 Hz is generated by all the interactions shown in Figure 10 (Line 1), so the interactions: 16p-49p, 31p-34p, 19p-49p, 29p-36p, 21p-44p, 26p-39p and 24p-41p (Considering only 50 space harmonics, from 1p to 50p). Comparing to the case of 64 bars (Table 6), the interactions in the machine with 65 bars are mostly between harmonics with relatively low amplitudes (Table A4) which explains the reduction of torque ripple thanks to this combination (N_{ph} = 5, N_{bar} = 65 and p_1 = 4).

Under the sequence "u = 3" (corresponds to the blue curve in Figure 11), the predominant pulsating component is at the frequency of 3211 Hz. Using the same approach (Appendix B), we determine that this frequency is generated by all the interactions in the last line of Figure 10, so: 32p-33p, 17p-48p, 18p-47p, 28p-37p, 27p-38p, 22p-43p, 23p-42p. These harmonics have also relatively low amplitudes (Table A4).

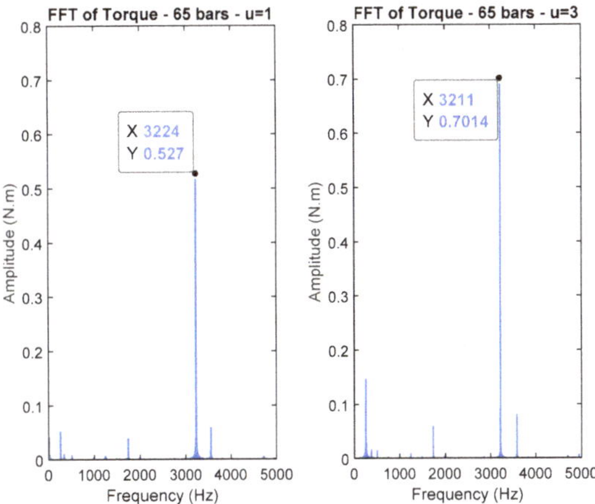

Figure 12. FFT analysis of torque curves for the IM with 65 bars, under sequences "u = 1" and "u = 3".

For five-phase induction machine, two α-β stator planes must be considered in modeling. The number of α-β rotor planes to be considered depends on the number of rotor bars. If there are interactions between harmonics with important amplitudes, the rotor planes containing these interactions must be considered to predict torque pulsations.

6. Conclusion

In order to be able to design multiphase induction drives which can be supplied with different harmonics of currents either for speed extension or for torque increase, the paper provides a mathematical approach to point out the excited α-β rotor planes which must be considered at two levels for the modelling. At first, during the electromagnetic design, the aim is to avoid wrong choices of bar numbers leading to torque pulsations. Secondly during the design of the vector control of the machine, it is necessary to use the simplest modeling of the rotor, with the minimum dimension, in order to define the vector control. The selection of the minimum number of rotor planes which contribute to the torque leads to a reduced-order modelling of the rotor cage. It has been shown that the rotor dimension, which is "N_{bar}" in the natural base model, can be much lower in the Concordia frame.

In this study, a general arithmetic tool (based on equations in Appendix B) was developed to predict, and consequently to avoid, torque ripples due to wrong combinations of phase number and rotor bar number. The result of the analysis is presented graphically in simple table where it is appearing clearly when interactions between space and time harmonics occur. This tool requires as inputs: the numbers of stator phases, rotor bars, pole-pairs (p_1), the stator frequency, the stator sequence "u" and the mechanical speed. It should be mentioned that this tool does not predict the pulsations due to slotting effect.

This arithmetic tool is validated on two different multiphase induction machines topologies (three and five phases) with different numbers of rotor bars. In the provided applications, the induced pulsating torque frequencies are predicted thanks to the arithmetic tool and validated by the FFT analysis of the steady-state torque determined by F-E simulations (Maxwell 2D software, with imposed sinusoidal stator currents).

Through the first application, it has been concluded that even for three-phase induction machines the simplified modeling of the rotor cage by only one equivalent two-phase circuit is not always relevant. Some numbers of bars lead to important interactions between space and time harmonics, and consequently important torque pulsations. To predict these pulsations, it is important to consider all the excited rotor planes (and not only the one containing the fundamental space harmonic). With a difference of one bar quite significant differences are observed and explained.

For five-phase induction machines, it is possible to use two independent sequences to supply the stator. The proposed mathematical approach permitted to model the rotor under both sequences, to predict precisely torque pulsations frequencies for two different numbers of bars. These results allow to explain the important difference in torque ripple, between the five-phase machines with 64 and 65 rotor bars, observed in a previous paper [22].

To design a multiphase induction machine the combination of phase number and rotor bar number must be carefully chosen, considering the different possible polarities related to the used supply sequences. In the literature [1,2], the reported rules regarding this choice are defined for three-phase machines and are not directly applicable to multiphase induction machines. Thanks to the arithmetic tool and the corresponding graphical representation proposed in this paper, the wrong combinations can be predicted and avoided thanks to very fast calculations.

Further work is ongoing regarding the identification of mutual inductance coefficients "M_{sr_v}" to predict precisely the amplitudes of torque pulsations and the mean torque under different supplying sequences for multiphase induction machines and under different levels of magnetic saturation.

Author Contributions: A.M. proposed the used mathematical approach, worked on the numerical applications, analyzed the results, proposed the structure of the paper and wrote it. E.S. improved the structure of the paper and its content and provided advanced theoretical elements. F.S. improved the structure of the paper and its content and helped in the analysis of results. H.Z. improved the content of the paper. All authors validated the final paper. All authors have read and agreed to the published version of the manuscript.

Conflicts of Interest: The authors declare no conflict of interest.

Nomenclature

N_{ph}	Number of stator phases
N_{bar}	Number of rotor bars
spp	Number of slots per pole and per phase
p	Number of pole-pairs
$\underline{V}_s$	Stator voltage vector, dimension N_{ph}
$\underline{I}_s$	Stator current vector, dimension N_{ph}
$\underline{I}_r$	Rotor current vector, dimension N_{bar}
$\underline{\Phi}_s$	Stator flux vector, dimension N_{ph}
$\underline{\Phi}_r$	Rotor flux vector, dimension N_{bar}
$\underline{R_s}$	Stator resistance matrix, dimension $N_{ph} \times N_{ph}$ (Diagonal matrix)
$\underline{R_r}$	Rotor resistance matrix, dimension $N_{bar} \times N_{bar}$
$\underline{L_{ss}}$	Stator inductance matrix, dimension $N_{ph} \times N_{ph}$
$\underline{L_{rr}}$	Rotor inductance matrix, dimension $N_{bar} \times N_{bar}$
$\underline{L_{sr}} = \underline{L_{rs}}'$	Mutual inductance matrix (between stator and rotor), dimension $N_{ph} \times N_{bar}$
θ	Mechanical angle
R_s	Stator phase resistance
R_b	Rotor bar resistance
R_e	Resistance of end-connection ring portion (connecting 2 adjacent bars)
L_{sij}	Mutual inductance between two stator phases "i" and "j"
L_{ls}	Leakage stator inductance
L_{rij}	Mutual inductance between two rotor phases "i" and "j"
L_b	Rotor bar leakage inductance
L_e	Leakage inductance of end-connection ring portion
I_{s_u}	Stator current amplitude (for the stator sequence "u")
f_{s_u}	Stator current frequency (for the stator sequence "u")
φ_{s_u}	Stator current phase (for the stator sequence "u")
I_{r_v}	Amplitude of induced rotor current harmonic "v"
f_{r_v}	Frequency of the induced rotor current harmonic "v"
φ_{r_v}	Phase of the induced rotor current harmonic "v"
Ω_{mec}	Mechanical speed

Appendix A

As seen in Figure 2, each space harmonic "k" is projected on a cell represented by $\underline{M_{k\,\alpha_{s,r}\beta_{s,r}}}(\theta)$ (Equation (18)), which can be a two-by-two matrix, a two-element vector or a scalar, depending on the rules described below:

- Rule 7: If $\mathrm{mod}(k, N_{ph}) \le \mathrm{floor}\left(\frac{N_{ph}}{2}\right)$ and $\mathrm{mod}(k.p, N_{bar}) \le \mathrm{floor}\left(\frac{N_{bar}}{2}\right)$:

$$\underline{M_{k\,\alpha_{s,r}\beta_{s,r}}}(\theta) = \frac{\sqrt{N_{ph}.N_{bar}}}{2}.M_{sr_k}\begin{pmatrix} \cos(-kp\theta + \varphi_{sr_k}) & \sin(-kp\theta + \varphi_{sr_k}) \\ -\sin(-kp\theta + \varphi_{sr_k}) & \cos(-kp\theta + \varphi_{sr_k}) \end{pmatrix}$$

- Rule 8: If $\mathrm{mod}(k, N_{ph}) \le \mathrm{floor}\left(\frac{N_{ph}}{2}\right)$ and $\mathrm{mod}(k.p, N_{bar}) > \mathrm{floor}\left(\frac{N_{bar}}{2}\right)$:

$$\underline{M_{k\,\alpha_{s,r}\beta_{s,r}}}(\theta) = \frac{\sqrt{N_{ph}.N_{bar}}}{2}.M_{sr_k}\begin{pmatrix} \cos(-kp\theta + \varphi_{sr_k}) & -\sin(-kp\theta + \varphi_{sr_k}) \\ -\sin(-kp\theta + \varphi_{sr_k}) & -\cos(-kp\theta + \varphi_{sr_k}) \end{pmatrix}$$

- Rule 9: If $\mathrm{mod}(k, N_{ph}) > \mathrm{floor}\left(\frac{N_{ph}}{2}\right)$ and $\mathrm{mod}(k.p, N_{bar}) \le \mathrm{floor}\left(\frac{N_{bar}}{2}\right)$:

$$\underline{M_{k\,\alpha_{s,r}\beta_{s,r}}}(\theta) = \frac{\sqrt{N_{ph}.N_{bar}}}{2}.M_{sr_k}\begin{pmatrix} \cos(-kp\theta + \varphi_{sr_k}) & \sin(-kp\theta + \varphi_{sr_k}) \\ \sin(-kp\theta + \varphi_{sr_k}) & -\cos(-kp\theta + \varphi_{sr_k}) \end{pmatrix}$$

- Rule 10: If $\mathrm{mod}(k, N_{ph}) > \mathrm{floor}\left(\frac{N_{ph}}{2}\right)$ and $\mathrm{mod}(k.p, N_{bar}) > \mathrm{floor}\left(\frac{N_{bar}}{2}\right)$:

$$\underline{M_{k\,\alpha_{s,r}\beta_{s,r}}}(\theta) = \frac{\sqrt{N_{ph}.N_{bar}}}{2}.M_{sr_k}\begin{pmatrix} \cos(-kp\theta + \varphi_{sr_k}) & -\sin(-kp\theta + \varphi_{sr_k}) \\ \sin(-kp\theta + \varphi_{sr_k}) & \cos(-kp\theta + \varphi_{sr_k}) \end{pmatrix}$$

- Rule 11: If $\mathrm{mod}(k, N_{ph}) = 0$ and $\mathrm{mod}(k.p, N_{bar}) \le \mathrm{floor}\left(\frac{N_{bar}}{2}\right)$:

$$\underline{M_{k\,\alpha_{s,r}\beta_{s,r}}}(\theta) = \sqrt{\frac{N_{ph}.N_{bar}}{2}}.M_{sr_k}\begin{pmatrix} \cos(-kp\theta + \varphi_{sr_k}) & \sin(-kp\theta + \varphi_{sr_k}) \end{pmatrix}$$

- Rule 12: If $\mathrm{mod}(k, N_{ph}) = 0$ and $\mathrm{mod}(k.p, N_{bar}) > \mathrm{floor}\left(\frac{N_{bar}}{2}\right)$:

$$\underline{M_{k\,\alpha_{s,r}\beta_{s,r}}}(\theta) = \sqrt{\frac{N_{ph}.N_{bar}}{2}}.M_{sr_k}\begin{pmatrix} \cos(-kp\theta + \varphi_{sr_k}) & -\sin(-kp\theta + \varphi_{sr_k}) \end{pmatrix}$$

- Rule 13 (Only when N_{ph} is even): If $\mathrm{mod}(k, N_{ph}) = \frac{N_{ph}}{2}$ and $\mathrm{mod}(k.p, N_{bar}) \le \mathrm{floor}\left(\frac{N_{bar}}{2}\right)$:

$$\underline{M_{k\,\alpha_{s,r}\beta_{s,r}}}(\theta) = \sqrt{\frac{N_{ph}.N_{bar}}{2}}.M_{sr_k}\begin{pmatrix} \cos(-kp\theta + \varphi_{sr_k}) & \sin(-kp\theta + \varphi_{sr_k}) \end{pmatrix}$$

- Rule 14 (Only when N_{ph} is even): If $\mathrm{mod}(k, N_{ph}) = \frac{N_{ph}}{2}$ and $\mathrm{mod}(k.p, N_{bar}) > \mathrm{floor}\left(\frac{N_{bar}}{2}\right)$:

$$\underline{M_{k\,\alpha_{s,r}\beta_{s,r}}}(\theta) = \sqrt{\frac{N_{ph}.N_{bar}}{2}}.M_{sr_v}\begin{pmatrix} \cos(-kp\theta + \varphi_{sr_k}) & -\sin(-kp\theta + \varphi_{sr_k}) \end{pmatrix}$$

- Rule 15: If $\mathrm{mod}(k, N_{ph}) \le \mathrm{floor}\left(\frac{N_{ph}}{2}\right)$ and $\mathrm{mod}(k.p, N_{bar}) = 0$:

$$\underline{M_{k\,\alpha_{s,r}\beta_{s,r}}}(\theta) = \sqrt{\frac{N_{ph}.N_{bar}}{2}}.M_{sr_k}\begin{pmatrix} \cos(-kp\theta + \varphi_{sr_k}) \\ -\sin(-kp\theta + \varphi_{sr_k}) \end{pmatrix}$$

- Rule 16: If $\mathrm{mod}\big(k, N_{ph}\big) > \mathrm{floor}\Big(\frac{N_{ph}}{2}\Big)$ and $\mathrm{mod}(k.p, N_{bar}) = 0$:

$$\underline{M_{k\,\alpha_{s,r}\beta_{s,r}}}(\theta) = \sqrt{\frac{N_{ph}.N_{bar}}{2}}.M_{sr_k}\begin{pmatrix} \cos(-kp\theta + \varphi_{sr_k}) \\ \sin(-kp\theta + \varphi_{sr_k}) \end{pmatrix}$$

- Rule 17 (Only when N_{bar} is even): If $\mathrm{mod}\big(k, N_{ph}\big) \leq \mathrm{floor}\Big(\frac{N_{ph}}{2}\Big)$ and $\mathrm{mod}(k.p, N_{bar}) = \frac{N_{bar}}{2}$:

$$\underline{M_{k\,\alpha_{s,r}\beta_{s,r}}}(\theta) = \sqrt{\frac{N_{ph}.N_{bar}}{2}}.M_{sr_k}\begin{pmatrix} \cos(-kp\theta + \varphi_{sr_k}) \\ -\sin(-kp\theta + \varphi_{sr_k}) \end{pmatrix}$$

- Rule 18 (Only when N_{bar} is even): If $\mathrm{mod}\big(k, N_{ph}\big) > \mathrm{floor}\Big(\frac{N_{ph}}{2}\Big)$ and $\mathrm{mod}(k.p, N_{bar}) = \frac{N_{bar}}{2}$:

$$\underline{M_{k\,\alpha_{s,r}\beta_{s,r}}}(\theta) = \sqrt{\frac{N_{ph}.N_{bar}}{2}}.M_{sr_k}\begin{pmatrix} \cos(-kp\theta + \varphi_{sr_k}) \\ \sin(-kp\theta + \varphi_{sr_k}) \end{pmatrix}$$

- Rule 19: If $\mathrm{mod}\big(k, N_{ph}\big) = 0$ and $\mathrm{mod}(k.p, N_{bar}) = 0$:

$$\underline{M_{k\,\alpha_{s,r}\beta_{s,r}}}(\theta) = \sqrt{N_{ph}.N_{bar}}.M_{sr_k}(\cos(-kp\theta + \varphi_{sr_k}))$$

- Rule 20 (Only when N_{bar} is even): If $\mathrm{mod}\big(k, N_{ph}\big) = 0$ and $\mathrm{mod}(k.p, N_{bar}) = \frac{N_{bar}}{2}$:

$$\underline{M_{k\,\alpha_{s,r}\beta_{s,r}}}(\theta) = \sqrt{N_{ph}.N_{bar}}.M_{sr_k}(\cos(-kp\theta + \varphi_{sr_k}))$$

- Rule 21 (Only when N_{ph} is even): If $\mathrm{mod}\big(k, N_{ph}\big) = \frac{N_{ph}}{2}$ and $\mathrm{mod}(k.p, N_{bar}) = 0$:

$$\underline{M_{k\,\alpha_{s,r}\beta_{s,r}}}(\theta) = \sqrt{N_{ph}.N_{bar}}.M_{sr_k}(\cos(-kp\theta + \varphi_{sr_k}))$$

- Rule 22 (Only when N_{ph} and N_{bar} are even): If $\mathrm{mod}\big(k, N_{ph}\big) = \frac{N_{ph}}{2}$ and $\mathrm{mod}(k.p, N_{bar}) = \frac{N_{bar}}{2}$:

$$\underline{M_{k\,\alpha_{s,r}\beta_{s,r}}}(\theta) = \sqrt{N_{ph}.N_{bar}}.M_{sr_k}(\cos(-kp\theta + \varphi_{sr_k}))$$

In the case where two, or more, harmonics are projected on the same stator and rotor planes, the matrices $\underline{M_{\alpha\beta_k}}$ related to these harmonics are superposed.

Appendix B

Under a stator sequence "u", the pulsating torque component due to the interaction between a space harmonic "v_1.p" and a rotor time harmonic "v_2.p" can be expressed, in the general form, as follows:

$$\begin{aligned} T_{v_1-v_2} = {} & \frac{N_{ph}.N_{bar}}{4}.p.|v_1|.I_{s_u}.I_{r_{v_2}}.(M_{sr})_{v_1}. \\ & \sin\Big(\delta_s(\omega_{s_u}.t + \varphi_{s_u}) + \delta_r\big(\big|\omega_{r_{v_2}}\big|.t + \varphi_{r_{v_2}}\big) + \delta_{sr}\big(-|v_1|.p.\Omega_{mec}.t + \varphi_{sr_{v_1}}\big)\Big) \end{aligned} \tag{27}$$

where: δ_s, δ_r and $\delta_{sr} = \pm 1$. These coefficients depend on the rules (Appendix A) related to the harmonics "v_1.p" and "v_2.p". The different cases of the coefficients δ_s, δ_r and δ_{sr} are shown in the Table A1 (Only rules 7 to 10 are considered).

Table A1. Pulsating torque components due to interactions of harmonics of rules 7, 8, 9 and 10.

	mod(u,N$_{ph}$)<floor($\frac{N_{ph}}{2}$)				**mod(u,N$_{ph}$)>floor($\frac{N_{ph}}{2}$)**			
Rule of "v$_2$.p" **Rule of "v$_1$.p"**	**Rule 7**	**Rule 8**	**Rule 9**	**Rule 10**	**Rule 7**	**Rule 8**	**Rule 9**	**Rule 10**
Rule 7	$\delta_s = -1$ $\delta_r = -1$ $\delta_{sr} = -1$	$\delta_s = 1$ $\delta_r = -1$ $\delta_{sr} = 1$	$\delta_s = -1$ $\delta_r = -1$ $\delta_{sr} = -1$	$\delta_s = 1$ $\delta_r = -1$ $\delta_{sr} = 1$	$\delta_s = -1$ $\delta_r = 1$ $\delta_{sr} = 1$	$\delta_s = 1$ $\delta_r = 1$ $\delta_{sr} = -1$	$\delta_s = -1$ $\delta_r = 1$ $\delta_{sr} = 1$	$\delta_s = 1$ $\delta_r = 1$ $\delta_{sr} = -1$
Rule 8	$\delta_s = 1$ $\delta_r = -1$ $\delta_{sr} = 1$	$\delta_s = -1$ $\delta_r = -1$ $\delta_{sr} = -1$	$\delta_s = 1$ $\delta_r = -1$ $\delta_{sr} = 1$	$\delta_s = -1$ $\delta_r = -1$ $\delta_{sr} = -1$	$\delta_s = 1$ $\delta_r = 1$ $\delta_{sr} = -1$	$\delta_s = -1$ $\delta_r = 1$ $\delta_{sr} = 1$	$\delta_s = 1$ $\delta_r = 1$ $\delta_{sr} = -1$	$\delta_s = -1$ $\delta_r = 1$ $\delta_{sr} = 1$
Rule 9	$\delta_s = -1$ $\delta_r = 1$ $\delta_{sr} = 1$	$\delta_s = 1$ $\delta_r = 1$ $\delta_{sr} = -1$	$\delta_s = -1$ $\delta_r = 1$ $\delta_{sr} = 1$	$\delta_s = 1$ $\delta_r = 1$ $\delta_{sr} = -1$	$\delta_s = -1$ $\delta_r = -1$ $\delta_{sr} = -1$	$\delta_s = 1$ $\delta_r = -1$ $\delta_{sr} = 1$	$\delta_s = -1$ $\delta_r = -1$ $\delta_{sr} = -1$	$\delta_s = 1$ $\delta_r = -1$ $\delta_{sr} = 1$
Rule 10	$\delta_s = 1$ $\delta_r = 1$ $\delta_{sr} = -1$	$\delta_s = -1$ $\delta_r = 1$ $\delta_{sr} = 1$	$\delta_s = 1$ $\delta_r = 1$ $\delta_{sr} = -1$	$\delta_s = -1$ $\delta_r = 1$ $\delta_{sr} = 1$	$\delta_s = 1$ $\delta_r = -1$ $\delta_{sr} = 1$	$\delta_s = -1$ $\delta_r = -1$ $\delta_{sr} = -1$	$\delta_s = 1$ $\delta_r = -1$ $\delta_{sr} = 1$	$\delta_s = -1$ $\delta_r = -1$ $\delta_{sr} = -1$

When a rotor zero-sequence is excited (rules 15, 16, 17 and 18 in Appendix A), the general form of the developed torque component due to the interaction of "v$_3$.p" and "v$_4$.p" (both exciting rotor zero-sequences) can be written as follows:

$$\begin{aligned} T_{v_3-v_4} = {} & \frac{N_{ph}.N_{bar}}{4}.p.|v_3|.I_{s_u}.I_{r_{v_4}}.(M_{sr})_{v_3}. \\ & \sin\left(\delta_{s1}(\omega_{s_u}.t + \varphi_{s_u}) + \delta_{r1}\left(\left|\omega_{r_{v_4}}\right|.t + \varphi_{r_{v_4}}\right) + \delta_{sr1}\left(-|v_3|.p.\Omega_{mec}.t + \varphi_{sr_{v_3}}\right)\right) \\ & + \sin\left(\delta_{s2}(\omega_{s_u}.t + \varphi_{s_u}) + \delta_{r2}\left(\left|\omega_{r_{v_4}}\right|.t + \varphi_{r_{v_4}}\right) + \delta_{sr2}\left(-|v_3|.p.\Omega_{mec}.t + \varphi_{sr_{v_3}}\right)\right) \end{aligned} \tag{28}$$

The different cases of the coefficients δ_{s1}, δ_{s2}, δ_{r1}, δ_{r2} and δ_{sr1}, δ_{sr2} are shown in the Table A2.

Table A2. Pulsating torque components due to interactions of harmonics of rules 15, 16, 17 and 18.

	mod(u,N$_{ph}$)<floor($\frac{N_{ph}}{2}$)				**mod(u,N$_{ph}$)>floor($\frac{N_{ph}}{2}$)**			
Rule of "v$_4$.p" **Rule of "v$_3$.p"**	**Rule 15**	**Rule 16**	**Rule 17**	**Rule 18**	**Rule 15**	**Rule 16**	**Rule 17**	**Rule 18**
Rule 15	$\delta_{s1} = 1$ $\delta_{r1} = -1$ $\delta_{sr1} = 1$ $\delta_{s2} = -1$ $\delta_{r2} = -1$ $\delta_{sr2} = -1$	$\delta_{s1} = 1$ $\delta_{r1} = -1$ $\delta_{sr1} = 1$ $\delta_{s2} = -1$ $\delta_{r2} = -1$ $\delta_{sr2} = -1$			$\delta_{s1} = 1$ $\delta_{r1} = 1$ $\delta_{sr1} = -1$ $\delta_{s2} = -1$ $\delta_{r2} = 1$ $\delta_{sr2} = 1$	$\delta_{s1} = 1$ $\delta_{r1} = 1$ $\delta_{sr1} = -1$ $\delta_{s2} = -1$ $\delta_{r2} = 1$ $\delta_{sr2} = 1$		
Rule 16	$\delta_{s1} = 1$ $\delta_{r1} = 1$ $\delta_{sr1} = -1$ $\delta_{s2} = -1$ $\delta_{r2} = 1$ $\delta_{sr2} = 1$	$\delta_{s1} = 1$ $\delta_{r1} = 1$ $\delta_{sr1} = -1$ $\delta_{s2} = -1$ $\delta_{r2} = 1$ $\delta_{sr2} = 1$			$\delta_{s1} = 1$ $\delta_{r1} = -1$ $\delta_{sr1} = 1$ $\delta_{s2} = -1$ $\delta_{r2} = -1$ $\delta_{sr2} = -1$	$\delta_{s1} = 1$ $\delta_{r1} = -1$ $\delta_{sr1} = 1$ $\delta_{s2} = -1$ $\delta_{r2} = -1$ $\delta_{sr2} = -1$		
Rule 17			$\delta_{s1} = 1$ $\delta_{r1} = -1$ $\delta_{sr1} = 1$ $\delta_{s2} = -1$ $\delta_{r2} = -1$ $\delta_{sr2} = -1$	$\delta_{s1} = 1$ $\delta_{r1} = -1$ $\delta_{sr1} = 1$ $\delta_{s2} = -1$ $\delta_{r2} = -1$ $\delta_{sr2} = -1$			$\delta_{s1} = 1$ $\delta_{r1} = 1$ $\delta_{sr1} = -1$ $\delta_{s2} = -1$ $\delta_{r2} = 1$ $\delta_{sr2} = 1$	$\delta_{s1} = 1$ $\delta_{r1} = 1$ $\delta_{sr1} = -1$ $\delta_{s2} = -1$ $\delta_{r2} = 1$ $\delta_{sr2} = 1$
Rule 18			$\delta_{s1} = 1$ $\delta_{r1} = 1$ $\delta_{sr1} = -1$ $\delta_{s2} = -1$ $\delta_{r2} = 1$ $\delta_{sr2} = 1$	$\delta_{s1} = 1$ $\delta_{r1} = 1$ $\delta_{sr1} = -1$ $\delta_{s2} = -1$ $\delta_{r2} = 1$ $\delta_{sr2} = 1$			$\delta_{s1} = 1$ $\delta_{r1} = -1$ $\delta_{sr1} = 1$ $\delta_{s2} = -1$ $\delta_{r2} = -1$ $\delta_{sr2} = -1$	$\delta_{s1} = 1$ $\delta_{r1} = -1$ $\delta_{sr1} = 1$ $\delta_{s2} = -1$ $\delta_{r2} = -1$ $\delta_{sr2} = -1$

For the rotor zero-sequence harmonics, the interaction between a space harmonic "v.p" and the time harmonic (of rotor current) of the same range "v.p" generates a constant torque and a pulsating component in the same time.

Appendix C

The three-phase induction machine has an overlapping distributed winding, its geometry is shown in Figure A1.

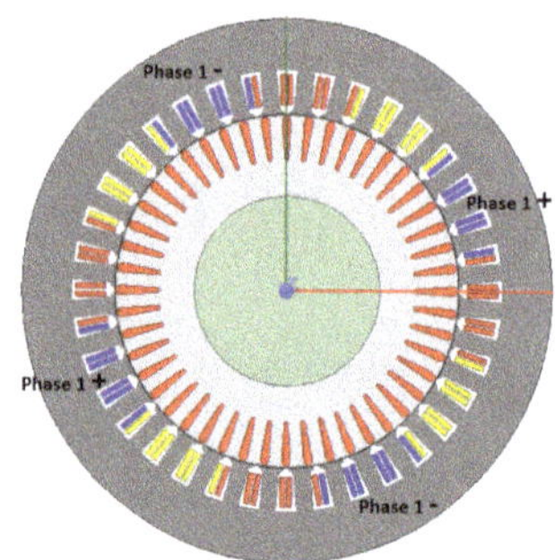

Figure A1. Geometry of the three-phase induction machine.

The parameters of this machine are:

- Number of phases, $N_{ph} = 3$;
- Number of stator slots, $N_s = 36$;
- Number of rotor bars (2 compared cases), $N_{bar} = 48, 49$;
- Number of pole-pairs under first current harmonic injection, $p_1 = 2$;
- Rated stator current, $I_s = 325\ A_{peak}$;
- Rated torque, T = 72 N.m.

The Table A3 shows the harmonic winding factors for the 50 first space harmonics. The winding factor "k_v" divided by its range "v" reflects the importance of the harmonic in torque production. In fact, the coefficient M_{sr_v} is proportional to the factor "k_v/v" (in the column 3). The lines containing the 15 most important space harmonics are colored in gray in the table.

Table A3. Harmonic winding factors of the three-phase winding.

v	k_v	k_v/v	v	k_v	k_v/v	v	k_v	k_v/v	v	k_v	k_v/v	v	k_v	k_v/v
1	0.945	0.945	11	0.061	0.006	21	0.577	0.027	31	0.140	0.005	41	0.140	0.003
2	0	0	12	0	0	22	0	0	32	0	0	42	0	0
3	0.577	0.192	13	0.140	0.011	23	0.140	0.006	33	0.577	0.017	43	0.061	0.001
4	0	0	14	0	0	24	0	0	34	0	0	44	0	0
5	0.140	0.028	15	0.577	0.038	25	0.061	0.002	35	0.945	0.027	45	0	0
6	0	0	16	0	0	26	0	0	36	0	0	46	0	0
7	0.061	0.009	17	0.945	0.056	27	0	0	37	0.945	0.026	47	0.061	0.001
8	0	0	18	0	0	28	0	0	38	0	0	48	0	0
9	0	0	19	0.945	0.050	29	0.061	0.002	39	0.577	0.015	49	0.140	0.003
10	0	0	20	0	0	30	0	0	40	0	0	50	0	0

Appendix D

The five-phase induction machine has a tooth-concentrated winding, its geometry is shown in Figure A2. The parameters of this machine are:

- Number of phases, $N_{ph} = 5$;
- Number of stator slots, $N_s = 20$;
- Number of rotor bars (2 compared cases), $N_{bar} = 64, 65$;
- Number of pole-pairs under first current harmonic injection, $p_1 = 4$;
- Rated stator current, $I_s = 400\ A_{peak}$;
- Rated torque, T = 250 N.m.

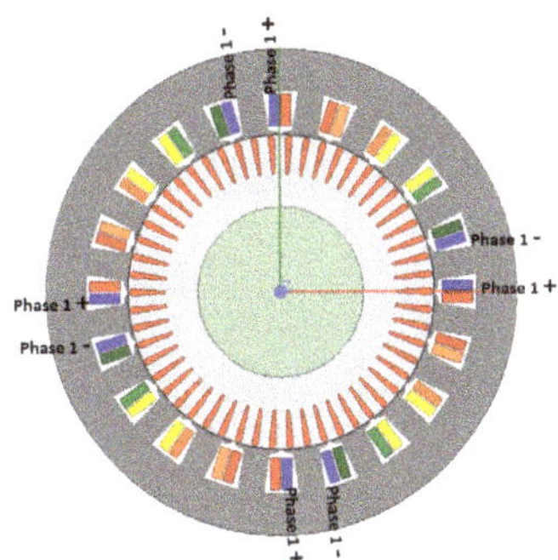

Figure A2. Geometry of the five-phase induction machine with double-layer tooth concentrated winding.

The Table A4 shows the harmonic winding factors for the 50 first space harmonics. The winding factor "k_v" divided by its range "v" reflect the importance of the harmonic in torque production. In fact, the coefficient M_{sr_v} is proportional to the factor "k_v/v" (in the column 3). The lines containing the 20 most important space harmonics are colored in gray in the table.

Table A4. Harmonic winding factors of the five-phase winding.

v	k_v	k_v/v	v	k_v	k_v/v	v	k_v	k_v/v	v	k_v	k_v/v	v	k_v	k_v/v
1	0.59	0.59	11	0.59	0.05	21	0.59	0.03	31	0.59	0.02	41	0.59	0.01
2	0.95	0.48	12	0.95	0.08	22	0.95	0.04	32	0.95	0.03	42	0.95	0.02
3	0.95	0.32	13	0.95	0.07	23	0.95	0.04	33	0.95	0.03	43	0.95	0.02
4	0.59	0.15	14	0.59	0.04	24	0.59	0.02	34	0.59	0.02	44	0.59	0.01
5	0	0	15	0	0	25	0	0	35	0	0	45	0	0
6	0.59	0.10	16	0.59	0.04	26	0.59	0.02	36	0.59	0.02	46	0.59	0.01
7	0.95	0.14	17	0.95	0.06	27	0.95	0.04	37	0.95	0.03	47	0.95	0.02
8	0.95	0.12	18	0.95	0.05	28	0.95	0.03	38	0.95	0.03	48	0.95	0.02
9	0.59	0.07	19	0.59	0.03	29	0.59	0.02	39	0.59	0.02	49	0.59	0.01
10	0	0	20	0	0	30	0	0	40	0	0	50	0	0

References

1. Boldea, I.; Nasar, S.A. *The Induction Machine Handbook*; CRC Press: Boca Raton, FL, USA, 2002.
2. Pyrhönen, J.; Jokinen, T.; Hrabovcovà, V. *Design of Rotating Electrical Machines*; Wiley: Hoboken, NJ, USA, 2008.
3. Lipo, T.A. *Introduction to AC Machine Design*; Wiley: Hoboken, NJ, USA, 2017.
4. Kim, Y.; Koo, B.K.; Nam, K. Induction Motor Design Strategy for Wide Constant Power Speed Range. *IEEE Trans. Ind. Electron.* **2018**, *66*, 8372–8381. [CrossRef]
5. Guan, Y.; Zhu, Z.; Afinowi, I.; Mipo, J. Influence of machine design parameters on flux- weakening performance of induction machine for electrical vehicle application. *IET Electr. Syst. Transp.* **2015**, *5*, 43–52. [CrossRef]
6. Abdel-Khalik, A.S.; Gadoue, S.M.; Masoud, M.I.; Wiliams, B.W. Optimum flux distribution with harmonic injection for a multiphase induction machine using genetic algorithms. *IEEE Trans. Energy Convers.* **2011**, *26*, 501–512. [CrossRef]
7. Mengoni, M.; Zarri, L.; Tani, A.; Parsa, L.; Serra, G.; Casadei, D. High-torque-density control of multiphase induction motor drives operating over a wide speed range. *IEEE Trans. Ind. Electron.* **2015**, *62*, 814–825. [CrossRef]
8. Pereira, L.A.; Haffner, S.; Pereira, L.F.A.; da Rosa, R.S. Torque capability of high phase induction machines with sinusoidal and trapezoidal airgap field under steady state. In Proceedings of the 39th Annual Conference of the IEEE Industrial Electronics Society, Vienna, Austria, 10–13 November 2013; pp. 3183–3188.

9. Kong, W.; Huang, J.; Qu, R.; Kang, M.; Yang, J. Nonsinusoidal Power Supply Analysis for Concentrated-Full-Pitch-Winding Multiphase Induction Motor. *IEEE Trans. Ind. Electron.* **2016**, *63*, 574–582. [CrossRef]
10. Dajaku, G.; Bachheibl, F.; Patzak, A.; Gerling, D. Intelligent Stator Cage Winding for Automotive Traction Electric Machines. In Proceedings of the EVS28 International Electric Vehicle Symposium and Exhibition, Goyang, Korea, 3–6 May 2015; pp. 1–8.
11. Yang, J.; Hu, H.; Huang, J. Electronic pole changing technique of multiphase induction motor based on vector control. *Int. Trans. Electr. Energy Syst.* **2012**, *23*, 901–913.
12. Ge, B.; Sun, D.; Wu, W.; Peng, F.Z. Winding Design, Modeling, and Control for Pole-Phase Modulation Induction Motors. *IEEE Trans. Magn.* **2013**, *49*, 898–911. [CrossRef]
13. Gautam, A.; Ojo, J.O. Variable Speed Multiphase Induction Machine Using Pole Phase Modulation Principle. In Proceedings of the 38th Annual Conference on IEEE Industrial Electronics Society, Montreal, QC, Canada, 25–28 October 2012; pp. 3659–3665.
14. Magill, M.P.; Member, S.; Krein, P.T. A Dynamic Pole-Phase Modulation Induction Machine Model. In Proceedings of the 2015 IEEE International Electric Machines & Drives Conference (IEMDC), Coeur d'Alene, ID, USA, 10–13 May 2015; pp. 13–19.
15. Grigore-Müler, O.; Barbelian, M. The simulation of a multi-phase induction motor drive. In Proceedings of the 12th International Conference on Optimization of Electrical and Electronic Equipment, Basov, Romania, 20–22 May 2010; pp. 297–306.
16. Liu, Z.; Wu, J.; Hao, L. Coordinated and fault-tolerant control of tandem 1five-phase induction motors in ship propulsion system. *IET Electr. Power Appl.* **2017**, *12*, 91–97. [CrossRef]
17. Patzak, A.; Bachheibl, F.; Baumgardt, A.; Dajaku, G.; Moros, O.; Gerling, D. Driving range evaluation of a multi-phase drive for low voltage high power electric vehicles. In Proceedings of the 2015 International Conference on Sustainable Mobility Applications, Renewables and Technology (SMART), Kuwait City, Kuwait, 23–25 November 2015; pp. 297–306.
18. Kron, G. Rules to Predetermine Crawling, Vibration, Noise and Hooks in the Speed-Torque Curve Induction Motor Slot Combinations. *Trans. Am. Inst. Electr. Eng.* **1931**, *50*, 757–767. [CrossRef]
19. Huang, J.; Kang, M.; Yang, J.; Jiang, H.; Liu, D. Multiphase Machine Theory and Its Applications. In Proceedings of the 2008 International Conference on Electrical Machines and Systems, Wuhan, China, 17–20 October 2008; pp. 1–7.
20. Duran, M.J.; Salas, F.; Arahal, M.R. Bifurcation Analysis of Five-Phase Induction Motor Drives With Third Harmonic Injection. *IEEE Trans. Ind. Electron.* **2008**, *55*, 2006–2014. [CrossRef]
21. Abdel-Khalik, A.S.; Masoud, M.I.; Ahmed, S.; Massoud, A.M. Effect of current harmonic injection on constant rotor volume multiphase induction machine stators: A comparative study. *IEEE Trans. Ind. Appl.* **2012**, *48*, 2002–2013. [CrossRef]
22. Mekahlia, A.; Semail, E.; Scuiller, F.; Hamiti, T.; Benlamine, R. Effect of Rotor Bar Number on Performance of Five-Phase Induction Machine for Traction. In Proceedings of the 2018 XIII International Conference on Electrical Machines (ICEM), Alexandroupoli, Greece, 3–6 September 2018; pp. 185–190.
23. Marfoli, A.; Sala, G.; Papini, L.; Bolognesi, P.; Gerada, C. Torque Ripple Investigation in Squirrel Cage Induction Machines. In Proceedings of the 2019 IEEE International Electric Machines & Drives Conference (IEMDC), San Diego, CA, USA, 12–15 May 2019; pp. 140–146.
24. White, D.C.; Woodson, H.H. *Electromechanical Energy Conversion*; Wiley: Hoboken, NJ, USA, 1959.
25. Gautam, A.; Ojo, O.; Ramezani, M.; Momoh, O. Computation of equivalent circuit parameters of nine-phase induction motor in different operating modes. In Proceedings of the IEEE Energy Conversion Congress and Exposition (ECCE), Raleigh, NC, USA, 15–20 September 2012; pp. 142–149.
26. Magill, M.P.; Member, S.; Krein, P.T.; Haran, K.S.; Selection, A.P.W. Equivalent Circuit Model for Pole-Phase Modulation Induction Machines. In Proceedings of the 2015 IEEE International Electric Machines & Drives Conference (IEMDC), Coeur d'Alene, ID, USA, 10–13 May 2015; pp. 293–299.
27. Fudeh, H.R.; Ong, C.M. Modeling and analysis of induction machines containing space harmonics: Part I: Modeling and Transformation. *IEEE Trans. Power Appar. Syst.* **1983**, *PAS-102*, 2608–2615. [CrossRef]
28. Toliyat, H.A.; Lipo, T.A.; White, J.C. Analysis of a Concentrated Winding Induction Machine for Adjustable Speed Drive Applications, Part I: Motor Analysis. *IEEE Trans. Energy Convers.* **1991**, *6*, 679–683. [CrossRef]

29. Muñoz, A.R. Analysis and Control of a Dual Stator Winding Squirrel Cage Induction Machine Drive. Ph.D. Thesis, University of Wisconsin, Madison, WI, USA, 1999.
30. Scuiller, F. Predicting the space harmonics generated by symmetrical multi-phase windings. In Proceedings of the 2019 IEEE International Electric Machines & Drives Conference (IEMDC), San Diego, CA, USA, 12–15 May 2019; pp. 1348–1355.

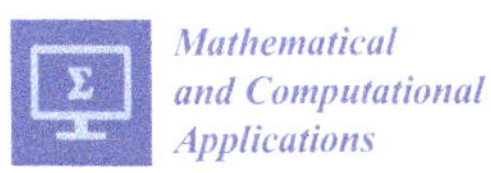
Mathematical and Computational Applications

Article

2D Hybrid Steady-State Magnetic Field Model for Linear Induction Motors

Samuil R. Aleksandrov *, Timo T. Overboom and Elena A. Lomonova

Department of Electrical Engineering, Electromechanics and Power Electronics, Eindhoven University of Technology, 5612 AZ Eindhoven, The Netherlands

* Correspondence: alexandrov.samuil@gmail.com

Received: 29 June 2019; Accepted: 24 July 2019; Published: 25 July 2019

Abstract: This paper presents a 2D hybrid steady-state magnetic field model, capable of accurately modeling the electromagnetic behavior in a linear induction motor, including primary slotting, finite yoke length, and longitudinal end-effects by primary motion. This model integrates a complex harmonic modeling technique with a discretized magnetic equivalent circuit model. The Fourier model is applied to regions with homogeneous material properties, e.g., air regions and the track of the motor, while the magnetic equivalent circuit (MEC) approach is used for the regions containing non-homogeneous material properties, e.g., the primary of the linear induction motor (LIM). By only meshing the domains containing highly-permeable materials, the computational effort is reduced in comparison with the finite element method (FEM). The model is applied to a double-layer single-sided LIM, and the resulting thrust and normal forces show an excellent agreement with respect to finite element analysis and measurement data.

Keywords: linear induction motors; complex harmonic modeling; hybrid analytical modeling; 2D steady-state models

1. Introduction

Linear induction motors (LIM) are widely used in long-stroke linear motion systems because of their inexpensive and robust construction. To obtain an optimal design, comprehensive methods able to predict the magnetic field distribution inside the electromagnetic structures play a crucial role. To allow extensive exploration of the design space, numerical methods such as the finite element method (FEM) are not preferable, as these models are computationally expensive.

In the literature, semi-analytical or hybrid methods are discussed, intending to reduce the needed computational efforts, while ensuring comparable accuracy to the numerical methods. However, all modeling techniques require certain assumptions, which limit their flexibility. In [1], an equivalent-circuit model of the LIM was proposed, determining the motor output thrust and vertical forces, while accounting for the longitudinal end-effects as a result of primary movement with respect to the secondary. In [2], an equivalent-circuit model for a high-speed industrial transportation LIM was presented, where the dynamic longitudinal and the transverse end-effects were accounted for by correction factors. In [3], an optimized end-effect equivalent-circuit model for LIM was presented, allowing modeling of partially-filled end-slots. However, equivalent circuit models are not suitable for design purposes, as their components need to be determined from measurements or magnetic field modeling [4].

In [5–7], magnetic field models for rotating and linear induction motors, using a two-dimensional field description by Fourier series, were presented. Although these models allow obtaining the magnetic field distribution inside the air gap, they do not include the magnetic field distribution inside the primary yoke or slots. In [5], the magnetic field distribution into a solid rotor was predicted, while considering the stator slots and tooth-tips and assuming an infinite permeability inside the stator yoke. In [6], the current carrying primary coils were replaced with infinitely-thin current sheets, and the primary slotting was accounted for by the use of Carter's coefficient. In [7], correction factors for the longitudinal end-effect and also for the primary core losses were presented. A semi-analytical model for LIM, based on harmonic modeling, was presented in [8]. The field inside the primary slotting was calculated, assuming an infinitely-permeable core, but the longitudinal end-effects of the motor and the magnetic field distribution in the primary yoke were neglected.

In [9,10], the primary core of a synchronous permanent magnet motor was successfully included in the field analysis. In [11], these models were extended to include saturation of the highly-permeable materials. Hybrid models combine the benefits of the magnetic equivalent circuit (MEC) method [12] and harmonic modeling [13]. However, these models have only been derived for magnetostatic fields, thus neglecting eddy-current effects.

As an alternative to the aforementioned magnetostatic hybrid techniques, this study presents a steady-state hybrid semi-analytical model, which combines an MEC-based description of the domains containing highly-permeable materials, e.g., the primary of the LIM, with complex Fourier modeling applied to the conductive medium of the secondary plate and surrounding air regions. This model allows modeling of the full primary core of the LIM, including longitudinal end-effects and the electromagnetic field in the primary yoke and slotting, while also accounting for the primary velocity. Including the velocity terms to the field solution allows time-stepping to be avoided, thus saving time, when compared to FEA.

The electromagnetic problem that is investigated in this paper is a linear induction motor (LIM) topology with a moving primary and an infinitely-long flat secondary (Figure 1). The primary contains a rewound laminated core from a Tecnotion TL-15 linear synchronous permanent magnet motor with double-layer three-phase distributed winding [14]. This topology is used to validate the model with static measurements.

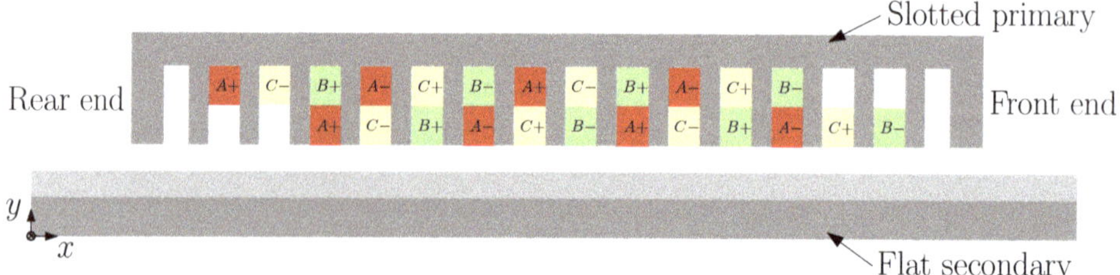

Figure 1. Analyzed linear induction motor (LIM) topology.

In this paper, a generalized description of the modeling methodology is presented in Section 2. Afterwards, the introduced model is applied to a double-layer single-sided LIM and validated with respect to a 2D steady-state FEA simulation and measurement data for the same topology, and the results are discussed in Section 3. Finally, the conclusions are presented in Section 4.

2. Modeling Methodology

To apply the hybrid modeling technique to an electromagnetic problem like the LIM, the topology was represented in the 2D Cartesian coordinate system and is divided into orthogonal regions, as depicted in Figure 2. The complex harmonic modeling was applied to Regions I, III, IV, V, and VI, as these regions contained only homogeneous, isotropic, and linear materials. As the primary of the LIM (Region II) contained different materials along the x-direction and y-direction, it was modeled using the mesh-based MEC formulation. Using the complex harmonic formulation required periodicity in the longitudinal direction. As the secondary of the LIM was considered infinitely long, and the finite length of the primary was included in the analysis to account for the longitudinal end-effects, while the periodicity in x-direction was ensured by adding air at the front and rear end of the primary yoke, defining the periodical length τ_{per} for the whole problem.

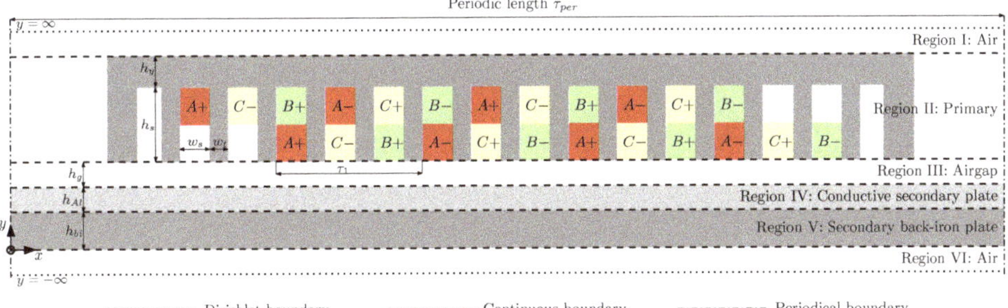

Figure 2. LIM topology: division into regions, dimensions, and considered boundary conditions.

The currents flowing into the three-phase distributed winding were as follows:

$$I_{phA}(t) = I_p e^{j2\pi f t}, \quad (1)$$

$$I_{phB}(t) = I_p e^{-j\frac{2\pi}{3}} e^{j2\pi f t}, \quad (2)$$

$$I_{phC}(t) = I_p e^{j\frac{2\pi}{3}} e^{j2\pi f t}, \quad (3)$$

where f is the synchronous frequency, I_p is the peak current, and t is the instance of time.

2.1. Complex Harmonic Modeling

The complex harmonic modeling technique is based on the analytical solution of the magnetic vector potential A_z, which was explained in detail in [15]. To derive the steady-state solution for the magnetic vector potential, time variation has to be included to account for the induced currents in the conductive regions. Their relation to the vector potential is expressed by the diffusion equation:

$$\frac{\partial^2 A_z}{\partial x^2} + \frac{\partial^2 A_z}{\partial y^2} = \mu_0 \mu_r \sigma \frac{\partial A_z}{\partial t}, \quad (4)$$

where σ is the conductivity of the considered region, μ_0 is the relative permeability of the free space, and μ_r is the relative permeability of the material in that region. The general form of the solution to the magnetic vector potential is obtained in complex form:

$$A_z(x,y,t) = \sum_{n=-\infty}^{\infty} (a_n e^{\sqrt{\lambda_n^2} y} + b_n e^{-\sqrt{\lambda_n^2} y}) e^{j\omega_n x} e^{j(2\pi f t + \omega_n v t)}, \quad (5)$$

where:

$$\lambda_n^2 = \omega_n^2 + j\mu_0 \mu_r \sigma (2\pi f + \omega_n v), \quad (6)$$

$$\omega_n = \frac{2n\pi}{\tau_{per}}. \quad (7)$$

In (5), ω_n is the spatial frequency for the nth space harmonic, v is the considered steady-state velocity of the primary with respect to the secondary, and a_n and b_n are the unknown coefficients for each harmonic, obtained from the applied boundary conditions explained in the following section.

The resulting flux density distributions for the tangential and normal direction were obtained as follows:

$$B_x(x,y,t) = \frac{\partial A_z(x,y,t)}{\partial y} = \sum_{n=-\infty}^{\infty} \left[\lambda_n (a_n e^{\lambda_n y} - b_n e^{-\lambda_n y}) \right] e^{j\omega_n x} e^{j(2\pi f t + \omega_n v t)}, \tag{8}$$

$$B_y(x,y,t) = -\frac{\partial A_z(x,y,t)}{\partial x} = -j \sum_{n=-\infty}^{\infty} \left[\omega_n (a_n e^{\lambda_n y} + b_n e^{-\lambda_n y}) \right] e^{j\omega_n x} e^{j(2\pi f t + \omega_n v t)}. \tag{9}$$

2.2. MEC

The primary of the LIM contained non-homogeneous material properties along the x- and y-direction, and for that reason, it was modeled using the MEC formulation. The region was discretized into L layers along the y-direction, each containing K rectangular elements along the x-direction, forming a mesh of $M = L \times K$ elements, as illustrated in Figure 3.

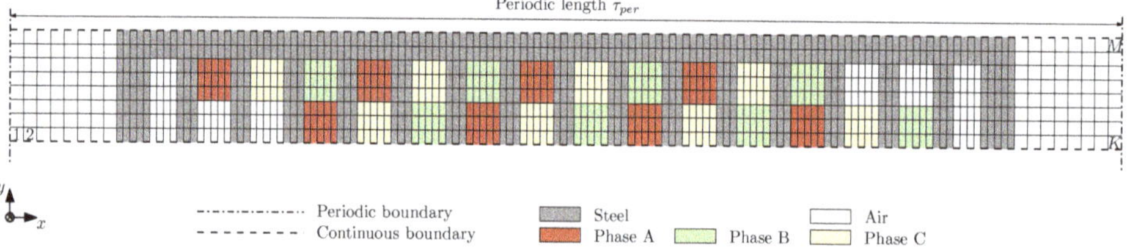

Figure 3. Discretization of the magnetic equivalent circuit (MEC) region.

Each MEC-element encompassed one potential node, $\psi(l,k)$, as is shown in Figure 4, and time dependency was accounted for by adapting the following expression:

$$\psi(l,k,t) = \psi(l,k)e^{j2\pi f t}, \tag{10}$$

where $\psi(l,k)$ is the complex value for each potential node, which is obtained after solving the set of linear equations, formed from the applied boundary conditions.

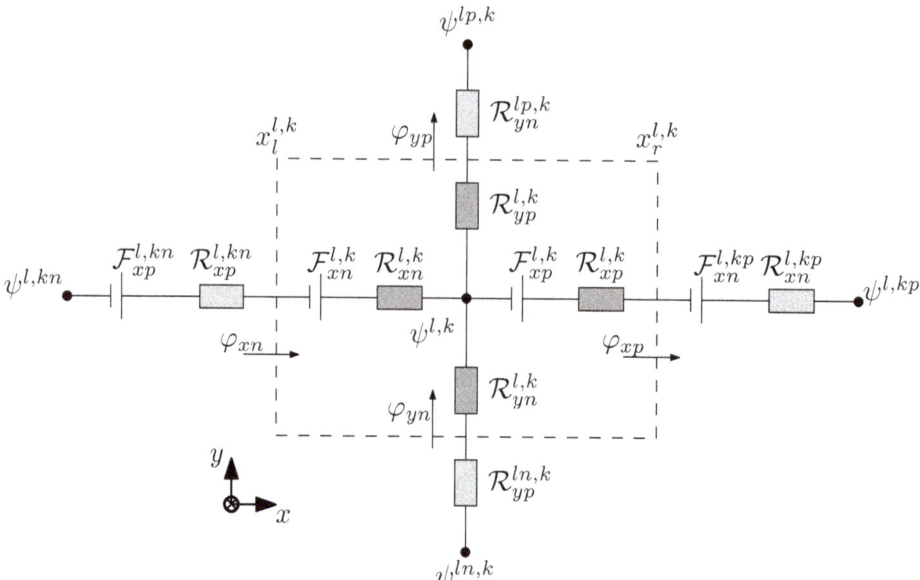

Figure 4. Single MEC element.

The reluctances for each MEC-element were defined by its dimensions and material properties:

$$\mathcal{R}_{xp}(l,k) = \mathcal{R}_{xn}(l,k) = \frac{l_x(l,k)}{2\mu_0\mu_r(l,k)S_{zy}(l,k)}, \tag{11}$$

$$\mathcal{R}_{yp}(l,k) = \mathcal{R}_{yn}(l,k) = \frac{l_y(l,k)}{2\mu_0\mu_r(l,k)S_{xz}(l,k)}, \tag{12}$$

where $l_x(l,k)$ and $l_y(l,k)$ are the lengths of each MEC-element in the x- and y-direction (Figure 4), $x_l^{l,k}$ and $x_r^{l,k}$ are the left and right coordinates of the MEC element, while $S_{zx}(l,k)$ and $S_{yz}(l,k)$ are the cross-sectional areas parallel to the zx- and yz-planes, respectively. The values, assigned to μ_r, depend on each element's location in the xy-plane, and as a consequence, the material each element encloses.

The magnetic equivalence of Kirchoff's current law was applied to each MEC-element. All magnetic flux entering one potential node ($\psi(l,k,t)$) should be equal to the magnetic flux leaving this node:

$$\varphi_{xn}(l,k,t) + \varphi_{yn}(l,k,t) = \varphi_{xp}(l,k,t) + \varphi_{yp}(l,k,t), \tag{13}$$

where:

$$\varphi_{xp}(l,k,t) = \frac{\psi(l,kp,t) - \psi(l,k,t)}{\mathcal{R}_{xp}(l,k) + \mathcal{R}_{xn}(l,kp)} + \frac{\mathcal{F}_{xp}(l,k,t) + \mathcal{F}_{xn}(l,kp,t)}{\mathcal{R}_{xp}(l,k) + \mathcal{R}_{xn}(l,kp)}, \tag{14}$$

$$\varphi_{xn}(l,k,t) = \frac{\psi(l,k,t) - \psi(l,kn,t)}{\mathcal{R}_{xn}(l,k) + \mathcal{R}_{xp}(l,kn)} + \frac{\mathcal{F}_{xn}(l,k,t) + \mathcal{F}_{xp}(l,kn,t)}{\mathcal{R}_{xn}(l,k) + \mathcal{R}_{xp}(l,kn)}, \tag{15}$$

$$\varphi_{yp}(l,k,t) = \frac{\psi(lp,k,t) - \psi(l,k,t)}{\mathcal{R}_{yp}(l,k) + \mathcal{R}_{yn}(lp,k)}, \tag{16}$$

$$\varphi_{yn}(l,k,t) = \frac{\psi(l,k,t) - \psi(ln,k,t)}{\mathcal{R}_{yn}(l,k) + \mathcal{R}_{yp}(ln,k)}, \tag{17}$$

where kp, kn, lp, and ln represent the indices of neighboring potential nodes.

To allow coupling with the complex harmonic regions, periodicity in the x-direction is fulfilled by linking the last element of each layer with the first element from the same layer.

The MMFsource terms, present in the primary of the LIM, included only coil excitations represented by MMF-sources:

$$\mathcal{F}_x(l,k,t) = \zeta\frac{N_t I_{ph}(t)}{2K_c}, \tag{18}$$

where I_{ph} is the complex phase current according to (1)–(3), N_t is the number of turns for a single coil, K_c is the number of MEC elements in the x-direction for a single coil, and ζ represents the scaling factor to account for the distribution of the MMF sources along the y-direction. In Figure 5, the magnitude variation of the MMF-sources in the top and bottom layer coils is depicted. The magnitude was maximal in the yoke elements, as the formed magnetic path through the air gap enclosed the whole area of the coil, while in the slot elements, the magnitude of the MMF-sources was proportional to the enclosed coil area. In case coils were present in both layers, superposition of both MMF-sources was applied.

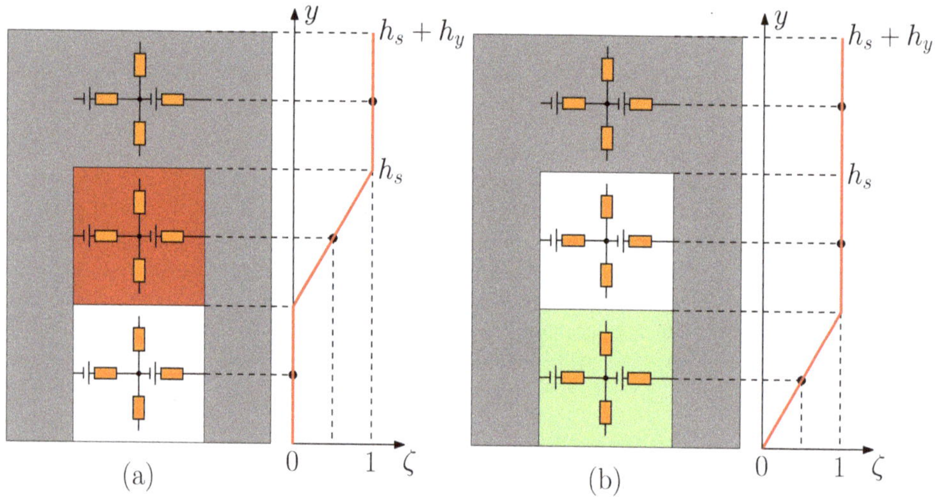

Figure 5. MMF-source distribution inside a single slot for (**a**) the top winding layer and (**b**) the bottom winding layer.

2.3. Boundary Conditions

To obtain the unknown coefficients for both harmonic and MEC regions, a set of linear equations, accounting the boundary conditions between every two adjacent regions, was solved. The presence of air above the primary and beneath the secondary of the investigated topology allowed the top and bottom boundaries of LIM to be extended to infinity, and thus, the Dirichlet boundary condition, forcing all field components to vanish, applied. To ensure continuity between every two neighboring Fourier regions i and j (e.g., Regions III and IV), the tangential components of the magnetic field strength and the normal components of the flux density, obtained by (8) and (9) for both regions, were equated. Analogically, the continuous boundary condition applied also on the border between each Fourier and MEC region (e.g., Regions I and II). While the obtained expressions for the magnetic fields in the harmonic regions were defined for the full periodical section of the analyzed problem, each expression for the MEC region was associated with a single mesh-element. A detailed explanation of the boundary conditions for both normal and tangential field components was given in [15], considering trigonometric harmonic solutions.

The main differences introduced by the complex harmonic solution, presented in this paper, were in the expressions of the coupled normal and tangential field components between each adjacent Fourier and MEC region.

Adapting (13), the coupled flux in the normal direction at the bottom and top of the MEC region took the form of:

$$\varphi_{xn}(1,k,t) + \varphi_{yn}^{HM}(1,k,t) = \varphi_{xp}(1,k,t) + \varphi_{yp}(1,k,t), \tag{19}$$

$$\varphi_{xn}(L,k,t) + \varphi_{yn}(L,k,t) = \varphi_{xp}(L,k,t) + \varphi_{yp}^{HM}(L,k,t), \tag{20}$$

where for the bottom layer of the MEC:

$$\begin{aligned} \varphi_{yn}^{HM}(1,k,t) &= L_s \int_{x_l(1,k)-vt}^{x_r(1,k)-vt} B_y^{HM}(x,y_{BC})\mathrm{d}x, \\ &= L_s \sum_{n=-\infty}^{\infty} (a_n e^{\lambda_n y_{BC}} + b_n e^{-\lambda_n y_{BC}})(e^{j\omega_n x_l(1,k)} - e^{j\omega_n x_r(1,k)})e^{j2\pi ft}, \end{aligned} \tag{21}$$

for $1 \leq k \leq K$, where y_{BC} is the y-coordinate between the regions, to which the boundary condition applies, and L_s is the depth of the domain. For the top layer of MEC, $\varphi_{yp}^{HM}(L,k,t)$ was derived analogically.

Substituting the derived equation for B_x from the adjacent Fourier region (8), the boundary condition for the tangential field strength used the the constitute relation $B = \mu_0\mu_r H$ and took the form of:

$$\frac{1}{\mu_0\mu_r^{HM}}B_x^{HM}(x,y_{BC},t) = \sum_{k=1}^{K}\frac{1}{\mu_0\mu_r^{MEC}(l,k)}B_x^{MEC}(x,y_{BC},t), \tag{22}$$

where μ_r^{HM} represents the homogeneous relative permeability within the Fourier region, while $\mu_r^{MEC}(l,k)$ is the relative permeability per element of the MEC-region. The magnetic flux density, calculated for the top or bottom layer of the MEC-region, was considered to be constant within a single element [9]. To allow coupling with the neighboring Fourier regions, the right-hand side of (22) was modified as:

$$B_x^{MEC}(x,y_{BC},t) = \sum_{n=-\infty}^{\infty}\frac{2}{\tau_{per}}\sum_{k=1}^{K}\int_{x_l(1,k)-vt}^{x_r(1,k)-vt} B_x^{MEC}(l,k,t)e^{-j\omega_n x}\mathrm{d}x, \tag{23}$$

where:

$$B_x^{MEC}(l,k,t) = \sum_{k=1}^{K}\frac{\varphi_{xn}(l,k,t)+\varphi_{xp}(l,k,t)}{2S_{zy}(l,k)} \tag{24}$$

is the average tangential flux density per element and $l = L$ or $l = 1$ for the top or bottom layer, respectively.

2.4. Force Calculation

The output thrust and normal forces acting on the primary were derived from the Maxwell stress tensor evaluated inside the air gap [16]. Taking into account the derived equations for B_x and B_y for Region III, the analytical force equations took the form of:

$$\begin{aligned} F_x &= -\frac{L_s}{\mu_0}\int_0^{\tau_{per}}\left[B_{xn}(x,y,t)B_{yn}^*(x,y,t)\right]dx \\ &= -\frac{jL_s\tau_{per}}{2\mu_0}\sum_{n=-\infty}^{\infty}\left[\lambda_n\omega_n(a_ne^{\lambda_n y}-b_ne^{-\lambda_n y})(a_n^*e^{\lambda_n^* y}+b_n^*e^{-\lambda_n^* y})\right] \end{aligned} \tag{25}$$

and:

$$\begin{aligned} F_y &= -\frac{L_s}{2\mu_0}\int_0^{\tau_{per}}\left[B_{xn}(x,y,t)B_{xn}^*(x,y,t)-B_{yn}(x,y,t)B_{yn}^*(x,y,t)\right]dx \\ &= -\frac{L_s\tau_{per}}{4\mu_0}\sum_{n=-\infty}^{\infty}\Big[\lambda_n\lambda_n^*(a_ne^{\lambda_n y}-b_ne^{-\lambda_n y})(a_n^*e^{\lambda_n^* y}-b_n^*e^{-\lambda_n^* y}) \\ &\quad -\omega_n^2(a_ne^{\lambda_n y}+b_ne^{-\lambda_n y})(a_n^*e^{\lambda_n^* y}+b_n^*e^{-\lambda_n^* y})\Big], \end{aligned} \tag{26}$$

where * is the complex conjugate.

2.5. Joule Losses' Calculation

The conduction losses inside the secondary can be calculated, using the Poynting vector, applied in the air gap (Region III) [17]:

$$P_{joule,sec} = -\frac{L_s}{2}\Re\int_0^{\tau_{per}} E_z(x,y,t)H_x^*(x,y,t)\,dx$$
$$= -\frac{L_s\tau_{per}}{j2\mu_0}\sum_{n=-\infty}^{\infty}\left[(\lambda_n^*\omega+\lambda_n^*\omega_n v)(a_n e^{\lambda_n y}+b_n e^{-\lambda_n y})(a_n^* e^{\lambda_n^* y}+b_n^* e^{-\lambda_n^* y})\right], \tag{27}$$

where the following expressions were used:

$$E_z(x,y,t) = -\frac{\partial A_z(x,y,t)}{\partial t}, \tag{28}$$
$$H_x^*(x,y,t) = \frac{1}{\mu_0}B_x^*(x,y,t). \tag{29}$$

3. Results and Model Validation

To validate the presented 2D complex hybrid steady-state model, 2D finite element analysis (FEA) was performed on the same topology. Table 1 contains the dimensions and design parameters used for both simulations. For the complex hybrid model, $N = 100$ harmonics and $K = 576$ elements in $L = 53$ layers were used, in order to generate a dense enough mesh, able to model the magnetic field in the primary of the motor and in the surrounding air accurately. The periodic length τ_{per} for both the complex hybrid model and FEA was selected to be even times the fundamental pitch of one periodical section of the primary τ_1 (in this case, $\tau_{per} = 12 \times \tau_1$). The conductivity of the aluminum plate was reduced accordingly, to take into consideration the transverse end-effects of the investigated motor [18].

Considering velocity $v = 0$ m/s, peak current $I_p = 10$ A, and synchronous frequency $f = 100$ Hz, the resulting magnetic flux density in the normal (Figure 6) and in longitudinal direction (Figure 7) was plotted against the steady-state FEA solution, showing excellent correspondence. The output thrust force, normal force, and Joule losses were calculated using (25)–(27), respectively, and predicted within 1.5%, 1.7%, and 3.1% when compared to FEA.

Table 1. Parameters of the double-layer single-sided LIM.

Parameter	Symbol	Value	Unit
Number of phases	N_p	3	-
Number of poles	$2p$	6	-
Number of slots	z_1	16	-
Number of turns per coil	N_t	57	-
Stack width	L_s	50	mm
Fundamental pitch of the primary	τ_1	12	mm
Primary tooth width	w_t	6	mm
Primary slot width	w_s	10	mm
Primary slot height	h_s	20	mm
Primary yoke height	h_y	6.5	mm
Air gap length	h_g	2.7	mm
Thickness of the aluminum plate	h_{Al}	2	mm
Thickness of the back-iron plate	h_{bi}	8	mm
Conductivity of aluminum	σ_{Al}	17×10^6	Sm^{-1}
Conductivity of iron	σ_{Fe}	4.5×10^6	Sm^{-1}
Relative permeability iron	μ_r	1000	mm

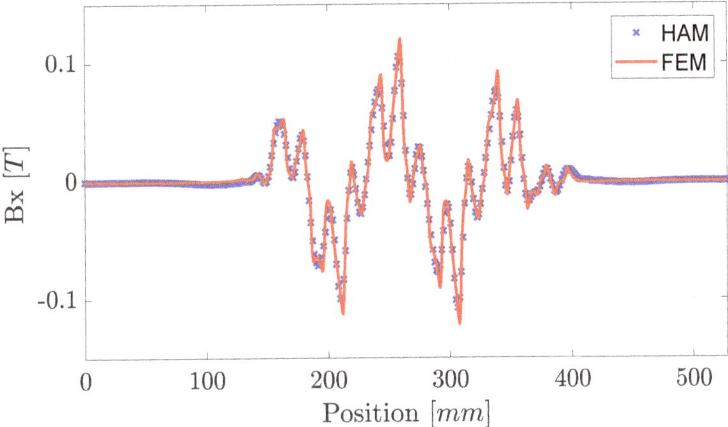

Figure 6. Magnetic flux density in normal direction in the middle of the air gap ($I_p = 10$ A, $v = 0$ m/s, $f = 100$ Hz).

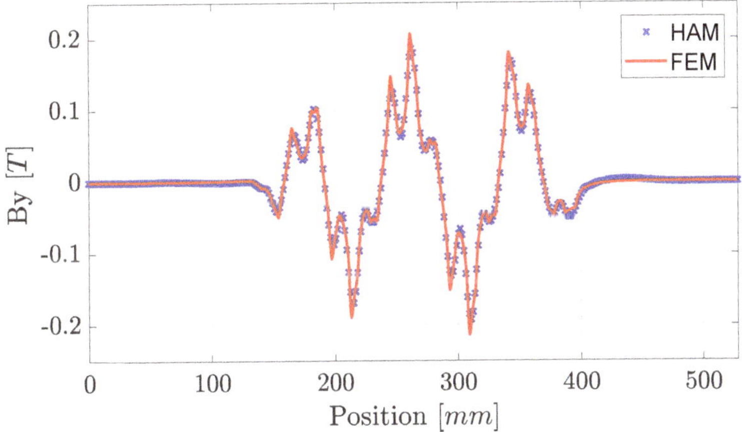

Figure 7. Magnetic flux density in longitudinal direction in the middle of the air gap ($I_p = 10$ A, $v = 0$ m/s, $f = 100$ Hz).

In addition, the thrust and normal forces, obtained at different synchronous frequencies from both the presented model and FEA, were validated by static measurements. As shown in Figure 8, the secondary of the LIM was mounted on a moving translator, while the mechanical construction on top of it held the primary and a six-axis load cell [14]. The thrust and normal force were measured with a fixed secondary and primary ($v = 0$ m/s). For $I_p = 10$ A, the results are shown in Figure 9.

Figure 8. Measurement setup [14].

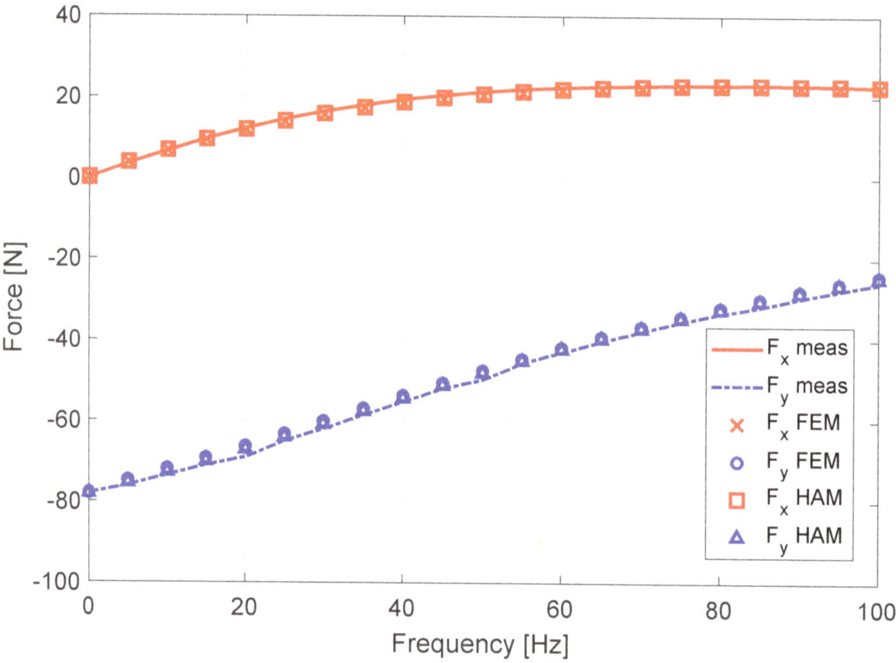

Figure 9. Normal and thrust force for different frequency in comparison with steady-state measurement data ($I_p = 10$ A, $v = 0$ m/s).

Having the resulting field represented by complex Fourier series allowed obtaining the contribution of each harmonic to the propulsion force. Including the end-effects of the primary provided information on the full harmonic spectrum contribution, as shown on Figure 10. As the velocity of the motor was accounted for in this steady-state hybrid model, two simulations at the same fundamental slip frequency:

$$f_{slip} = f - \frac{v}{2\tau_1} \tag{30}$$

were performed:

- Case 1: velocity $v = 0$ m/s at frequency $f = 50$ Hz,
- Case 2: velocity $v = 10$ m/s at frequency $f = 154.17$ Hz.

In Case 1, the eddy-currents generated in the static secondary were interacting with the traveling wave, produced by the three-phase primary winding, and thus, propulsion force in the positive x-direction was acting on the primary. The fundamental motor harmonic, contributing to the propulsion force generation, was defined by one periodical section of the primary. Due to the definition of this electromagnetic problem, where $x_p = 12\tau$, the fundamental motor harmonic was equal to the sixth field harmonic in the Fourier series, and analogically, the fifth motor harmonic was equal to the thirtieth field harmonic in the Fourier series, as can be clearly seen in Figure 10. Additional thrust force contributions from the fifth and seventh field harmonics were caused by the end-effects, and the total thrust force in Case 1 was $F_x = 20.4$ N.

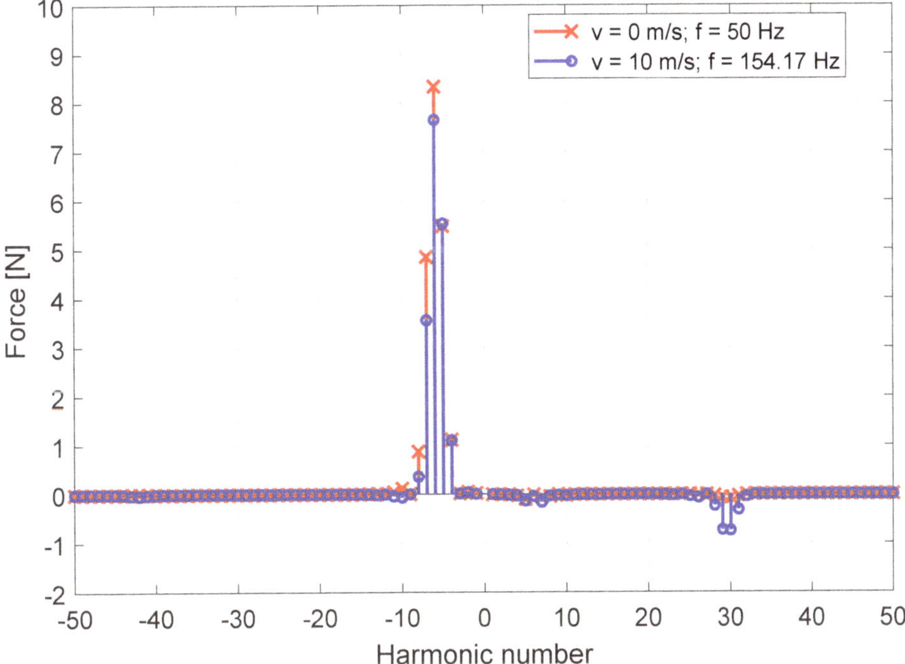

Figure 10. Harmonics contributing to the propulsion force at $v = 0$ m/s; $f = 50$ Hz and $v = 10$ m/s; $f = 154.1667$ Hz.

As the velocity was accounted for in Case 2, new conductive material, unaffected by the induced magnetic field, was constantly seen by the front end of the primary, while there were still trailing eddy-currents in the conductive plate, behind the rear end of the motor. This effect caused the fifth motor harmonic (thirtieth field harmonic) to oppose the fundamental motor harmonic (sixth field harmonic), which was additionally reduced by the longitudinal end-effects, and thus, the generated thrust force was decreased to $F_x = 15.8$ N, as seen in Figure 10.

4. Conclusions

In this paper, a 2D hybrid steady-state magnetic field model that included the full primary of a double-layer single-sided linear induction motor, thus accounting for the longitudinal end-effects, was presented. Compared to finite elements analysis, the model showed excellent correspondence of the magnetic field distribution inside the LIM. The velocity of the primary, with respect to the secondary, and the resulting longitudinal end-effects were accounted for in the solution of the magnetic field. The obtained thrust and normal forces for different fundamental slip frequencies had a discrepancy within 1.7% compared to FEA and were verified by static measurements. Future research will focus on derivation of secondary parameters and adding saturation effects to the model, which will allow the implementation for different motor topologies.

Author Contributions: The theory presented in this paper was developed by S.R.A. The analysis of the results was performed in cooperation with T.T.O. and E.A.L. The paper was written by S.R.A., and contributions and improvements to the content were made by T.T.O. and E.A.L.

Conflicts of Interest: The authors declare no conflict of interest.

References

1. Duncan, J. Linear induction motor-equivalent-circuit model. *IEE Proc. B Electr. Power Appl.* **1983**, *130*, 51–57. [CrossRef]
2. Lu, J.; Ma, W. Research on End Effect of Linear Induction Machine for High-Speed Industrial Transportation. *IEEE Trans. Plasma Sci.* **2011**, *39*, 116–120. [CrossRef]
3. Hu, Y.; Cosic, A.; Östlund, S.; Hui, Z. Design and Optimization Procedure of a Single-Sided Linear Induction Motor Applied to an Articulated Funiculator. In Proceedings of the 2016 IEEE 8th International Power Electronics and Motion Control Conference (IPEMC-ECCE Asia), Hefei, China, 22–26 May 2016; pp. 3096–3102. [CrossRef]
4. Woronowicz, K.; Safaee, A. A novel linear induction motor equivalent-circuit with optimized end-effect model including partially-filled end slots. In Proceedings of the 2014 IEEE Transportation Electrification Conference and Expo (ITEC), Dearborn, MI, USA, 15–18 June 2014; pp. 1–5. [CrossRef]
5. Boughrara, K.; Dubas, F.; Ibtiouen, R. 2-D Analytical Prediction of Eddy Currents, Circuit Model Parameters, and Steady-State Performances in Solid Rotor Induction Motors. *IEEE Trans. Magn.* **2014**, *50*, 1–14. [CrossRef]
6. Gieras, J.F.; Eastham, A.R.; Dawson, G.E. Performance calculation for single-sided linear induction motors with a solid steel reaction plate under constant current excitation. *IEE Proc. B Electr. Power Appl.* **1985**, *132*, 185–194. [CrossRef]
7. Boldea, I.; Babescu, M. Multilayer approach to the analysis of single-sided linear induction motors. *Proc. Inst. Electr. Eng.* **1978**, *125*, 283–287. [CrossRef]
8. Overboom, T.; Smeets, J.; Jansen, J.; Lomonova, E. Semi-analytical modeling of a linear induction motor including primary slotting. In Proceedings of the 15th International Symposium on Electromagnetic Fields in Mechatronics (ISEF), Funchal, Portugal, 1–3 September 2011; pp. 1–8.
9. Pluk, K.J.W.; Jansen, J.W.; Lomonova, E.A. Hybrid Analytical Modeling: Fourier Modeling Combined With Mesh-Based Magnetic Equivalent Circuits. *IEEE Trans. Magn.* **2015**, *51*, 1–12. [CrossRef]
10. Ouagued, S.; Diriye, A.A.; Amara, Y.; Barakat, G. A General Framework Based on a Hybrid Analytical Model for the Analysis and Design of Permanent Magnet Machines. *IEEE Trans. Magn.* **2015**, *51*, 1–4. [CrossRef]
11. Bao, J.; Gysen, B.L.J.; Lomonova, E.A. Hybrid Analytical Modeling of Saturated Linear and Rotary Electrical Machines: Integration of Fourier Modeling and Magnetic Equivalent Circuits. *IEEE Trans. Magn.* **2018**, *54*, 1–5. [CrossRef]
12. Ostovic, V. *Dynamics of Saturated Electric Machines*; Springer: Berlin, Germany, 1989.
13. Gysen, B.L.J.; Meessen, K.J.; Paulides, J.J.H.; Lomonova, E.A. General Formulation of the Electromagnetic Field Distribution in Machines and Devices Using Fourier Analysis. *IEEE Trans. Magn.* **2010**, *46*, 39–52. [CrossRef]

14. Overboom, T.; Smeets, J.; Jansen, J.; Lomonova, E. Decoupled control of thrust and normal force in a double-layer single-sided linear induction motor. *Mechatronics* **2013**, *23*, 213–221. [CrossRef]
15. Aleksandrov, S.R.; Overboom, T.T.; Lomonova, E.A. Design Optimization and Performance Comparison of Two Linear Motor Topologies With PM-Less Tracks. *IEEE Trans. Magn.* **2018**, *54*, 1–8. [CrossRef]
16. Amrhein, M.; Krein, P.T. Force Calculation in 3-D Magnetic Equivalent Circuit Networks with a Maxwell Stress Tensor. *IEEE Trans. Energy Convers.* **2009**, *24*, 587–593. [CrossRef]
17. Poynting, J.H. On the transfer of energy in the electromagnetic field. *Philos. Trans. R. Soc. Lond.* **1884**, *175*, 343–361. [CrossRef]
18. Bolton, H. Transverse edge effect in sheet-rotor induction motors. *Proc. Inst. Electr. Eng.* **1969**, *116*, 725–731. [CrossRef]

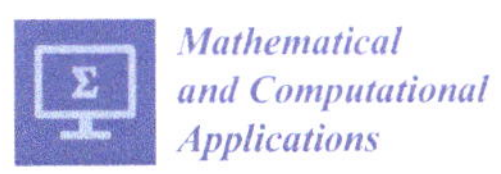

Article

Combining the Magnetic Equivalent Circuit and Maxwell–Fourier Method for Eddy-Current Loss Calculation

Youcef Benmessaoud, Frédéric Dubas * and Mickael Hilairet

Département ENERGIE, FEMTO-ST, CNRS, Univ. Bourgogne Franche-Comté, F90000 Belfort, France; youcef.benmessaoud@femto-st.fr (Y.B.); mickael.hilairet@univ-fcomte.fr (M.H.)

* Correspondence: frederic.dubas@univ-fcomte.fr; Tel.: +33-384-583-648

Received: 27 March 2019; Accepted: 2 June 2019; Published: 4 June 2019

Abstract: In this paper, a hybrid model in Cartesian coordinates combining a two-dimensional (2-D) generic magnetic equivalent circuit (MEC) with a 2-D analytical model based on the Maxwell–Fourier method (i.e., the formal resolution of Maxwell's equations by using the separation of variables method and the Fourier's series) is developed. This model coupling has been applied to a U-cored static electromagnetic device. The main objective is to compute the magnetic field behavior in massive conductive parts (e.g., aluminum, magnets, copper, iron) considering the skin effect (i.e., with the eddy-current reaction field) and to predict the eddy-current losses. The magnetic field distribution for various models is validated with 2-D and three-dimensional (3-D) finite-element analysis (FEA). The study is also focused on the discretization influence of 2-D generic MEC on the eddy-current loss calculation in conductive regions. Experimental tests and 3-D FEA have been compared with the proposed approach on massive conductive parts in aluminum. For an operating point, the computation time is divided by ~4.6 with respect to 3-D FEA.

Keywords: eddy-current losses; experiment; hybrid model; magnetic equivalent circuit; numerical; Maxwell–Fourier method

1. Introduction

1.1. Context of This Paper

Political and economic issues are one of the main drawback of permanent-magnet (PM) synchronous machines (PMSMs) due to the presence of rare-earth PMs. Indeed, economic dependency constitutes a strong objective for industrial electronics companies and that is the reason why industry and academia conduct research on PM-less machines (e.g., synchronous or switched-reluctance machines, induction machines) [1]. However, today, PMSMs are one of the most competitive machines for their high electromagnetic performances, massive torque, high efficiency, and low torque ripple [2,3]. Nevertheless, the speed variation leads to variable magnetic fields constituted of: (i) temporal harmonics due to the current waveform (e.g., sinusoidal, six-step rectangular, pulse-width modulation currents, etc.), and (ii) spatial harmonics, both the stator slotting permeance and the magnetomotive force (MMF) distribution [4,5]. Consequently, eddy-currents appear inside the PM volume, which contributes to supplement losses, namely eddy-current losses.

At high-speed or high-frequency, PM losses can be important [4]. The electrical conductivity can also be affected at high temperatures, leading to a loss increase and faulty conditions due to the PM. Therefore, the study of this phenomenon is required to predict the PM eddy-current losses in order to improve the design procedure in electromagnetic devices. Different formulations have been developed in order to estimate these eddy-current losses, such as: (i) semi-analytical methods based on the electrical equivalent circuit (EEC) and/or magnetic equivalent circuit (MEC) [6], (ii) analytical methods based on the formal resolution of Maxwell's equations [4,7–10], and (iii) the numerical hybrid method based on the 3-D finite-element analysis (FEA) and the 3-D finite-difference method [11]. In [12,13], the model enables consideration of both spatial and temporal harmonics using a resistance-limited magnetic potential vector to formulate the system resolution. In [14], the eddy-currents induced by the magnetic field variation are computed by solving the Maxwell's equations. The losses are calculated by using the integral volume accordingly with the eddy-current calculated previously. Also, eddy-currents could be considered as additional induced terms in the Ampere laws [15]. The armature reaction can be modeled by MMF sources [16], or by including hysteresis and eddy-current coefficients [17,18].

In [19,20], the eddy-currents are obtained by additional capacitors in the MEC. Coupling an EEC with a MEC or a 2-D solution of Maxwell's equations is also used in [21]. The authors conclude that the armature reaction does not contribute significantly to the increase of the PM eddy-current losses. The advantage of this method consists of the separation of the magnetic phenomenon from those of the EEC responsible for the magnetic field reaction. In [22], a magnetic inductance can also be incorporated in the MEC where the eddy-currents can be modeled by short single coil encircling iron elements with a resistance. In [23,24], an EEC taking into account to the magnetic field reaction has been considered. The local quantities can be derived from MEC or FEA. Using the EEC, the authors consider the eddy-currents by incorporating a magnetic inductance in a regular MEC. This was applied to a static electromagnetic device as well as a PMSM. A similar approach based on a coupling model between a MEC and an EEC is used in [25], where the 3-D MEC serves to compute the flux crossing perpendicularly the section of the massive conductive part and gives the induced voltage, while the EEC is used to compute the eddy-currents losses [26]. It is explained in [27] how to incorporate the eddy-current losses in 3-D FEA. It can also be found in [28] a 3-D approach method using a magnetic conductance. An eddy-current loss estimation made by using an EEC is detailed in [29] to modelize a PMSM, where the resolution of equation systems are done simultaneously with multi-slice 2-D FEA. A hybrid method for the eddy-current loss calculation was proposed in [30]. It combines a 2-D transient FEA to obtain the magnetic field distribution in PMs and a 2-D analytical model to determine resulting eddy-current losses. The FEA output is used as the data input of the analytical method. The results are validated by 3-D transient FEA computations and by experimental measurements.

1.2. Objectives of This Paper

The major drawbacks of the previously cited papers are linked to the high computational time and depend strongly on the FEA. A model coupling (or a hybrid model), combining an analytical model based on the Maxwell–Fourier method (i.e., the formal resolution of Maxwell's equations by using the separation of variables method and the Fourier's series) in massive conductive parts (e.g., aluminum, PMs, copper, iron) with a generic MEC, appeared as a promising solution [31]. MECs are largely used in modeling with a greater or lesser time depending on the fineness applied to the reluctances network. Analytical methods are also well-known for their short computation time. Therefore, unlike [30], it is necessary to combine these two models in order to find a good compromise between the computation time and the model accuracy. Hence, the scientific objective of this paper is to describe this type of hybrid model by validating it with numerical and experimental results.

The 2-D generic MEC determine the magnetic flux density distribution in massive conductive parts without the skin effect (i.e., without the eddy-current reaction field). The 2-D analytical model based on the Maxwell–Fourier method calculates the magnetic field distribution in massive conductive parts considering the skin effect as well as the resultant eddy-current density. The boundary conditions (BCs) imposed on the 2-D analytical model are equivalent to the magnetic field obtained from the MEC. Therefore, the 2-D analytical model will be applied across different layers in the y-axis of massive conductive parts. From local quantities with the skin effect, the 3-D eddy-current loss distribution in massive conductive parts can be observed. It is interesting to note that special attention should be paid to BCs of the 2-D analytical model. Frequently, only the middle component of the magnetic field (i.e., by assuming a uniform magnetic field) is taken to calculate the eddy-current losses [7,30]. In this work, the BC number influence is accounted for by applying three different limit conditions. Moreover, this paper contributes to the study of the discretization influence in order to reduce the computer time consumed for optimization of the design and takes into account to the eddy-current losses in thermal design in a more accurate manner.

This paper is organized as follows. First, Section 2 describes the U-cored static electromagnetic device used to validate the proposed approach with 3-D FEA and experimental results. Secondly, the model coupling is exposed by describing the 2-D generic MEC and the 2-D analytical model based on the Maxwell–Fourier method. The magnetic field distribution for various models is validated with 2-D and 3-D FEA [32]. The mathematic formulation as well as the experimental and 3-D numerical validation of eddy-current losses in massive conductive parts in aluminum are given in Section 3. The discretization influence is also discussed in the same section.

2. Model Coupling: 2-D Generic MEC/Maxwell–Fourier

2.1. U-Cored Static Electromagnetic Device

The 2-D view of the U-cored static electromagnetic device is shown in Figure 1. It is constituted of a mobile armature that allows the insertion of massive aluminum conductive parts of various thicknesses. Two coils having N_t series turns are connected in parallel. The electromagnetic device is supplied with a sinusoidal voltage. The magnetic circuit is not saturated with the voltage levels. Therefore, the current waveform is purely sinusoidal with a maximum amplitude of $I_{\max}$. The current direction in the conductor is defined by ⊗ for the forward conductor and ⊙ for return conductor. The U-cored static electromagnetic device as well as the experimental tests have been presented in [33]. The geometrical and physical parameters are detailed respectively in Tables 1 and 2.

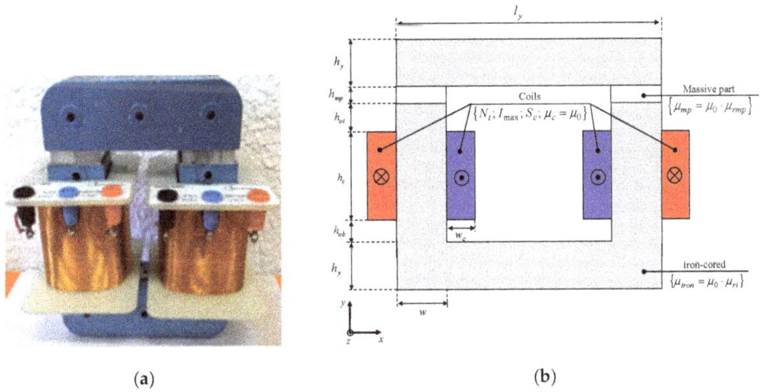

Figure 1. U-cored static electromagnetic device: (**a**) Experimental test [33], and (**b**) geometrical parameters (see Table 1 for the various parameters).

Table 1. Geometrical parameters.

Parameters, Symbols (Units)	Values
Depth, d (mm)	43
Width, w (mm)	43
Coil height and width, $\{h_c;\ w_c\}$ (mm)	{77; 10}
Coil section, $S_c = h_c \cdot w_c$ (mm^2)	770
Yoke height and length, $\{h_y;\ l_y\}$ (mm)	{43; 150}
Thickness of massive part, h_{mp} (mm)	6 or 10
Height of overhang top and low, $\{h_{ot}; h_{ob}\}$ (mm)	{19; 4}

Table 2. Physical parameters.

Parameters, Symbols (Units)	Values
Electrical frequency, f (Hz)	50
Maximal current, I_{max} (A)	0 to 8.2
Number of turns, N_t (-)	500
Relative permeability of massive parts in aluminum, μ_{rmp} (-)	1
Electrical conductivity of massive parts in aluminum, σ_{mp} (S/m)	38.46×10^6
Vacuum permeability, μ_0 (H/m)	$4\pi \times 10^{-7}$
Relative permeability of iron core, μ_{n} (-)	1500

2.2. Proposed Approach

The approach consists of combining two models, viz.: (i) a 2-D generic MEC, and (ii) a 2-D analytical model based on the Maxwell–Fourier method (i.e., the formal resolution of Maxwell's equations by using the separation of variables method and the Fourier's series). The 2-D generic MEC gives us the ability to determine the magnetic flux density distribution in massive conductive parts without the skin effect, while the 2-D analytical model provides the local quantities with the skin effect. By applying the Poynting vector, the eddy-current losses in massive conductive parts across a closed surface can be determined. Figure 2 shows the principle of model coupling.

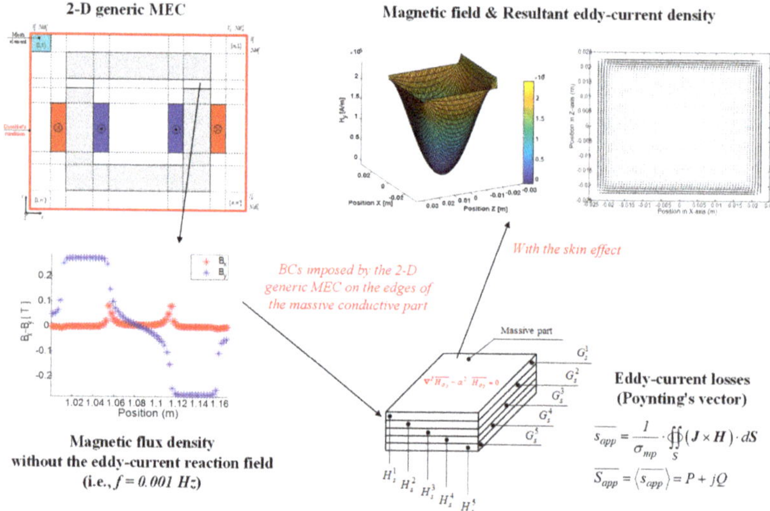

Figure 2. Principle of model coupling.

It can be noticed that the input data of the 2-D generic MEC are the geometrical and physical parameters (see Tables 1 and 2) as well as the discretization vectors in both directions (i.e., x- and y-axis). The output data of the 2-D generic MEC (i.e., the magnetic flux density without the skin effect) will be used as the input data of the 2-D analytical model, where the main variables are the skin depth of the massive conductive part and the spatial harmonics number.

2.3. 2-D Generic MEC

2.3.1. General Assumptions

The 2-D generic MEC is based on the following simplifying assumptions:

- The saturation and hysteresis effects are neglected;

- The end-effects in the z-axis are neglected (i.e., the semi-analytical is assumed to be in 2-D);
- The eddy-current effects in all materials (e.g., the massive parts, the copper, the iron) are neglected (i.e., the electrical conductivities are assumed to be null);
- The magnetic materials are considered as isotropic;
- The mechanical stress on the nonlinear *B(H)* curve is ignored;
- Since the magnetic circuit is not saturated, the magnetic permeability is supposedly constant, corresponding to the linear zone of the nonlinear *B(H)* curve.

2.3.2. Automatic Mesh

In an (x, y) coordinate system, Figure 3 represents the generalized discretization of a U-cored static electromagnetic device for the development of 2-D generic MEC [34,35]. The device is inserted in an infinite box whose outer edges respect the Dirichlet's conditions. It can be divided into $n = 9$ zones in the x-axis and $n' = 8$ zones in the y-axis. The intersection of these zones in both axes gives rise to mesh elements $\{j, i\}$, having the same magnetic permeability, of size $l_i^x \times l_j^y$ with $i = 1, \ldots, n$ and $j = 1, \ldots, n'$. So, the total number of mesh elements is equal to $n \times n' = 72$. The mesh elements $\{j, i\}$ can be discretized one or several bidirectional (BD) blocks from $\{Nd_j^y, Nd_i^x\}$ which are respectively the vectors (of dimension $n' \times 1$ and $n \times 1$) of discretization number in the y- and x-axis for the zone j and i (see Figure 3a).

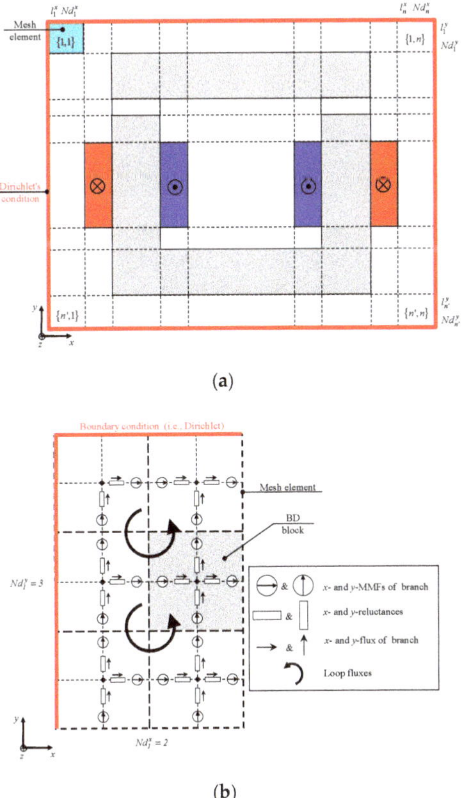

Figure 3. Generalized discretization of a U-cored static electromagnetic device: (**a**) Automatic mesh, and (**b**) discretization of a mesh element (e.g., for the mesh element $\{1, 1\}$).

The mesh elements are so composed of BD blocks depending on the discretization chosen by the designer. Figure 3b describes an example of discretization for the mesh element {1, 1} (see the mesh element in sky blue color in Figure 3a) with $Nd_1^x = 2$ and $Nd_1^y = 3$, where the number of BD blocks is equal to $Nd_1^x \times Nd_1^y = 6$. The BD blocks, connected between them by the loop fluxes ψ and giving to the magnetic flux the possibility to flow in both directions, are described by a middle-point related to (except for the outer edges of the device due to Dirichlet's conditions):

- 4 branch MMFs (i.e., two x-MMFs and two y-MMFs);
- and 4 magnetic reluctances (i.e., two x-reluctances and two y-reluctances) crossed by branch fluxes φ.

In general, the number of loop fluxes ψ is given by

$$N_\psi = N_\psi^x \cdot N_\psi^y, \tag{1a}$$

$$N_\psi^x = p - 1 \text{ with } p = \sum_{i=1}^{n} Nd_i^x, \tag{1b}$$

$$N_\psi^y = m - 1 \text{ with } m = \sum_{j=1}^{n'} Nd_j^y. \tag{1c}$$

The number of magnetic reluctances (or branch fluxes and MMFs) is defined by

$$N = N^x + N^y \tag{2a}$$

$$N^x = m \cdot P \text{ with } P = 2N_\psi^x, \tag{2b}$$

$$N^y = p \cdot M \text{ with } M = 2N_\psi^y. \tag{2c}$$

It is interesting to note that the number of BD blocks can be given by $N_{BD} = p \cdot m$.

One should notice that the accuracy and the computational time of 2-D generic MEC rise by increasing the number of BD blocks in each mesh element.

2.3.3. Matrix Formulation

Using the Maxwell's equations as well as the magnetic material equations, the 2-D generic MEC (where the loop fluxes ψ are the unknowns) can be governed by

$$[F] - [\chi] \cdot [\Re] \cdot [\chi]^T \cdot [\psi] = 0, \tag{3a}$$

$$[F] = [\chi] \cdot [MMF], \tag{3b}$$

in which

- $[\psi]$ is the loop fluxes vector (of dimension $N_\psi \times 1$);
- $[F]$ is the loop MMFs vector (of dimension $N_\psi \times 1$);
- $[MMF]$ is the branch MMFs vector (of dimension $N \times 1$) defined by

$$[MMF] = \begin{bmatrix} [MMF^x] \\ [MMF^y] \end{bmatrix}. \tag{4}$$

The branch MMFs vectors $[MMF^x]$ and $[MMF^y]$ in the x- and y-axis (of dimension $N^x \times 1$ and $N^y \times 1$) are given by

$$[MMF^{\bullet}] = \begin{bmatrix} MMF^{\bullet}_{\{1,1\}} \\ \vdots \\ MMF^{\bullet}_{\{1,n\}} \\ \vdots \\ MMF^{\bullet}_{\{n',1\}} \\ \vdots \\ MMF^{\bullet}_{\{n',n\}} \end{bmatrix}, \tag{5}$$

with

$$MMF^{x}_{\{j,i\}} = \begin{cases} [Z]_{(2\cdot Nd^x_i-1)\cdot Nd^y_j,1} \text{ for } i = \left| \begin{matrix} 1 \\ n \end{matrix} \right. \quad \forall j \\ otherwise\ [Z]_{2\cdot Nd^x_i\cdot Nd^y_j,1} \end{cases} \tag{6a}$$

$$MMF^{y}_{\{j,i\}} = \begin{cases} [Z]_{(2\cdot Nd^y_j-1)\cdot Nd^x_i,1} \text{ for } j = \left| \begin{matrix} 1 \\ n' \end{matrix} \right. \quad \forall i \\ f_{MMF}(N_t, I_{\max}) \cdot [O]_{2\cdot Nd^x_i\cdot Nd^y_j,1} \text{ for } \left| \begin{matrix} j = 5 \\ i = 2\,to\,4\ \wedge\ 6\,to\,8 \end{matrix} \right. \\ otherwise\ [Z]_{2\cdot Nd^x_i\cdot Nd^y_j,1} \end{cases} \tag{6b}$$

where $[Z]_{\bullet,*}$ is the zeros matrix of dimension $\bullet \times *$, $[O]_{\bullet,*}$ is the ones matrix of dimension $\bullet \times *$, and $f_{MMF}(N_t, I_{\max})$ is a MMF function explained in [35]. Figure 4a represents the waveform of this function at $t = 0\,\text{s}$ corresponding to $I_{\max}$. The MMF curve for a coil is defined by a trapezoidal waveform whose Ampere-turns change linearly from 0 to $N_t \cdot I_{\max}$ for the forward conductor ⊗ and from $N_t \cdot I_{\max}$ to 0 for the return conductor ⊙. Figure 4b describes an example of MMF values according to the discretization number for the mesh element with $Nd^y_5 = 2$ and $Nd^x_2 = 2$. All BD blocks in the y-axis have the same values of MMFs. The MMF values differ with the discretization number according to the MMF slope in the conductor.

• $[\Re]$ the diagonal matrix of magnetic reluctances (of dimension $N \times N$) defined by

$$[\Re] = \begin{bmatrix} [\Re^x] & 0 \\ 0 & [\Re^y] \end{bmatrix}, \tag{7}$$

The diagonal matrices of magnetic reluctances $[\Re^x]$ and $[\Re^y]$ in the x- and y-axis (of dimension $N^x \times N^x$ and $N^y \times N^y$) are given by

$$[\Re] = \begin{bmatrix} [\Re^x] & 0 \\ 0 & [\Re^y] \end{bmatrix}, \tag{8}$$

with

$$\left[\Re^x_{\{j,i\}}\right] = \Re^x_{j,i} \cdot \begin{cases} [O]_{Nd^y_j,2\cdot Nd^x_i-1} \text{ for } i = \left| \begin{matrix} 1 \\ n \end{matrix} \right. \quad \forall j \\ otherwise\ [O]_{Nd^y_j,2\cdot Nd^x_i} \end{cases} \tag{9a}$$

$$\left[\Re^y_{\{j,i\}}\right] = \Re^y_{j,i} \cdot \begin{cases} [O]_{2\cdot Nd^y_j-1,Nd^x_i} \text{ for } j = \left| \begin{matrix} 1 \\ n' \end{matrix} \right. \quad \forall i \\ otherwise\ [O]_{2\cdot Nd^y_j,Nd^x_i} \end{cases}, \tag{9b}$$

where

$$\Re^x_{j,i} = L^x_i / \left(\mu_{j,i} \cdot S^x_j\right), \tag{10a}$$

$$\Re^y_{j,i} = L^y_i / \left(\mu_{j,i} \cdot S^y_j\right), \tag{10b}$$

in which $\mu_{j,i}$ is the absolute magnetic permeability of mesh elements $\{j,i\}$ defined by

$$\mu_{j,i} = \mu_0 \cdot \begin{cases} \mu_{ri} \text{ for} \left| \begin{array}{l} \{2 \wedge 7, 3\, to\, 7\} \\ \{4\, to\, 6, 3 \wedge 7\} \end{array} \right. \\ \mu_{rmp} \text{ for } \{3, 3 \wedge 7\} \\ otherwise\ 1 \end{cases} . \tag{11}$$

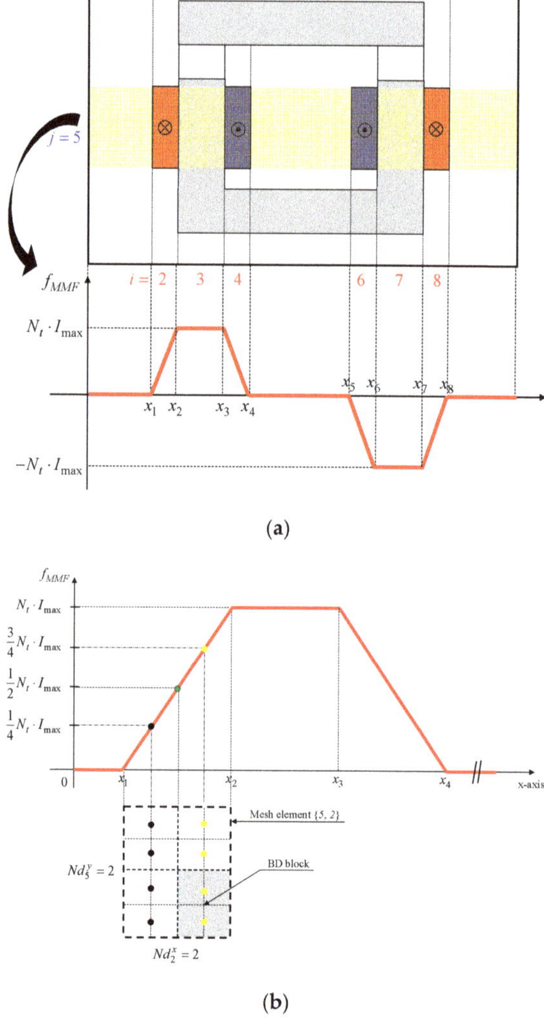

Figure 4. The magnetomotive force (MMF) function $f_{MMF}(N_t, I_{\max})$: (**a**) Waveform, and (**b**) MMF values according to the discretization number (e.g., for the mesh element {5, 2} with $Nd_5^y = 2$ and $Nd_2^x = 2$).

The lengths (viz., L_i^x and L_j^y) and sections (viz., S_j^x and S_i^y) of magnetic reluctances in the x- and y-axis are given in Table 3.

Table 3. Lengths and sections of magnetic reluctances.

	Length (m)	Section (m^2)
x-axis	$L_i^x = \frac{l_i^x}{2 \cdot Nd_i^x}$	$S_j^x = \frac{d \cdot l_j^y}{Nd_j^y}$
y-axis	$L_j^y = \frac{l_j^y}{2 \cdot Nd_j^y}$	$S_i^y = \frac{d \cdot l_i^x}{Nd_i^x}$

- $[\chi]$ is the topological (or incidence) matrix (of dimension $N_\psi \times N$) defined by

$$[\chi] = \begin{bmatrix} [\chi^x] & [\chi^y] \end{bmatrix}, \tag{12}$$

where $[\chi^x]$ and $[\chi^y]$ are respectively the topological matrices in the x- and y-axis (of dimension $N_\psi \times N^x$ and $N_\psi \times N^y$). The elements $[\chi]_{k,k'}$ are then equal to [36]:

$$[\chi]_{k,k'} = \begin{cases} \pm 1 & if\ \phi_{k'} \in \psi_k^{\pm} \\ 0 & if\ \phi_{k'} \notin \left(\psi_k^+ \cup \psi_k^-\right) \end{cases}, \tag{13}$$

with ψ_k^+—branch and loop fluxes have the same direction, and —branch and loop fluxes have opposite directions. Therefore, the topological matrices $[\chi^x]$ and $[\chi^y]$ are given by

$$[\chi^x] = [Y]_{N_\psi^y} \otimes \left([I]_{N_\psi^x, N_\psi^x} \otimes [O]_{1,2}\right), \tag{14a}$$

$$[\chi^y] = [I]_{N_\psi^y, N_\psi^y} \otimes \begin{bmatrix} [Y]_{N_\psi^x} & [Y]_{N_\psi^x} \end{bmatrix}, \tag{14b}$$

where $[I]_{*,*}$ is the identity matrix of dimension $* \times *$, $\otimes$ is the Kronecker's product, and

$$[Y]_* = \begin{array}{c} \begin{matrix} 1 & 2 & \cdots & \cdots & *+1 \end{matrix} \\ \begin{bmatrix} -1 & 1 & & & \\ & -1 & 1 & & \\ & & \ddots & \ddots & \\ & & & -1 & 1 \end{bmatrix} \end{array} \begin{matrix} 1 \\ 2 \\ \vdots \\ * \end{matrix}. \tag{15}$$

2.3.4. Problem Solving

To solve the Cramer's system (3), a numerical matrix inversion is required for the calculation of $[\psi]$, viz., $[\psi] = [A]^{-1}[F]$ with $[A] = [\chi] \cdot [\Re] \cdot [\chi]^T$. For a saturated system, it is interesting to note that (3) can be solved iteratively with a constant relative magnetic permeability μ_{ri} according to the nonlinear *B(H)* curve at each iteration by using the fixed-point iteration method. The flowchart of the nonlinear system solving is detailed in [37].

Knowing $[\psi]$, the branch fluxes vector $[\phi]$ (of dimension $N \times 1$) and the magnetic flux densities vector $[B]$ (of dimension $N \times 1$) are respectively defined by

$$[\phi] = [\chi]^T \cdot [\psi], \tag{16}$$

$$[B] = [\phi]/[S], \tag{17}$$

with the reluctances surface vector (of dimension $N \times 1$) in the various BD blocks given by

$$[S] = \begin{bmatrix} [S^x] \\ [S^y] \end{bmatrix} \tag{18}$$

The reluctances surface vectors $[S^x]$ and $[S^y]$ in the x- and y-axis (of dimension $N^x \times 1$ and $N^y \times 1$) are given by

$$[S^\bullet] = \begin{bmatrix} S^\bullet_{\{1,1\}} \\ \vdots \\ S^\bullet_{\{1,n\}} \\ \vdots \\ S^\bullet_{\{n',1\}} \\ \vdots \\ S^\bullet_{\{n',n\}} \end{bmatrix}, \tag{19}$$

with

$$S^x_{\{j,i\}} = S^x_{j,i} \cdot \begin{cases} [Z]_{(2 \cdot Nd^x_i - 1) \cdot Nd^y_j, 1} \text{ for } i = \left| \begin{matrix} 1 \\ n \end{matrix} \right. & \forall j \\ otherwise\ [Z]_{2 \cdot Nd^x_i \cdot Nd^y_j, 1} \end{cases} \tag{20a}$$

$$S^y_{\{j,i\}} = S^y_{j,i} \cdot \begin{cases} [Z]_{(2 \cdot Nd^y_j - 1) \cdot Nd^x_i, 1} \text{ for } j = \left| \begin{matrix} 1 \\ n' \end{matrix} \right. & \forall i \\ otherwise\ [Z]_{2 \cdot Nd^x_i \cdot Nd^y_j, 1} \end{cases} \tag{20b}$$

2.3.5. Comparing with 2-D FEA

The validation of 2-D generic MEC has been realized by Cedrat's Flux2D software package (i.e., an advanced FE method based numeric field analysis program) [32]. The parameters of a U-cored static electromagnetic device have been sent to a 2-D FEA pre-processor in the application *"Magneto Static 2-D"*. The 2-D FEA is done with the same assumptions as the 2-D generic MEC (see Section 2.3.1). It has been implemented in Matlab® by using the sparse matrix/vectors. The discretization in the x- and y-axis have been considered as follows

$$Nd^x = \begin{bmatrix} 1 & 4 & \boldsymbol{k} & 4 & 14 & 4 & \boldsymbol{k} & 4 & 1 \end{bmatrix}, \tag{21a}$$

$$Nd^y = \begin{bmatrix} 1 & 10 & \boldsymbol{k} & 5 & 6 & 6 & 10 & 1 \end{bmatrix}, \tag{21b}$$

where $\boldsymbol{k} = 2;\ 6;\ 10;\ 14;\ 18;\ 22;\ 24$ is the discretization number in massive conductive parts.

Consequently, for the high discretization (i.e., $\boldsymbol{k} = 24$), (3) is composed of N_{BD} = 5040 BD blocks, N_ψ = 4898 loop fluxes, and $N = 19,874$ branch fluxes, which is much smaller than the 2-D FEA mesh having 38,897 nodes, 2081 line elements, and 19,288 surface elements of the second order (viz., the triangles number of system). Figure 5 shows the consumption time for the 2-D generic MEC versus k. By using the high discretization (i.e., $\boldsymbol{k} = 24$) in the 2-D generic MEC, the computation time is the same for both modeling methods, viz., ~4 s.

The validation paths of $\boldsymbol{B} = \{B_x;\ B_y;\ 0\}$ for the comparison are given in Figure 6. The waveforms of B_x and B_y are represented on various paths in Figures 7–9 for $I_{\max}$ = 7.78A at $t = 0$ s and $h_{mp} = 6$ mm. The dotted lines represent the components of $\boldsymbol{B}$ calculated by the 2-D FEA and the circles correspond to 2-D generic MEC. It can be seen that a very good agreement is obtained for the components of $\boldsymbol{B}$ whatever the paths. Figures 7a and 9b confirm that the electromagnetic device is not saturated with a maximum level of $\boldsymbol{B}$ equal to 1 T. In Figure 8a, it is interesting to note that the level of B_x on the edges of massive conductive parts are not the same, which is due to the electromagnetic device structure. Indeed, the magnetic leakages are more important inside than outside. It should be noted that B_x in massive conductive parts is considered negligible in relation to B_y (see Figure 8). Moreover, Figure 10 presents a zoom of B_y in the left massive conductive part for the various paths (viz., Path_{mp1} to Path_{mp4}) between x_1 and x_4 (see Figure 6). Due to leakage fluxes, the levels of B_y are different at the edges and in the middle of the massive conductive part whatever the path in the y-axis. These various magnetic flux densities will be used as BCs in the 2-D analytical model based on the Maxwell–Fourier method, which calculate the magnetic field distribution considering the skin effect as well as the resultant eddy-current density. It is interesting to note that the path number depends on the discretization number in the y-axis, so the influence of the discretization number will be discussed in Section 3 in the eddy-current loss calculation.

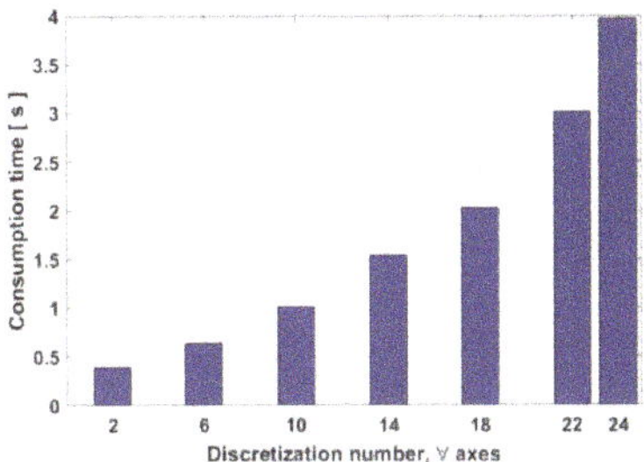

Figure 5. Consumption time for the 2-D generic magnetic equivalent circuit (MEC) according to k (viz., the discretization number in massive conductive parts).

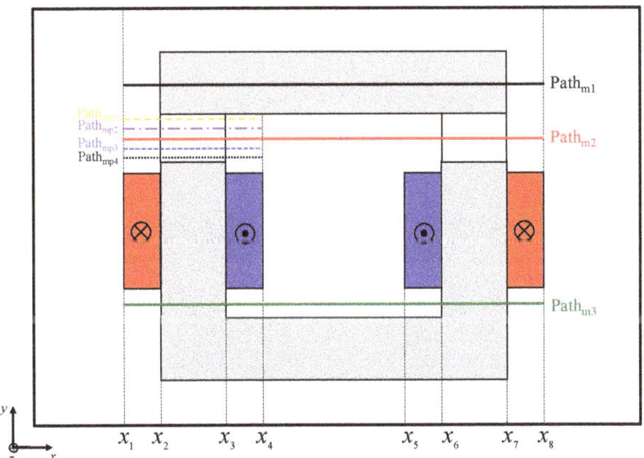

Figure 6. Paths of magnetic flux density validation for the comparison.

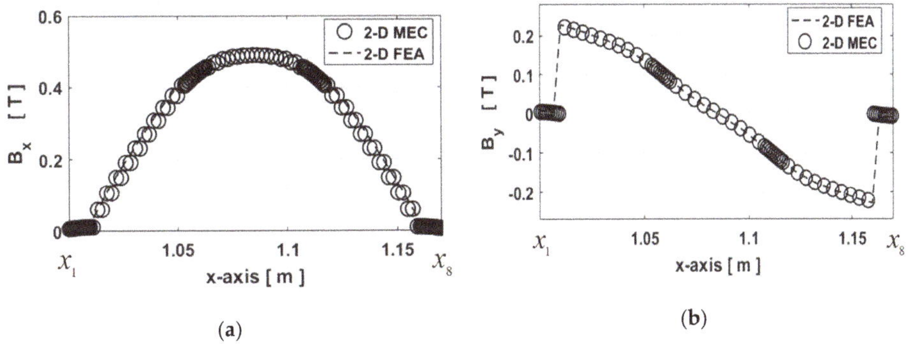

Figure 7. Waveform of B for Path$_{m1}$ with $I_{max} = 7.78$A at $t = 0$s and $h_{mp} = 6$mm: (**a**) x- and (**b**) y-component.

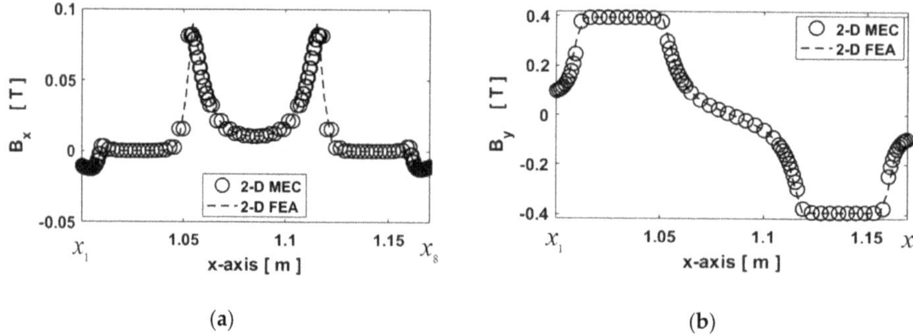

(a) (b)

Figure 8. Waveform of $\boldsymbol{B}$ for Path_{m2} with $I_{\max} = 7.78\,A$ at $t = 0\,\text{s}$ and $h_{mp} = 6\,\text{mm}$: (**a**) x- and (**b**) y-component.

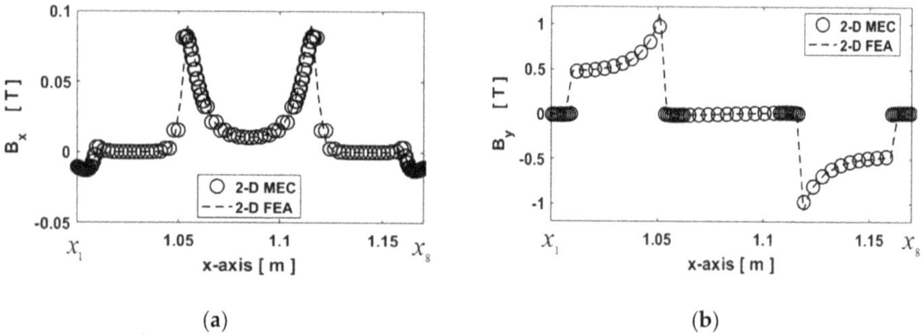

(a) (b)

Figure 9. Waveform of $\boldsymbol{B}$ for Path_{m3} with $I_{\max} = 7.78\,A$ at $t = 0\,\text{s}$ and $h_{mp} = 6\,\text{mm}$: (**a**) x- and (**b**) y-component.

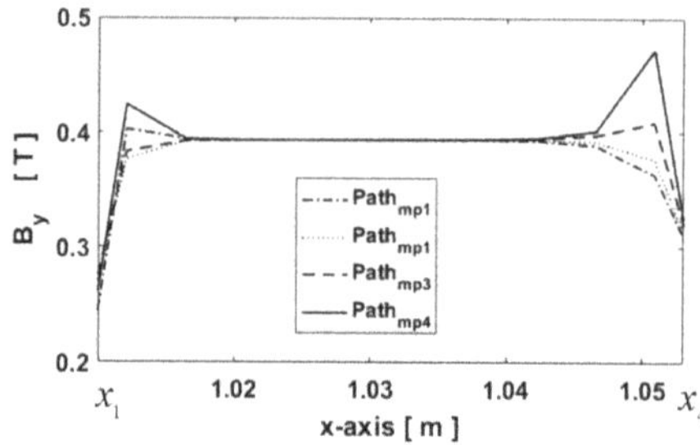

Figure 10. The y-component of $\boldsymbol{B}$ in the left massive conductive part (between x_1 and x_4) for various paths with $I_{\max} = 7.78\,A$ at $t = 0\,\text{s}$ and $h_{mp} = 6\,\text{mm}$.

2.4. 2-D Maxwell–Fourier

2.4.1. General Assumptions

The 2-D analytical model based on the Maxwell–Fourier method is defined by the following simplifying assumptions:

- The massive conductive parts are excited by the magnetostatic magnetic field from the 2-D generic MEC which is assumed normal to the xz-plane;
- Since the magnetic circuit is not saturated (see Figures 7a and 9b), the excitation magnetic field varies sinusoidally in time which is similar to the power supply source;

- The resultant eddy-current density in massive conductive parts has two components, i.e., $\boldsymbol{J}=\{J_x; 0; J_z\}$;
- The relative magnetic permeability and electrical conductivity of massive conductive parts (i.e., μ_{mp} and σ_{mp}) are assumed to be constant.

It is interesting to note that the 2-D analytical model, for calculating the magnetic field distribution with the skin effect as well as the resultant eddy-current density, will be applied across different layers in the y-axis of massive conductive parts (viz., the mesh elements {3,3} and {3,7} in Figure 3a). These different layers depend on the discretization Nd_3^y of BD blocks in massive conductive parts.

2.4.2. Governing Partial Differential Equations (PDEs) in Cartesian Coordinates

By assuming that the term $\partial \boldsymbol{D}/\partial t$ (with $\boldsymbol{D}$ as the displacement field vector) is negligible in comparison with the resultant eddy-current density $\boldsymbol{J}$, the Maxwell's equations are represented by

$$\nabla \times \boldsymbol{H} = \boldsymbol{J} \cdot \text{implying } \nabla \bullet \boldsymbol{J} = 0 \text{ (Maxwell – Ampère)}, \tag{22a}$$

$$\nabla \times \boldsymbol{E} = -\partial \boldsymbol{B}/\partial t \text{ (Maxwell – Faraday)}, \tag{22b}$$

$$\nabla \bullet \boldsymbol{B} = 0 \text{ (Maxwell – Thomson)}, \tag{22c}$$

where $\boldsymbol{E}$ is the electrical field vector.

In a conductor, $\boldsymbol{E}$ is linked to $\boldsymbol{J}$ by

$$\boldsymbol{J} = \sigma \cdot \boldsymbol{E} \text{ (Ohm's law)}, \tag{23}$$

where σ is the electrical conductivity.

The field vectors $\boldsymbol{B}$ and $\boldsymbol{H}$ are coupled by

$$\boldsymbol{B} = \mu \cdot \boldsymbol{H} + \mu_0 \cdot \boldsymbol{M}_r \text{ (Magnetic material equation)}, \tag{24}$$

where $\boldsymbol{M}_r$ is the remnant magnetization vector (with $\boldsymbol{M}_r \neq 0$ for the PMs or $\boldsymbol{M}_r = 0$ for the other materials).

Inside a linear magnetic or nonmagnetic material of constant electrical conductivity without electromagnetic sources (i.e., $\boldsymbol{M}_r = 0$), the magnetodynamic PDEs in terms of $\boldsymbol{H}$ can be defined by

$$\nabla^2 \boldsymbol{H} - \mu \cdot \sigma \cdot \frac{\partial \boldsymbol{H}}{\partial t} = 0 \text{ (Diffusion equation)}. \tag{25}$$

From the general assumptions, and using the complex notation, the magnetic field $\boldsymbol{H} = \left\{0; H_{\sigma y}; 0\right\}$ inside the massive conductive part considering the skin effect can be written as

$$H_{\sigma y} = \Re\left\{\overline{H_{\sigma y}} \cdot e^{j \cdot \omega \cdot t}\right\} \tag{26}$$

where $j = \sqrt{-1}$ and $\omega = 2\pi \cdot f$ is the electrical pulse.

Therefore, (25) becomes

$$\nabla^2 \overline{H_{\sigma y}} - \alpha^2 \cdot \overline{H_{\sigma y}} = 0 \text{ with } \alpha^2 = j \cdot \mu_{mp} \cdot \sigma_{mp} \cdot \omega, \tag{27}$$

which is the complex Helmholtz's equation.

In (x, z) coordinate system, the distribution of the magnetic field inside the massive conductive part considering the skin effect is then governed by

$$\frac{\partial^2 \overline{H_{\sigma y}}}{\partial x^2} + \frac{\partial^2 \overline{H_{\sigma y}}}{\partial z^2} - \alpha^2 \cdot \overline{H_{\sigma y}} = 0 \tag{28}$$

2.4.3. Definition of BCs

The BCs at the edges of massive conductive parts for the 2-D analytical model are equivalent to the magnetostatic magnetic fields of 2-D generic MEC (see Section 2.3). Usually, BCs are considered homogeneous at the edges, which are often equal to the excitation magnetic field value in the middle of the massive conductive part as [7,30]. From the 2-D generic MEC simulations, the magnetic field levels are different at the edges and in the middle of the massive conductive part, whatever the path in the y-axis (see Figure 10). Hence, BCs in the 2-D analytical model are considered as non-homogeneous. Figure 11 represents the BCs at the edges of massive conductive parts in a (x, z) coordinate system and $\forall l$, where M_s^l, L_s^l, and R_s^l are respectively the magnetostatic magnetic field values in the middle, at the left edge, and at the right edge of massive conductive parts. The index $l = 1, \ldots, 2 \cdot Nd_3^y$ is the path in the y-axis (or the parallel path in the x-axis). Figure 12 shows the value locations of M_s^l, L_s^l, and R_s^l in the massive conductive part from the 2-D generic MEC in a (x, y) coordinate system.

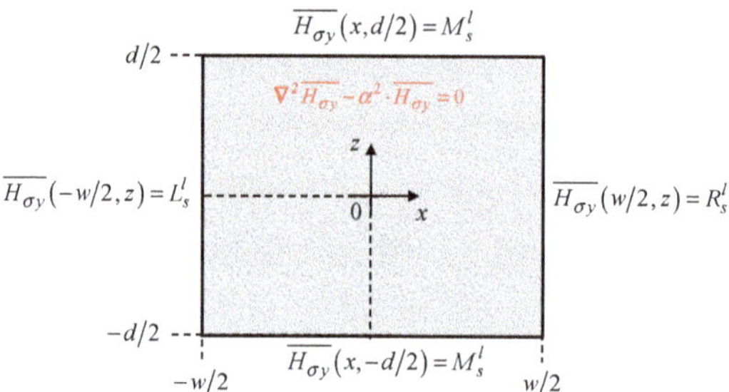

Figure 11. Boundary conditions (BCs) at edges of massive conductive parts in a (x, z) coordinate system and $\forall l$.

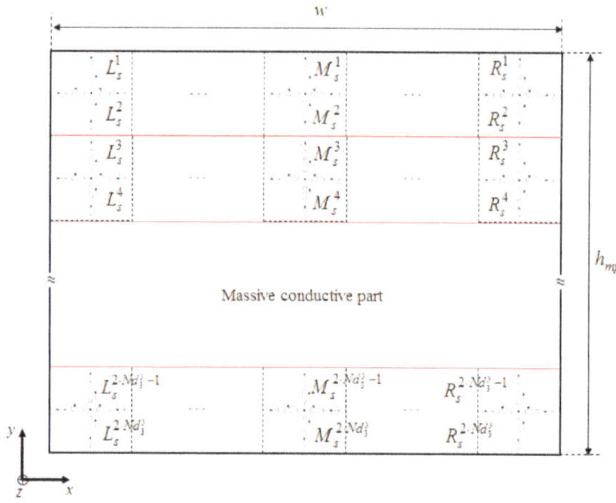

Figure 12. The values location of M_s^l, L_s^l, and R_s^l in the massive conductive part from the 2-D generic MEC in a (x, y) coordinate system.

2.4.4. General Solution of the Magnetic Field

Using the separation of variables method, the 2-D general solution of $H_{\sigma y}$ in both directions (i.e., x- and y-edges) can be written as a Fourier's series, $\forall l$,

$$\overline{H_{\sigma y}} = \sum_{h=0}^{\infty} \begin{bmatrix} \overline{c_{xh}^l} \cdot \mathrm{ch}(\chi_h \cdot z) \\ \cdots + \overline{d_{xh}^l} \cdot \mathrm{sh}(\chi_h \cdot z) \end{bmatrix} \cdot \begin{bmatrix} \overline{e_{xh}^l} \cdot \cos(\beta_h \cdot x) \\ \cdots + \overline{f_{xh}^l} \cdot \sin(\beta_h \cdot x) \end{bmatrix} + \sum_{k=0}^{\infty} \begin{bmatrix} \overline{c_{zk}^l} \cdot \cos(\lambda_k \cdot z) \\ \cdots + \overline{d_{zk}^l} \cdot \sin(\lambda_k \cdot z) \end{bmatrix} \cdot \begin{bmatrix} \overline{e_{zk}^l} \cdot \mathrm{ch}(\delta_k \cdot x) \\ \cdots + \overline{f_{zk}^l} \cdot \mathrm{sh}(\delta_k \cdot x) \end{bmatrix}, \quad (29a)$$

where $\overline{c_{xh}^{l}} \sim \overline{f_{xh}^{l}}$ and $\overline{c_{zk}^{l}} \sim \overline{f_{zk}^{l}}$ are the integration constants, β_h and λ_k are the periodicity of $\overline{H_{\sigma y}}$ in the x- and z-axis, h and k are the spatial harmonic orders, and

$$\chi_h = \sqrt{\alpha^2 + \beta_h{}^2}, \tag{29b}$$

$$\delta_k = \sqrt{\alpha^2 + \lambda_k{}^2}. \tag{29c}$$

The coefficients $\overline{c_{xh}^{l}} \sim \overline{f_{xh}^{l}}$ and $\overline{c_{zk}^{l}} \sim \overline{f_{zk}^{l}}$ are determined by applying the BCs illustrated in Figure 9. Therefore, (29a) becomes, $\forall l$,

$$\overline{H_{\sigma y}} = H_s^l \cdot \left\{ \frac{\mathrm{ch}(\alpha \cdot z)}{\mathrm{ch}\left(\alpha \cdot \frac{d}{2}\right)} + \sum_{k=1,3,\ldots}^{\infty} \left[\overline{e_{zk}^{l}} \cdot \frac{\mathrm{ch}(\delta_k \cdot x)}{\mathrm{ch}\left(\delta_k \cdot \frac{w}{2}\right)} + \overline{f_{zk}^{l}} \cdot \frac{\mathrm{sh}(\delta_k \cdot x)}{\mathrm{sh}\left(\delta_k \cdot \frac{w}{2}\right)} \right] \cdot \sin\mathrm{c}\left(\lambda_k \cdot \frac{d}{2}\right) \cdot \cos(\lambda_k \cdot z) \right\}, \tag{30a}$$

$$\overline{e_{zk}^{l}} = \left[\frac{R_s^l + L_s^l}{M_s^l} - 2\left(\frac{\lambda_k}{\delta_k}\right)^2 \right], \tag{30b}$$

$$\overline{f_{zk}^{l}} = \frac{R_s^l - L_s^l}{M_s^l}, \tag{30c}$$

with $\lambda_k = k\pi/d$.

It should be noted that if $M_s^l = L_s^l = R_s^l$ then (30) is identical to the relation provided in [7]. Moreover, when $\alpha = 0$ (viz., σ_{mp} = 0 S/m and/or $f \cong 0^+$ Hz) the field distribution is equivalent to the excitation magnetic field.

2.4.5. Resultant Eddy-Current Density

From the general assumptions, and using the complex notation, the components of resultant eddy-current $\boldsymbol{J} = \{J_x; 0; J_z\}$ in massive conductive parts can be written as

$$J_x = \Re\left\{\overline{J_x} \cdot e^{j \cdot \omega \cdot t}\right\}. \tag{31a}$$

$$J_z = \Re\left\{\overline{J_z} \cdot e^{j \cdot \omega \cdot t}\right\}. \tag{31b}$$

Using $\boldsymbol{J} = \nabla \times \boldsymbol{H}$, the complex components of $\boldsymbol{J}$ in Cartesian coordinates (x,z) can be deduced by

$$\overline{J_x} = \frac{\partial \overline{H_{\sigma y}}}{\partial z}, \tag{32a}$$

$$\overline{J_z} = \frac{\partial \overline{H_{\sigma y}}}{\partial x}, \tag{32b}$$

which leads to

$$\overline{J_x} = H_s^l \cdot \left\{ -\alpha \cdot \frac{\mathrm{sh}(\alpha \cdot z)}{\mathrm{ch}\left(\alpha \cdot \frac{d}{2}\right)} + \sum_{k=1,3,\ldots} \lambda_k \cdot \left[\overline{e_{zk}^{l}} \cdot \frac{\mathrm{ch}(\delta_k \cdot x)}{\mathrm{ch}\left(\delta_k \cdot \frac{w}{2}\right)} + \overline{f_{zk}^{l}} \cdot \frac{\mathrm{sh}(\delta_k \cdot x)}{\mathrm{sh}\left(\delta_k \cdot \frac{w}{2}\right)} \right] \cdot \sin\mathrm{c}\left(\lambda_k \cdot \frac{d}{2}\right) \cdot \sin(\lambda_k \cdot z) \right\} \tag{33a}$$

$$\overline{J_z} = H_s^l \cdot \sum_{k=1,3,\ldots} \delta_k \cdot \left[\overline{e_{zk}^{l}} \cdot \frac{\mathrm{sh}(\delta_k \cdot x)}{\mathrm{ch}\left(\delta_k \cdot \frac{w}{2}\right)} + \overline{f_{zk}^{l}} \cdot \frac{\mathrm{ch}(\delta_k \cdot x)}{\mathrm{sh}\left(\delta_k \cdot \frac{w}{2}\right)} \right] \cdot \sin\mathrm{c}\left(\lambda_k \cdot \frac{d}{2}\right) \cdot \cos(\lambda_k \cdot z) \tag{33b}$$

2.4.6. Comparing with 3-D FEA

The validation of the 2-D analytical model based on the Maxwell–Fourier method has been realized using Cedrat's Flux3D software package by using the application *"Harmonic State 3-D"* [32].

The analytical solution of $H_{\sigma y}$ and $\boldsymbol{J} = \{J_x; 0; J_z\}$ have been computed with a finite number of spatial harmonic term $2 \cdot K_{\max} - 1 = 241$. The 3-D FEA mesh consists of 32,169 surface and 93,031 volume elements of second order. Figure 13 shows the consumption time for the hybrid model versus **k** for only one operating point. By using the high discretization (i.e., $\boldsymbol{k}$ = 24) in the 2-D generic MEC for the BCs of the 2-D analytical model, the computation time for the hybrid model is greatly reduced as short as 6.5 s, whereas the 3-D FEA requires as much as 30 s. The proposed approach can thus reduce the computation time by approximately 4.6-fold compared to 3-D FEA.

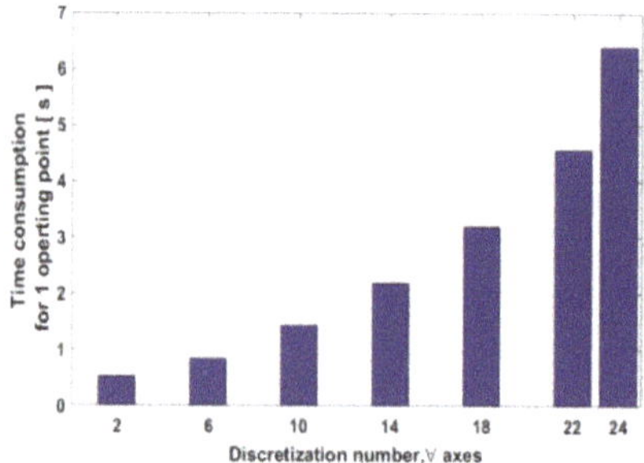

Figure 13. Consumption time for the hybrid model according to $\boldsymbol{k}$ (viz., the discretization number in massive conductive parts) for only one operating point.

Figure 14 shows the magnetic field distribution considering the skin effect on a 2-D grid parallel to the xz-plane located in the middle of the massive conductive part in the y-axis. The waveforms of $H_{\sigma y}$ have been calculated with $I_{\max}$ = 7.78A at t = 0 s and h_{mp} = 6 mm for two values of electrical frequency (viz., 50 Hz and 1600 Hz). Also for the same conditions, the evolution of the resultant eddy-current density is given in Figure 15. There is a very good agreement of the results given by analytic and numeric calculation. The electrical frequency effect on the behavior of $H_{\sigma y}$ and $\boldsymbol{J}$ can be clearly seen. In these figures, it can be seen that the skin effect appears slightly at 50 Hz, contrary to 1600 Hz where the massive conductive part act as a barrier to the crossing flux. The error order is less than 16% for $H_{\sigma y}$ (viz., 12% at 50 Hz and 16% at 1600 Hz) and less than 4% for $\boldsymbol{J}$ (viz., 1% at 50 Hz and 4% for 1600 Hz).

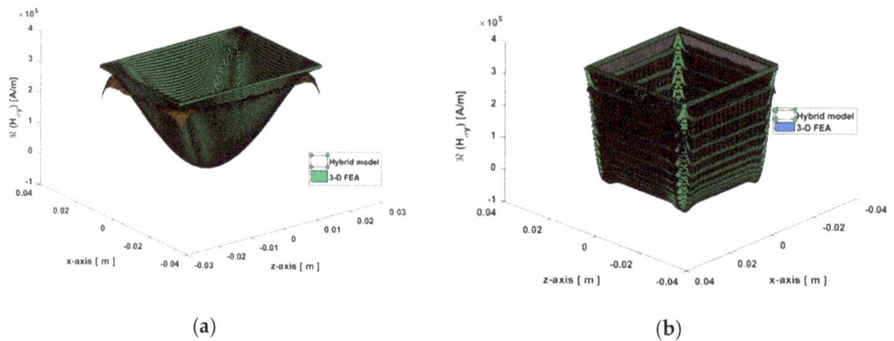

Figure 14. Evolution of $H_{\sigma y}$ in the massive conductive part with $I_{\max} = 7.78\,A$ at $t = 0$ s and $h_{mp} = 6$ mm for: (**a**) $f = 50$ Hz, and (**b**) $f = 1600$ Hz.

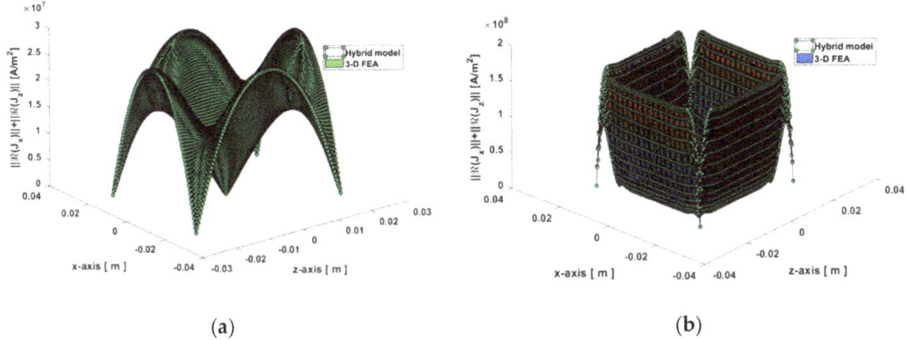

(a) (b)

Figure 15. Evolution of J in the massive conductive part with $I_{\max} = 7.78\,A$ at $t = 0$ s and $h_{mp} = 6$ mm for: (**a**) $f = 50$ Hz, and (**b**) $f = 1600$ Hz.

3. 3-D Eddy-Current Loss Calculation

3.1. Mathematic Formulation

The instantaneous density of power flow $\prod$ at a point is defined by the Poynting vector [7]

$$\prod = \boldsymbol{E} \times \boldsymbol{H}. \tag{34}$$

Using $\boldsymbol{J} = \sigma \cdot \boldsymbol{E} = \nabla \times \boldsymbol{H}$, this power across a closed surface, in terms de complex vectors, is given by the instantaneous apparent power

$$s_{app} = \oiint_S \prod \cdot d\boldsymbol{S} = \frac{1}{\sigma_{mp}} \cdot \oiint_S (\boldsymbol{J} \times \boldsymbol{H}) \cdot d\boldsymbol{S} = p + \boldsymbol{j} \cdot q \tag{35}$$

The real part of the surface integral of the complex Poynting vector gives the instantaneous ohmic losses, and the imaginary part the instantaneous magnetic energy.

From BCs at the edges of massive conductive parts (see Figure 10a), the average of $\overline{s_{app}}$ over an electrical cycle $T = 2\pi/\omega$ can be defined by, $\forall l$,

$$\overline{S_{app}} = \left\langle \overline{s_{app}} \right\rangle = \frac{h^l}{2\sigma_{mp}} \cdot \left\{ \int_{-w/2}^{w/2} 2 \cdot \overline{J_x} \cdot \overline{H_{\sigma y}}^* \Big|_{z=-d/2} \cdot dx - \int_{-d/2}^{d/2} \left(\overline{J_z} \cdot \overline{H_{\sigma y}}^* \Big|_{x=-w/2} - \overline{J_z} \cdot \overline{H_{\sigma y}}^* \Big|_{x=w/2} \right) \cdot dz \right\} \tag{36}$$

where $h^l = h_{mp} / \left(2 \cdot Nd_3^y\right)$ is the layers thickness in the y-axis of the massive conductive part.

After the development, by substituting (30) and (33a) into (36), the average density of power flow is then given by, $\forall l$,

$$\overline{S_{app}} = P^l + \boldsymbol{j} \cdot Q^l = \frac{h^l \cdot \left(H_s^l\right)^2}{\sigma_{mp}} \cdot \left\{ w \cdot \alpha \cdot \frac{\mathrm{sh}\left(\alpha \cdot \frac{d}{2}\right)}{\mathrm{ch}\left(\alpha \cdot \frac{d}{2}\right)} + \sum_{k=1,3,\ldots} \frac{2\delta_k}{d \cdot (\lambda_k)^2} \cdot \left[\begin{array}{c} \left(\overline{e_{zk}^l}\right)^2 \cdot \frac{\mathrm{sh}\left(\delta_k \cdot \frac{w}{2}\right)}{\mathrm{ch}\left(\delta_k \cdot \frac{w}{2}\right)} \\ \cdots + \left(\overline{f_{zk}^l}\right)^2 \cdot \frac{\mathrm{ch}\left(\delta_k \cdot \frac{w}{2}\right)}{\mathrm{sh}\left(\delta_k \cdot \frac{w}{2}\right)} \end{array} \right] \right\} \tag{37}$$

It should be noted that $P = \sum_i P^l$ only represents the 3-D eddy-current losses in massive conductive parts.

3.2. Experimental and Numerical Validations

In what follow, the analytical results are obtained by applying only one BC (viz., $M_s^l = L_s^l = R_s^l$) and also by applying three different BCs. The discretization impact of 2-D generic MEC in massive conductive parts is also discussed. The 3-D eddy-current loss results given by the model coupling are compared with those obtained by 3-D FEA and experimental tests. The experimental and numerical validations were performed at $f = 50$ Hz for two thicknesses in aluminum, viz., 6 mm and 10 mm.

3.2.1. Experimental Acquisition [33]

The eddy-current losses are calculated by using the separation of losses method. Firstly, the active power of the U-cored static electromagnetic device without the massive conductive parts is measured, then the active power after the insertion of massive conductive parts is measured. The difference between these two active powers gives the eddy-current losses created by the sinusoidal variation of the magnetic field in massive conductive parts.

3.2.2. Validation of Model Coupling with $M_s^l = L_s^l = R_s^l$

Figure 16 represent the evolution of P according to I_{max} when only the medium BC is applied over the edges of the massive conductive part (i.e., $M_s^l = L_s^l = R_s^l$) for $h_{mp} = 6\,mm$ and $h_{mp} = 10\,mm$. The analytical results give a good agreement with 3-D FEA and experimental results.

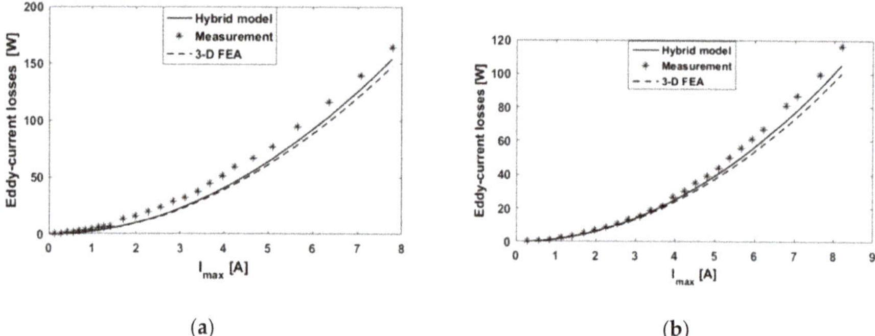

Figure 16. Validation of eddy-current losses (analytical, numerical, and experimental) in massive conductive parts versus $I_{\max}$ with $M_s^l = L_s^l = R_s^l$ for: (**a**) $h_{mp} = 6$ mm and (**b**) $h_{mp} = 10$ mm.

The difference between the analytical and experimental results can be linked to, $\forall h_{mp}$: (i) the experimental method, (ii) the electrical conductivity variation due to the temperature rise, and (iii) non-homogeneous BCs in the 2-D analytical model. For the experimental method, the use of analogue measurement instruments can affect the values obtained. In the 2-D analytical model developed, the electrical conductivity is assumed constant and invariant according to the temperature. In reality, the temperature variation influences the electrical conductivity values, and therefore the eddy-current losses in the massive conductive part due to the eddy-current reaction field. Then, it is interesting to note that the development of a magneto-thermal model would improve the error between the analytical and experimental results. Non-homogeneous BCs at the edges of the massive conductive part related to magnetic leakages (see Figure 10) affect to the distribution of the magnetic field $H_{\sigma y}$ inside the massive conductive part, and therefore the eddy-current losses. However, in [38], a 3-D generic MEC considering the skin effect would improve the volumic eddy-current loss calculation and observe the magnetic reaction field influence of the massive conductive parts on the magnetic circuit of the U-cored static electromagnetic device.

3.2.3. Validation of Model Coupling with $M_s^l \neq L_s^l \neq R_s^l$

To study the BCs influence, three different BCs at the edges of massive conductive parts are applied (i.e., $M_s^l \neq L_s^l \neq R_s^l$) (see Figure 11), and the results comparison are given in Figure 17. The results show a good agreement with experimental results compared to those obtained with $M_s^l = L_s^l = R_s^l$. The computation time is still acceptable even in high discretization.

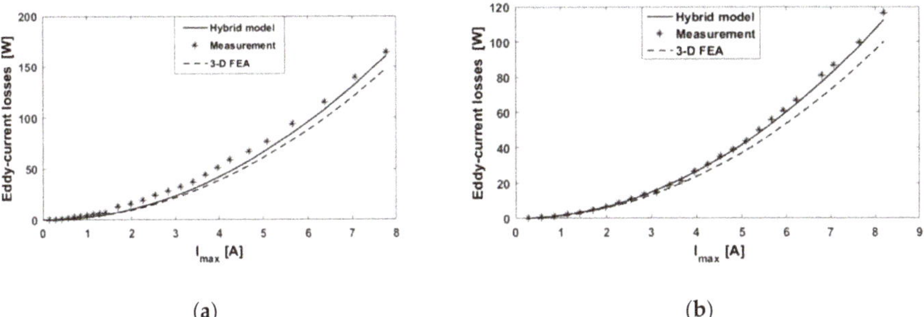

Figure 17. Validation of eddy-current losses (analytical, numerical, and experimental) in massive conductive parts versus $I_{\max}$ with $M_s^l \neq L_s^l \neq R_s^l$ for: (**a**) $h_{mp} = 6$ mm and (**b**) $h_{mp} = 10$ mm.

Figure 18 shows the normalized root-mean-square deviation (NRMSD) according to the discretization number in both axes (i.e., x- and y-axis). The NRMSD formulation, for all studied currents (see Table 2), is defined by

$$NRMSD = \frac{RMSD}{P_N^{meas} - P_1^{meas}} \text{ with } RMSD = \sqrt{\frac{\sum_{i=1}^{N} \left(P_i^{meas} - P_i^{anal}\right)^2}{N}} \tag{38}$$

where N is the total number of currents used in experimental measures, P_i^{meas} and P_i^{anal} are respectively the eddy-current losses obtained analytically and experimentally for the ith current. It can be remarked that NRMSD decreases by increasing the discretization number, $\forall h_{mp}$. With the discretization used in (21), NRMSD is equal to 4.5% for 6 mm and 3% for 10 mm.

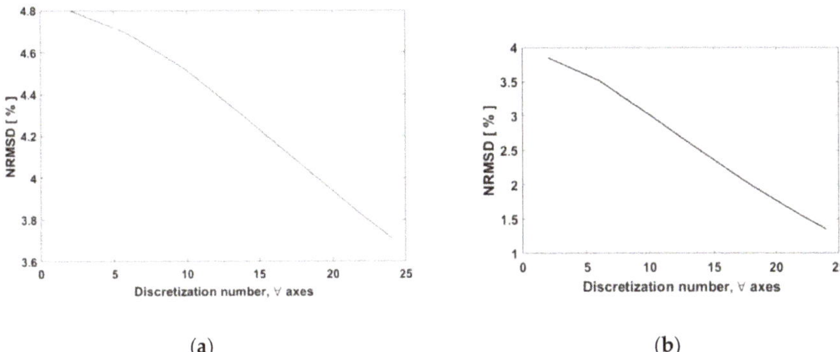

Figure 18. NRMSD according to the discretization number in both axes (i.e., x- and y-axis) for: (**a**) $h_{mp} = 6$ mm and (**b**) $h_{mp} = 10$ mm.

The eddy-current losses were calculated on different layers in the y-axis of massive conductive parts. Figure 19 shows the evolution of P^l at the top, middle, and bottom of the massive conductive part for h_{mp} = 6 mm with a discretization Nd_3^y. It is interesting to note that the eddy-current losses present a non-uniform distribution depending on the height of the massive conductive part. This can lead to a non-uniform temperature distribution on the massive part. According to the magnetic field distribution in the y-axis (see Figure 10), the level of eddy-current losses is higher at the bottom of the massive conductive part. This is due to the magnetic flux density which presents a high level compared to the medium layer.

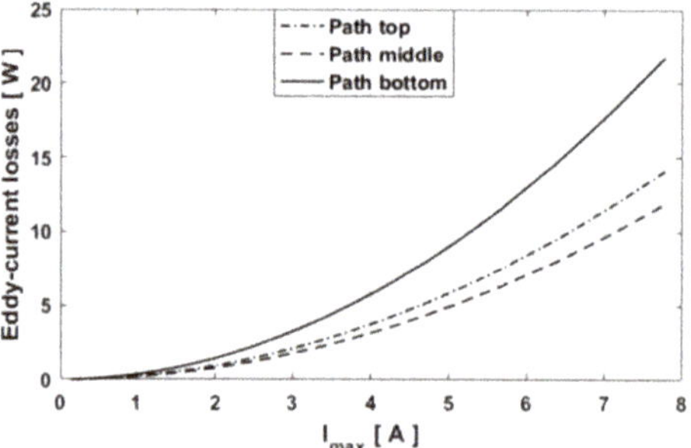

Figure 19. Evolution of the eddy-current losses in the massive conductive part for h_{mp} = 6 mm according to the layer in the y-axis.

4. Conclusions

In this work, a 3-D eddy-current losses model is developed, combining a 2-D generic MEC with a 2-D analytical model based on the Maxwell–Fourier method (i.e., the formal resolution of Maxwell's equations by using the separation of variables method and the Fourier's series). This model coupling has been applied to a U-cored static electromagnetic device [33]. The 2-D generic MEC determines the magnetic flux density distribution in massive conductive parts without the skin effect (i.e., without the eddy-current reaction field). The 2-D analytical model based on the Maxwell–Fourier method calculates the magnetic field distribution in massive conductive parts considering the skin effect as well as the resultant eddy-current density. BCs imposed on the 2-D analytical model are equivalent to the magnetic field of the MEC. The magnetic field distribution for various models is validated with 2-D and 3-D FEA. Experimental tests and 3-D FEA have been compared with the proposed approach for massive conductive parts in aluminum. For an operating point, the computation time is divided by 4.6 with respect to 3-D FEA. The study of the homogenous and non-homogenous BCs on the edges of massive conductive parts has been analyzed. Moreover, according to the 2-D MEC discretization, the model coupling is able to give more or less accurately the behavior of the magnetic field and eddy-current distribution in the massive conductive part.

The magnetic flux density repartition in different paths parallel to the x-axis present different variations. Consequently, the eddy-current losses present a non-uniform distribution over the massive conductive part. This can lead to predicting the temperature distribution over the conducting region while designing thermal components of the electromagnetic devices.

Furthermore, one advantage of this coupling model would be its exploitation in studies of PMSMs with(out) circumferential and/or axial PMs segmentation in order to reduce the computation time, which remains a major problem in this numerical method.

Author Contributions: Conceptualization, F.D.; Formal analysis, Y.B.; Methodology, F.D.; Project administration, F.D.; Supervision, F.D. and M.H.; Writing—original draft, Y.B.; Writing—review & editing, F.D.

Funding: This research received no external funding.

Acknowledgments: The authors would like to thank Boespflug, J., Bidoire, A., and Chetangny, P.K. for the experimental tests on the U-cored static electromagnetic device. This work was supported by RENAULT-SAS, Guyancourt, France. This scientific study is related to the project "Conception optimale des chaines de Traction

Electrique" (COCTEL) financed by the "Agence De l'Environnement et de la Maîtrise de l'Énergie" (ADEME) in the program "Véhicule du future des Investissements de l'Avenir".

Conflicts of Interest: The authors declare no conflict of interest.

References

1. Combe, M. La Chine, un quasi-monopole sur la production de terres rares. 24 January 2018. Available online: https://www.techniques-ingenieur.fr/actualite/articles/chine-monopole-production-terres-rares-51380/ (accessed on 4 June 2019).
2. Jang, J.; Ha, J.; Ohto, M.; Ide, K.; Sul, S. Analysis of permanent-magnet machine for sensorless control based on high-frequency signal injection. *IEEE Trans. Ind. Appl.* **2004**, *40*, 595–1604. [CrossRef]
3. Benlamine, R.; Dubas, F.; Esapnet, C.; Randi, S.A.; Lhotellier, D. Design of an axial-flux interior permanent-magnet synchronous motor for automotive application: performance comparison with electric motors used in EVS and HEVS. In Proceedings of the 2014 IEEE Vehicle Power and Propulsion Conference (VPPC), Coimbra, Portugal, 27–30 October 2014. [CrossRef]
4. Dubas, F.; Espanet, C.; Miraoui, A. Field diffusion equation in high-speed surface-mounted permanent magnet motors, parasitic eddy-current losses. In Proceedings of the 6th International Symposium on Advanced Electro Mechanical Motion Systems (ELECTROMOTION 2005), Lausanne, Switzerland, 27–29 September 2005.
5. Masmoudi, A.; Masmoudi, A. 3-D analytical model with the end effect dedicated to the prediction of PM eddy-current loss in FSPMMs. *IEEE Trans. Magn.* **2015**, *51*, 8103711. [CrossRef]
6. Li, W.; Jeong, Y.W.; Koh, C.S. An adaptive equivalent circuit modeling method for the eddy current-driven electromechanical system. *IEEE Trans. Magn.* **2010**, *46*, 1859–1862. [CrossRef]
7. Stoll, R.L. *The Analysis of Eddy Currents*; Clarendon Press: Oxford, UK, 1974.
8. Zhu, Z.Q.; Ng, K.; Schofield, N.; Howe, D. Improved analytical modeling of rotor eddy current loss in brushless machines equipped with surface mounted permanent magnets. *IEE Proc. Electr. Power Appl.* **2004**, *151*, 641–650. [CrossRef]
9. Dubas, F.; Rahideh, A. Two-dimensional analytical permanent-magnet eddy-current loss calculations in slotless PMSM equipped with surface-inset magnets. *IEEE Trans. Magn.* **2014**, *50*, 6300320. [CrossRef]
10. Pfister, P.-D.; Yin, X.; Fang, Y. Slotted permanent-magnet machines: general analytical model of magnetic fields, torque, eddy currents, and permanent-magnet power losses including the diffusion effect. *IEEE Trans. Magn.* **2016**, *52*, 8103013. [CrossRef]
11. Benlamine, R.; Dubas, F.; Randi, S.A.; Lhotellier, D.; Espanet, C. 3-D numerical hybrid method for pm eddy-current losses calculation: application to axial-flux PMSMS. *IEEE Trans. Magn.* **2015**, *51*, 8106110. [CrossRef]
12. Ishak, D.; Zhu, Z.Q.; Howe, D. Eddy-current loss in the rotor magnets of permanent magnet brushless machines having a fractional number of slots per pole. *IEEE Trans. Magn.* **2005**, *41*, 2462–2469. [CrossRef]
13. Wang, J.; Atallah, K.; Chin, R.; Arshad, W.M.; Lendenmann, H. Rotor eddy-current loss in permanent-magnet brushless AC machines. *IEEE Trans. Magn.* **2010**, *46*, 2701–2707. [CrossRef]
14. Sinha, G.; Prabhu, S.S. Analytical model for estimation of eddy current and power loss in conducting plate and its application. *Phys. Rev. Spec. Top. Accel. Beams* **2011**, *14*, 062401. [CrossRef]
15. Hur, J.; Toliyat, H.A.; Hong, J.-P. 3-D time-stepping analysis of induction motor by new equivalent magnetic circuit network-model. *IEEE Trans. Magn.* **2001**, *37*, 3225–3228. [CrossRef]
16. Mahyob, A.; Ould Elmoctar, M.Y.; Reghem, P.; Barakat, G. Induction machine modelling using permeance network method for dynamic simulation of air-gap eccentricity. In Proceedings of the 2007 European Conference on Power Electronics and Applications (EPE), Aalborg, Denmark, 2–5 September 2007. [CrossRef]
17. Nakamura, K.; Fujio, S.; Ichinokura, O. A method for calculating iron loss of an SR motor based on reluctance network analysis and comparison of symmetric and asymmetric excitation. *IEEE Trans. Magn.* **2006**, *42*, 3440–3442. [CrossRef]
18. Nakamura, K.; Fujio, S.; Ichinokura, O. A method for calculating iron loss of a switched reluctance motor based on reluctance network analysis. In Proceedings of the 12th International Power Electronics and Motion Control Conference (EPE PMC), Portoroz, Slovenia, 30 August–1 September 2006. [CrossRef]

19. Carpenter, M.J.; Macdonald, D.C. Circuit representation of inverter-fed synchronous motors. *IEEE Trans. Energy Conv.* **1989**, *4*, 531–537. [CrossRef]
20. Slemon, G.R. An equivalent circuit approach to analysis of synchronous machines with saliency and saturation. *IEEE Trans. Energy Conv.* **1990**, *5*, 538–545. [CrossRef]
21. Hemeida, A.; Sergeant, P.; Vansompel, H. Comparison of methods for permanent magnet eddy-current loss computations with and without reaction field considerations in axial flux PMSM. *IEEE Trans. Magn.* **2015**, *51*, 8106110. [CrossRef]
22. Busch, T.J.; Law, J.D.; Lipo, T.A. Magnetic circuit modeling of the field regulated reluctance machine. Part II: saturation modeling and results. *IEEE Trans. Energy Conv.* **1996**, *11*, 56–61. [CrossRef]
23. Yoshida, Y.; Nakamura, K.; Ichinokura, O. A method for calculating eddy current loss distribution based on electric and magnetic networks. *IEEE Trans. Magn.* **2011**, *47*, 4155–4158. [CrossRef]
24. Yoshida, Y.; Nakamura, K.; Ichinokura, O. Consideration of eddy current loss estimation in SPM motor based on electric and magnetic networks. *IEEE Trans. Magn.* **2012**, *48*, 3108–3111. [CrossRef]
25. Nakamura, K.; Hisada, S.; Arimatsu, K.; Ohinata, T.; Sakamoto, K.; Ichinokura, O. Iron loss calculation in a three-phase-laminated-core based on reluctance network analysis. *IEEE Trans Magn.* **2009**, *45*, 4781–4784. [CrossRef]
26. Bormann, D.; Tavakoli, H. Reluctance network treatment of skin and proximity effects in multi-conductor transmission lines. *IEEE Trans. Magn.* **2012**, *48*, 735–738. [CrossRef]
27. Demenko, A.; Sykulski, J.; Wojciechowski, R. Network representation of conducting regions in 3-D finite-element description of electrical machines. *IEEE Trans. Magn.* **2008**, *44*, 714–717. [CrossRef]
28. Demenko, A. Three dimensional eddy current calculation using reluctance-conductance network formed by means of FE method. *IEEE Trans. Magn.* **2000**, *36*, 741–745. [CrossRef]
29. Fu, W.N.; Liu, Z.J. Estimation of eddy-current loss in permanent magnets of electric motors using network-field coupled multi-slice time-stepping finite-element method. *IEEE Trans. Magn.* **2002**, *38*, 1225–1228. [CrossRef]
30. Gerlach, T.; Rabenstein, L.; Dietz, A.; Kremser, A.; Gerling, D. Determination of eddy current losses in permanent magnets of SPMSM with concentrated windings: A hybrid loss calculation method and experimental verification. In Proceedings of the 8th International Conference on Ecological Vehicles and Renewable Energies (EVER), Monte-Carlo, Monaco, Monaco, 10–12 April 2018. [CrossRef]
31. Mohammadi, S.; Mirsalim, M.; Vaez-Zadeh, S. Nonlinear modeling of eddy-current couplers. *IEEE Trans. Energy Conv.* **2014**, *29*, 224–231. [CrossRef]
32. Flux2D/3D. *General Operating Instructions*; Version 11.1.; Cedrat, S.A., Ed.; Electrical Engineering: Grenoble, France, 2013.
33. Chetangny, P.K.; Houndedako, S.; Vianou, A.; Espanet, C. Eddy-current loss in a conductive material inserted into a U-cored electromagnetic device. In Proceedings of the 2017 IEEE Vehicle Power and Propulsion Conference (VPPC), Belfort, France, 11–14 December 2017. [CrossRef]
34. Benmessaoud, Y.; Dubas, F.; Benlamine, R.; Hilairet, M. Three-dimensional automatic generation magnetic equivalent circuit using mesh-based formulation. In Proceedings of the 20th International Conference on Electrical Machines and Systems (ICEMS), Sydney, NSW, Australia, 11–14 August 2017. [CrossRef]
35. Benmessaoud, Y.; Dubas, F.; Benlamine, R.; Hilairet, M. Circuit équivalent magnétique générique tridimensionnel pour systèmes électromagnétiques. In Proceedings of the Symposium de Génie Électrique (SGE), Nancy, France, 3–5 July 2018.
36. Delale, A.; Albert, L.; Gerbaud, L.; Wurtz, F. Automatic generation of sizing models for the optimization of electromagnetic devices using reluctance networks. *IEEE Trans. Magn.* **2004**, *40*, 830–833. [CrossRef]
37. Benlamine, R.; Benmessaoud, Y.; Dubas, F.; Espanet, C. Nonlinear adaptive magnetic equivalent circuit of a radial-flux interior PM machine using air-gap sliding-line technic. In Proceedings of the 2017 IEEE Vehicle Power and Propulsion Conference (VPPC), Belfort, France, 11–14 December 2017. [CrossRef]
38. Benmessaoud, Y.; Belguerras, W.; Dubas, F.; Hilairet, M. 3-D generic magnetic equivalent circuit taking into account skin effect: Magnetic field and eddy-current losses. In Proceedings of the 13th international conference of the IMACS TC1 Committee (ELECTRIMACS 2019), Salerno, Italy, 21–23 May 2019.

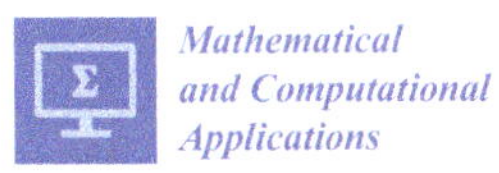
Mathematical and Computational Applications

Article

Investigation of Volumic Permanent-Magnet Eddy-Current Losses in Multi-Phase Synchronous Machines from Hybrid Multi-Layer Model

Youcef Benmessaoud, Daoud Ouamara *, Frédéric Dubas * and Mickael Hilairet

Département ENERGIE, FEMTO-ST, CNRS, Univ. Bourgogne Franche-Comté, F90000 Belfort, France; youcef.benmessaoud@femto-st.fr (Y.B.); mickael.hilairet@univ-fcomte.fr (M.H.)

* Correspondence: daoud.ouamara@gmail.com (D.O.); frederic.dubas@univ-fcomte.fr (F.D.)

Received: 5 October 2019; Accepted: 4 March 2020; Published: 4 March 2020

Abstract: This paper investigates the permanent-magnet (PM) eddy-current losses in multi-phase PM synchronous machines (PMSM) with concentric winding and surface-mounted PMs. A hybrid multi-layer model, combining a two-dimensional (2-D) generic magnetic equivalent circuit (MEC) with a 2-D analytical model based on the Maxwell–Fourier method (i.e., the formal resolution of Maxwell's equations by using the separation of variables method and the Fourier's series), performs the eddy-current loss calculations. First, the magnetic flux density was obtained from the 2-D generic MEC and then subjected to the Fast Fourier Transform (FFT). The semi-analytical model includes the automatic mesh of static/moving zones, the saturation effect and zones connection in accordance with rotor motion based on a new approach called "Air-gap sliding line technic". The results of the hybrid multi-layer model were compared with those obtained by three-dimensional (3-D) nonlinear finite-element analysis (FEA). The PM eddy-current losses were estimated on different paths for different segmentations as follow: (i) one segment (no segmentation), (ii) five axial segments, and (iii) two circumferential segments, where the non-uniformity loss distribution is shown. The top of PMs presents a higher quantity of losses compared to the bottom.

Keywords: eddy-current losses; hybrid model; magnetic equivalent circuit; Maxwell–Fourier method; multi-phase; segmentation; synchronous machines

---

1. Introduction

Currently, there is trend of the electrification of transport applications, where PMSM are increasingly being used. The reason is their good torque density, lower losses and higher power-to-mass ratio [1]. In addition to advantages associated to this topology, a multi-phase system allows continuous operation in degraded mode, which is a powerful asset in traction application [2,3].

One of the major concerns in PMSM is the PM demagnetization risk. This occur by temperature rise caused by the eddy-currents [4]. The high-order magnetic flux density harmonics, which are not synchronous with the rotor, cause those parasitic eddy-current losses in the PMs [5]. Eddy-current losses may be separated into two cases according to the operating point:

- No-load (without stator current): the reluctance variation due to the stator slot-opening [5,6];
- Load (with stator current): the spatio-temporal magnetomotive force (MMF) harmonics [7,8].

In design process, the PM eddy-current losses must be calculated. Several methods have been developed and categorized into three families viz.: (i) (Semi-)analytical methods based the formal resolution of Maxwell's equations [9,10], (ii) numerical methods [11,12], and (iii) hybrid methods [13,14]. An overview on the eddy-current loss calculation has been realized in [15], where several investigations have been chronologically listed. In [16], sources, calculation methods, reduction techniques, and thermal analysis of PM eddy-current losses were revised. A review of (semi-)analytical models based on the subdomain technique for PM eddy-current calculation was made by [17].

The fact that the eddy-currents flow is along the loops inside the PM volume leads to the classification of their calculation as a 3-D problem [18]. The 2-D calculation ignores end-effects which cause a large error in the PM eddy-current computation. To consider end-effects and obtain a more accurate calculation, a 3-D model must be used [19].

This paper investigates the PM eddy-current losses in massive conducting parts of a multi-phase PMSM from hybrid multi-layer model. The latter consists of combining two models, viz.: (i) a 2-D generic MEC, and (ii) a 2-D analytical model based on the Maxwell–Fourier method (i.e., the formal resolution of Maxwell's equations by using the separation of variables method and the Fourier's series). The 2-D generic MEC determines the magnetic flux density distribution in the massive conductive parts (e.g., the PMs,...) without the skin effect (i.e., without the eddy-current reaction field). Moreover, this model can give the distribution of other local/integral physical quantities in the electrical machine with the nonlinear of $B(H)$ curve. The 2-D analytical model based on the Maxwell–Fourier method calculates the magnetic field distribution in massive conductive parts considering the skin effect as well as the resultant eddy-current density. The boundary conditions (BCs) imposed on the 2-D analytical model are equivalent to the magnetic field obtained from the MEC. These BCs are considered homogeneous on the edges and are only equal to normal magnetic field in the top or middle of massive conductive parts. This is a strong hypothesis, because the magnetic field is naturally non-uniform. In our case, the model gives the ability to use both conditions (i.e., in the top or middle of massive conductive parts), allowing to see the impact on the eddy-current loss calculation.

Eddy-current losses are investigated on a multi-phase PMSM with a non-overlapping (i.e., concentrated) all-teeth wound winding (double-layer winding with left and right layer) having 5-phases/20-slots/16-poles [20].

The proposed approach is introduced in section 2 with a short focus of the 2-D generic MEC and its validation. Section 3 shows the obtained PM eddy-current losses on different paths for different segmentations. In the last section, the results of the hybrid multi-layer model are compared with those obtained by 3-D nonlinear FEA [21].

2. Hybrid Multi-Layer Model Description

The performed approach is composed of two main modules, viz.: (i) a 2-D generic MEC, and (ii) a 2-D analytical model based on the Maxwell–Fourier method. The first one allows the calculation of magnetic flux density $\boldsymbol{B}$ without the skin effect over all the PMSM. In the PM zones, the magnetic flux density is levied on different paths parallel in the tangential direction. Indeed, the magnetic flux density varies depending on the PM depth direction. These magnetic flux densities will be subjected to the FFT giving the amplitude as well as the frequency of different harmonic numbers. Furthermore, these outputs constitute the main input parameters to estimate the PM eddy-current losses by using the developed formal Maxwell model. Note that only the harmonics are going to be taken into account when calculating the eddy-current losses. The eddy-current losses due to the average value are null because the resultant eddy-current density in the PMs does not exist. The developed approach for the eddy-current loss calculation is explained in the flowchart shown in Figure 1.

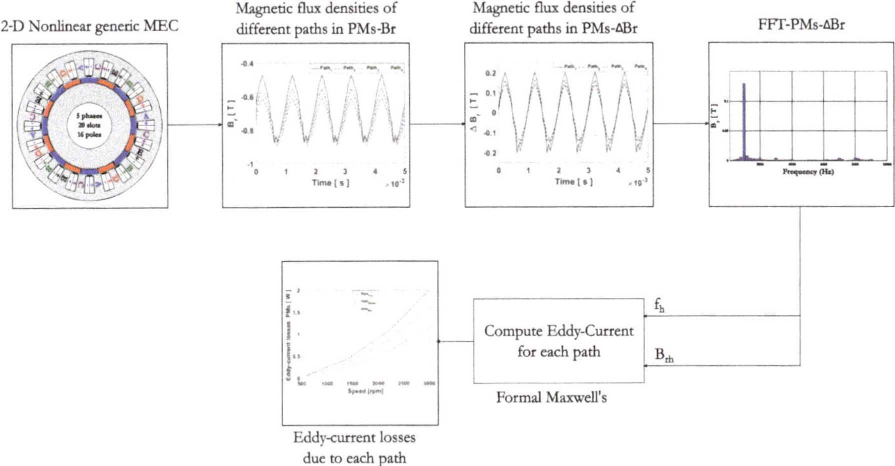

Figure 1. General flowchart of hybrid multi-layer model.

2.1. Geometrical and Physical Parameters of PMSM

The performed model was applied to a five-phase surface-mounted PMSM, where the geometrical and physical parameters are given, respectively, in Tables 1 and 2. Figure 2 gives the topology of the machine and the spatial distribution of various phases. The stator has a double-layer concentrated winding distribution (viz., the non-overlapping winding with all teeth wound), supplied by sinusoidal current waveform. The five-phases windings are star-connected. The PMs, radially magnetized, and the sheet metal are, respectively, *NdFeB* and *M270-35A*. The dashed lines represent the different paths used to proof this approach. Based on this path, the local quantities will be levied and furthermore compared to those of FEA [21].

Attention should be paid to Figure 3 which presents the different tangential paths located in both South and North PM. This path will be used to get the different values of the magnetic field along the symmetric axis of the PM.

Table 1. Geometrical data of the machine [20].

Designation	Symbol	Value	Unit
Slots number	Q_s	20	-
Phases number	m	5	-
Poles number	$2p$	16	-
Slot-opening/tooth-pitch	α_{so}	60	%
Isthmus-opening/slot	α_{is}	53	%
PM pole-arc/pole-pitch	α_a	72	%
Winding-opening/slot	α_w	92	%
Stator bore diameter	D_e	49.4	mm
External diameter	D_{es}	65.7	mm
Internal diameter	D_i	34.9	mm
Stator yoke height	h_{cs}	3	mm
Rotor yoke height	h_{cr}	3	mm
PM height	h_a	2.5	mm
Tooth height	h_d	4.8	mm
Machine length	l_m	100	mm

Table 2. Physical data of the machine.

Designation	Symbol	Value	Unit
Speed	N	3,000	rpm
Electromagnetic torque	T_{em}	30	Nm
Current	I	30	A

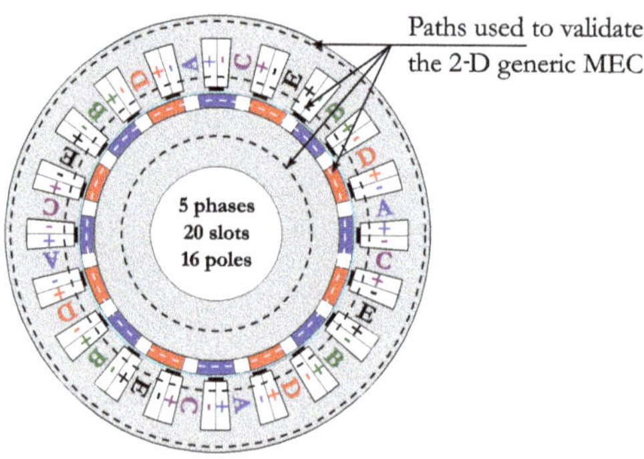

Figure 2. Description of the machine: Topology and winding distribution.

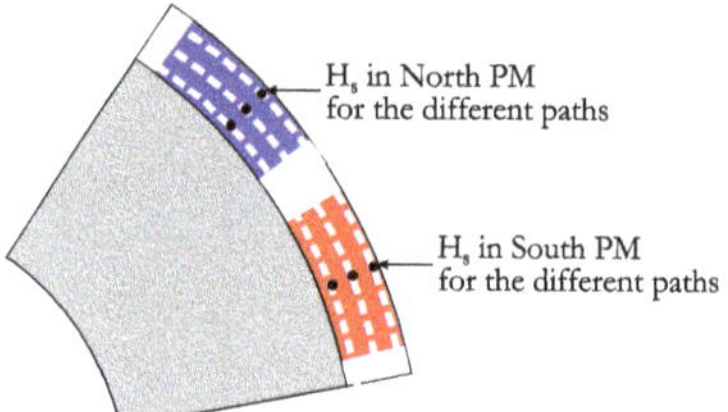

Figure 3. Tangential parallel paths in PMs.

2.2. 2-D Generic MEC

In this work, the approach proposed in [22] (viz., the generalized nonlinear adaptive MEC) was applied onto the 5-phases PMSM. Indeed, the 2-D MEC is based on the discretization principle, which allows bidirectional (BD) automatic mesh reluctance generation. In a general manner, the electrical machine can be discretized in a set of mesh elements that can be, in turn, discretized in a high number of BD blocks. As an example, the mesh element spotted by the grey color in Figure 4 can be discretized in a set of BD blocks, as shown in Figure 5.

The main advantages of such a model include of the flexibility, generic, compromise between computation time and precision. The semi-analytical model includes the automatic mesh of static/moving zones, the saturation effect and zones connection in accordance with rotor motion based on a new approach called "Air-gap sliding line technic". This technique was applied in [23–26] on different electrical machine configurations, viz., axial-flux interior PM machine [23,24], coaxial magnetic gear equipped with surface-mounted PMs [25], wound-rotor synchronous machine [27], and radial-flux interior PM machine [26]. This novel technique of connection between static/moving zones as well as the approach proposed in [22] is applied for the first time on a radial-flux surface-mounted PMSM having multi-phases for automotive application. Nevertheless, it should be noted that there are other techniques permitting to connect static/moving zones in electrical machines which have been overviewed in [22].

Comparing to [26], in our case, the stator tooth-pitch will be constituted by five radial zones, and four radial zones for the rotor pole pitch. For the tangential zones, the rotor pole pitch is consists of three zones, while the stator pitch consists of five zones. The discretization number in each zones is given by Table 3.

The developed model respects the flowchart given in Figure 6, where the different steps of the nonlinear system solving are presented. To account for the saturation effect, an iterative process is considered by using the fixed point iteration method. The nonlinear B(H) curve of the iron is then introduced by using the Marrocco's interpolation function. In Figure 6, μ_{ri} represents the initial relative permeability in the magnetic circuit and N_{sat} is the total iterations number for taking into account the saturation effect. The 2-D nonlinear adaptive MEC (where the loop fluxes ψ are the unknowns) can be expressed by

$$[\chi] \cdot [\Re] \cdot [\chi]^T \cdot [\psi] - [F] = 0 \quad \text{with} \quad [F] = [\chi] \cdot [MMF] \tag{1}$$

$$[\phi] = [\chi]^T \cdot [\psi] \tag{2}$$

where $[\Re]$ is the diagonal matrix of reluctances, $[\psi]$ is the loop fluxes vector, $[\chi]$ is the topological (or incidence) matrix, $[F]$ is the loop MMFs vector, and $[MMF]$ is the branch MMFs vector. These matrixes and vectors are dependent upon the discretization of the various zones. Their dimensions are invariable with the time (or the mechanical angular position between the rotor and the stator) [22]. Finally, dividing this later by the corresponding reluctance surfaces leads to the magnetic flux density in each magnetic branch.

Table 3. Discretization number.

Discretization/Zones	Stator	Rotor
Tangential	[2 1 7 1 2]	[2 7 2]
Radial	[1 2 3 2 3]	[3 6 4 1]

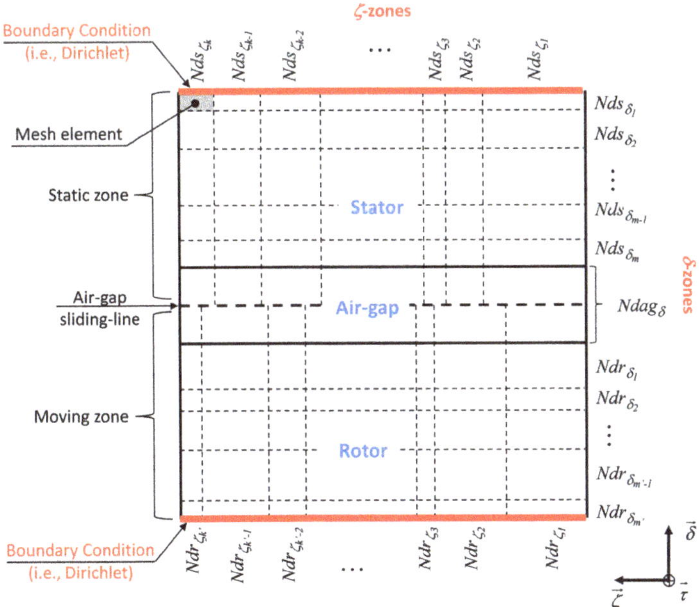

Figure 4. Mesh element discretization [22].

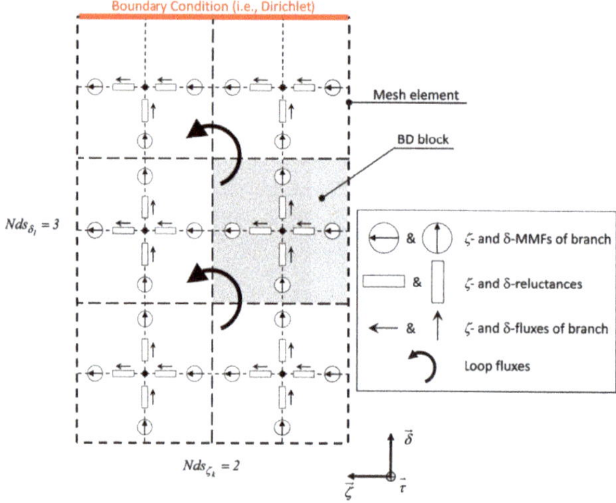

Figure 5. Components of mesh elements [22].

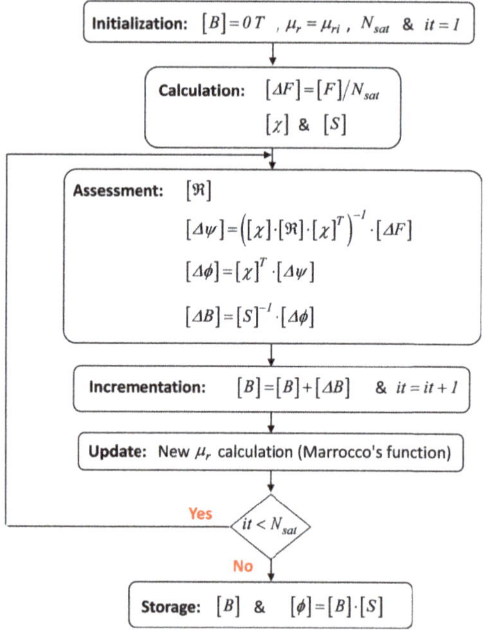

Figure 6. Flowchart of the iterative solving of the nonlinear system [22].

2.3. Validation of 2-D Generic MEC

The tangential and radial components of $\boldsymbol{B}$ at $t = 0\ sec$ are given in Figures 7–12 for the different path crossing , viz.: the stator yoke, stator teeth, stator air-gap, rotor air-gap, rotor PMs and rotor yoke. The components of $\boldsymbol{B}$ in the stator teeth [see Figure 8] are very saturated. The results are in agreement with 2-D nonlinear FEA, which confirms that the nonlinear solving method is accurate. The errors are due to the discretization of static and moving zones in both tangential and radial directions. For example, a sensibility study versus stator/rotor discretization on the computation time as well as on the iron losses determined from the magnetic flux density in the stator/rotor iron have been performed in [27]. The difference is mainly linked to the discretization for the rotor yoke where $Nd_3^r = 4$, which can be observed on the Figure 12a of the tangential component of $\boldsymbol{B}$ in the rotor yoke. The computation time is reduced by twice compared to the 2-D nonlinear FEA.

The simulation in time stepping leads to the results represented in Figures 13–15. The electromagnetic torque evolution at locked rotor is given in Figure 13; this help us to find the shift angle to obtain the maximal electromagnetic torque at synchronism shown in Figure 14. The results give a decent satisfaction. The average electromagnetic torque given by 2-D nonlinear FEA is equal to 32 Nm while the 2-D nonlinear adaptive MEC is slightly above 30 Nm. The error is equal to 6.25 %. This small difference can be linked to the approach used to take into account the magnetic saturation, the rotor motion as well as the air-gap, stator and rotor discretization.

As mentioned in Section 2, the performed model will be used to extract the local physical quantities in PMs in order to estimate the eddy-current losses. Moreover, it should be noted that B_θ in the PMs is considered negligible in relation to B_r (see Figure 11). The losses are then determined from the normal magnetic field in the PMs. The model can give the waveform of $\boldsymbol{B}$ according to the time or the rotor position. The simulations are done for the path crossing the middle of PMs for one periodic pattern as shown in Figure 15a. The results are in agreement with the 2-D nonlinear FEA. Figure 15b gives the magnetic flux density evolution in the middle of North and South PMs.

Note that the average value of $\boldsymbol{B}$ is assumed to be not considered for the eddy-current loss calculation. The evolution of ΔB_r for both middle of North and South PMs is shown in Figure 15b. It can be seen that the variation of $\boldsymbol{B}$ is practically identical. Thus, it can be concluded that the losses in North and South PMs are identical. Furthermore, only the results given for the South PM are given.

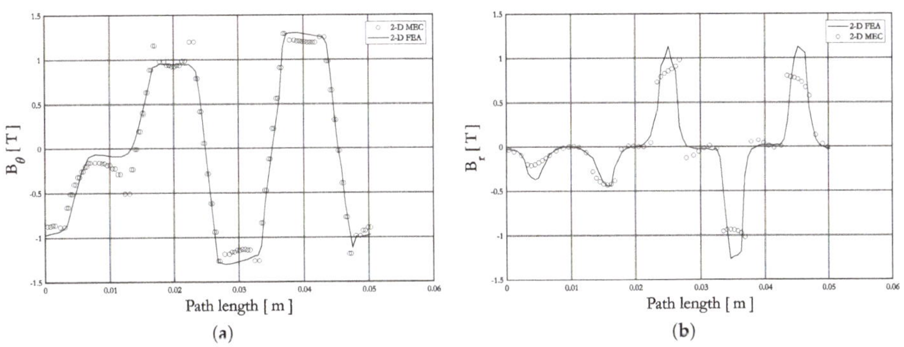

Figure 7. Waveform of $\boldsymbol{B}$ in the stator yoke: (**a**) tangential, and (**b**) radial component.

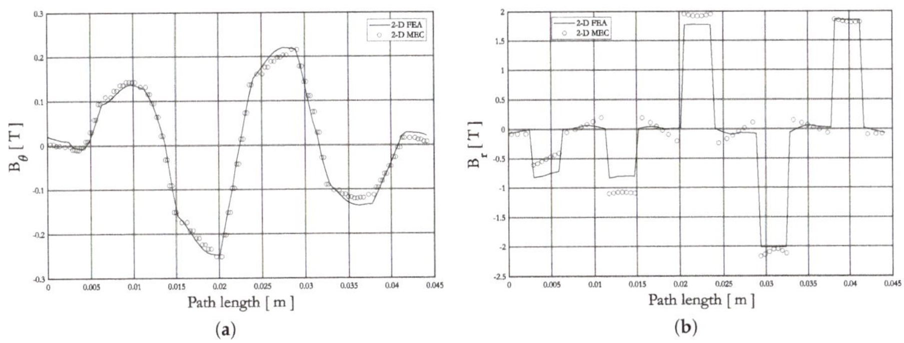

Figure 8. Waveform of $\boldsymbol{B}$ in the stator teeth: (**a**) tangential, and (**b**) radial component.

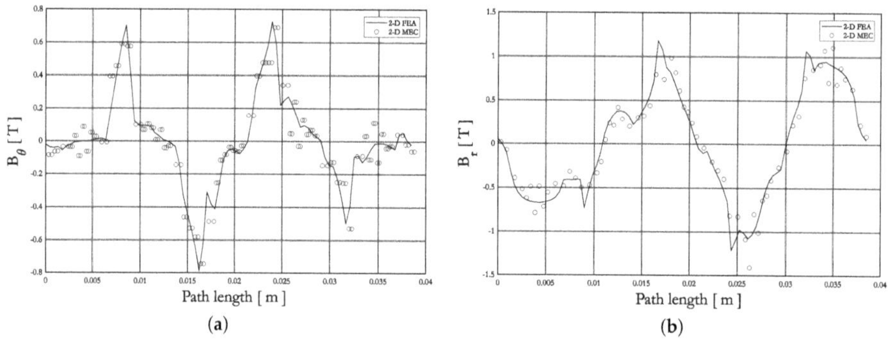

Figure 9. Waveform of B in the stator air-gap: (**a**) tangential, and (**b**) radial component.

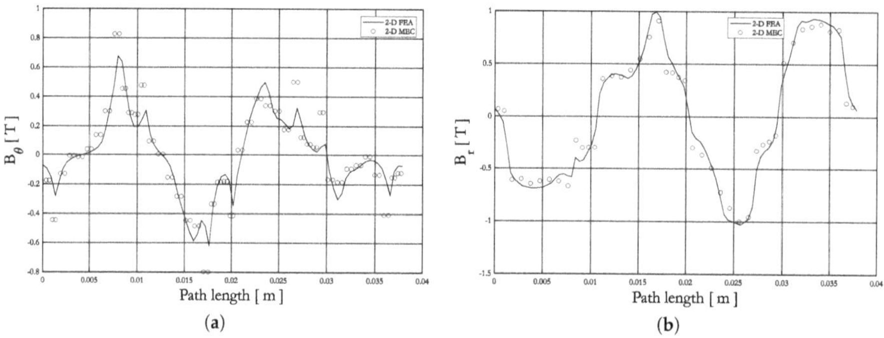

Figure 10. Waveform of B in the rotor air-gap: (**a**) tangential, and (**b**) radial component.

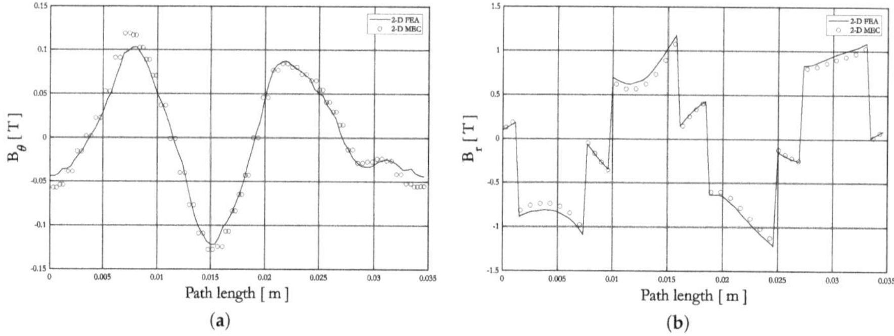

Figure 11. Waveform of B in the rotor PMs: (**a**) tangential, and (**b**) radial component.

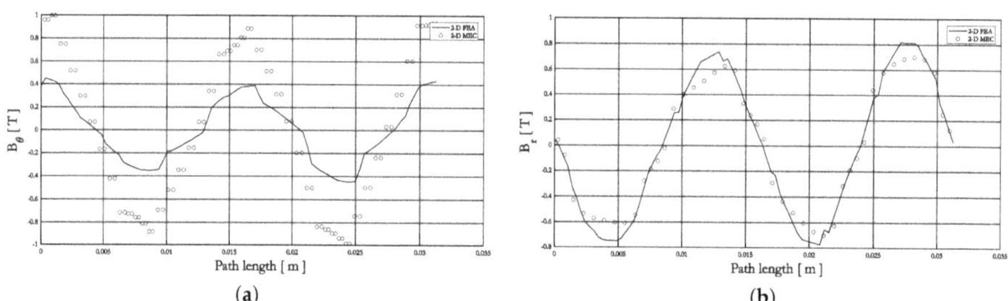

Figure 12. Waveform of B in the rotor yoke: (**a**) tangential, and (**b**) radial component.

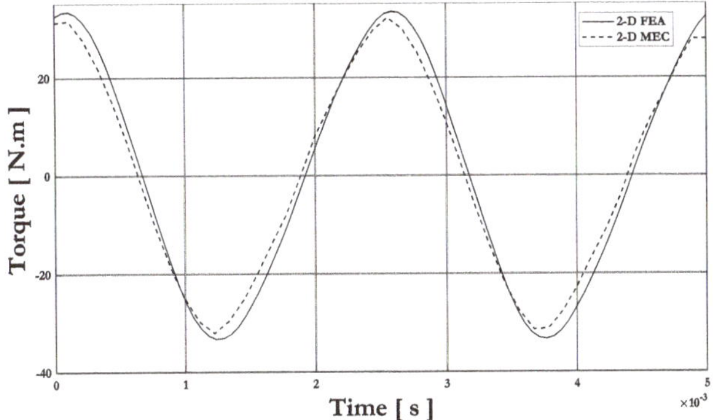

Figure 13. Electromagnetic torque at locked rotor.

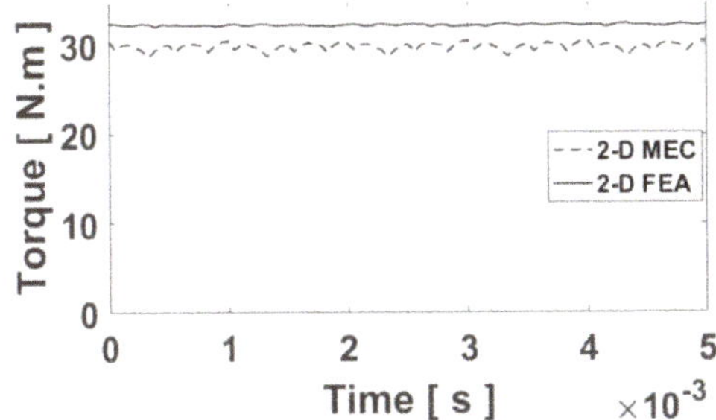

Figure 14. Electromagnetic torque at synchronism.

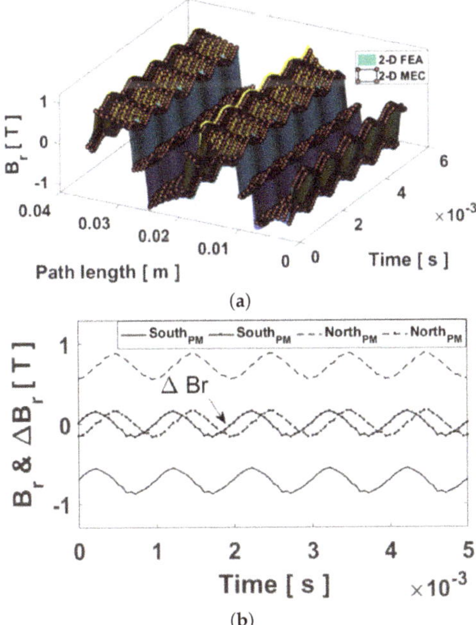

Figure 15. Evolution of B: (**a**) versus time for the path crossing the middle of PMs, and (**b**) in the middle of North and South PMs with(out) the average value.

2.4. Formal Resolution of Maxwell's Equations

2.4.1. General Solution of Magnetic Field with the Skin Effect

The magnetic field with the skin is determined from an analytical model in 2-D Cartesian coordinates (i.e., the PMs are assumed rectangular) based on the Maxwell–Fourier method (i.e., the formal resolution of Maxwell's equations by using the separation of variables method and the Fourier's series). This permits to estimate the resultant eddy-current density in PMs in both directions (i.e., axial and circumferential axis). The BCs imposed on the 2-D analytical model are equivalent to the magnetic field obtained from the MEC. These BCs are considered homogeneous on the edges of massive conductive parts, which are equal to H_s. This is a strong hypothesis because the magnetic field is naturally non-uniform. The value of magnetic field is defined as the normal magnetic field in the top or middle of PMs determined by 2-D generic MEC i.e., without the skin effect, viz., $H_s = B_r(0,0)/\mu_{mp}$. In our case, the model gives ability to use both conditions (i.e., in the top or middle of massive conductive parts), allowing to see the impact on the eddy-current loss calculation. Furthermore, the results will be given for different paths parallel to the tangential direction. The various BCs are shown in Figure 16.

In quasi-stationary approximation, inside a linear (non)magnetic material of electrical conductivity without electromagnetic sources, the partial differential equation in magnetodynamic in term of H^{mp} can be defined by [28]

$$\nabla^2 \boldsymbol{H}^{mp} - \mu_{mp} \cdot \sigma_{mp} \cdot \frac{\partial \boldsymbol{H}^{mp}}{\partial t} = 0 \quad \text{(Diffusion equation)} \tag{3}$$

Using the complex notation, the magnetic field $\boldsymbol{H}^{mp} = \left\{0;\ H_{\sigma y}^{mp};\ 0\right\}$ in massive conductive parts can be written as

$$H_{\sigma y}^{mp} = \Re\left\{\overline{H_{\sigma y}^{mp}} \cdot e^{j\omega t}\right\} \tag{4}$$

where $j = \sqrt{-1}$ and $\omega = 2\pi f$ is the electrical pulse.

Therefore, (3) becomes in 2-D Cartesian coordinates the complex Helmholtz's equation, viz.,

$$\frac{\partial^2 \overline{H_{\sigma y}^{mp}}}{\partial x^2} + \frac{\partial^2 \overline{H_{\sigma y}^{mp}}}{\partial z^2} - \alpha^2 \cdot \overline{H_{\sigma y}^{mp}} = 0 \tag{5}$$

where $\alpha^2 = j \cdot \mu_{mp} \cdot \sigma_{mp} \cdot \omega$.

By using the separation of variables method and by applying the BCs, the 2-D general solution of $H_{\sigma y}^{mp}$ in both directions (i.e., x- and z-edges) can be written as Fourier's series [28]

$$\overline{H_{\sigma y}^{mp}} = H_s \cdot \overline{f_\sigma(x,z)}, \tag{6a}$$

$$\overline{f_\sigma(x,z)} = \left\{ \frac{ch(\alpha \cdot z)}{ch\left(\alpha \cdot \frac{d}{2}\right)} + \sum_{k=1,3,\ldots}^{\infty} \overline{g_k^z} \cdot ch\left(\delta_k \cdot x\right) \cos\left(\lambda_k \cdot z\right) \right\}, \tag{6b}$$

$$\overline{g_k^z} = \overline{e_k^z} \cdot \frac{sinc\left(\lambda_k \cdot \frac{d}{2}\right)}{ch\left(\delta_k \cdot \frac{w}{2}\right)} \text{ with } \overline{e_k^z} = 2\left[1 - \left(\frac{\lambda_k}{\delta_k}\right)^2\right], \tag{6c}$$

where $\lambda_k = k\pi/d$ is the periodicity of $\overline{H_{\sigma y}^{mp}}$ in the z-axis, $\delta_k = \sqrt{\alpha^2 + \lambda_k^2}$, and k are the spatial harmonic orders.

It should be noted that $H_{\sigma y}^{mp}$ is assumed to be invariant in the y-axis according to the study path. Moreover, when $\alpha = 0$ (viz., $\sigma_{mp} = 0\ S/m$ and $f \cong 0^{+}Hz$) then $\overline{f_{\sigma}(x,z)} = 1$ thus giving $\overline{H_{\sigma y}^{mp}} = H_s$.

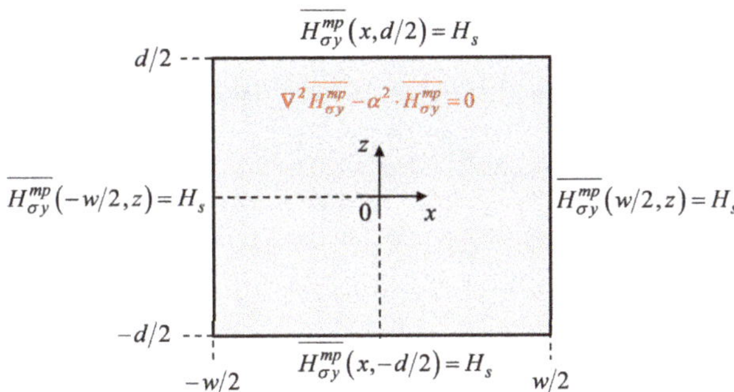

Figure 16. BCs at edges of PMs in (x, z) coordinate system.

2.4.2. Eddy-Current Loss Formulation

The instantaneous density of power flow $\prod$ at a point is defined by the Poynting's vector [28]

$$\prod = E \times H^{mp} \tag{7}$$

Using $J = \sigma \cdot E = \nabla \times H^{mp}$, this power across a closed surface, in terms of complex vectors, is given by the instantaneous apparent power

$$\overline{s_{app}} = \oiint_S \prod \cdot dS = \frac{1}{\sigma_{mp}} \cdot \oiint_S (J \times H^{mp}) \cdot dS = p + j \cdot q \tag{8}$$

The real part of the surface integral of the complex Poynting's vector gives the instantaneous ohmic losses and the imaginary part the instantaneous magnetic energy.

From BCs at edges of PMs (see Figure 16), the average of $\overline{s_{app}}$ over an electrical cycle $T = 2\pi/\omega$ can be defined by

$$\overline{S_{app}} = \langle \overline{s_{app}} \rangle = \frac{h^l}{\sigma_{mp}} \cdot \left\{ \int_{-w/2}^{w/2} \overline{J_x} \cdot \overline{H_{\sigma y}^{mp}}^* \bigg|_{z=-d/2} \cdot \mathrm{d}x - \int_{-d/2}^{d/2} \overline{J_z} \cdot \overline{H_{\sigma y}^{mp}}^* \bigg|_{x=-w/2} \cdot \mathrm{d}z \right\}, \tag{9}$$

where $h^l = h_a/\,(2 \cdot Nd_2^r)$ is the layers thickness in the r-axis of PMs with l the path number.

After development, (9) is then given by

$$\overline{S_{app}} = \frac{h^l \cdot (H_s)^2}{\sigma_{mp}} \cdot \left\{ w \cdot \alpha \cdot \frac{sh\left(\alpha \cdot \frac{d}{2}\right)}{ch\left(\alpha \cdot \frac{d}{2}\right)} + \sum_{k=1,3,\ldots}^{\infty} \frac{2\delta_k}{d} \cdot \left(\frac{\overline{e_k^z}}{\lambda_k}\right)^2 \cdot \frac{sh\left(\delta_k \cdot \frac{w}{2}\right)}{ch\left(\delta_k \cdot \frac{w}{2}\right)} \right\}, \tag{10a}$$

$$\overline{S_{app}} = P + \mathrm{j} \cdot Q \tag{10b}$$

It should be noted that $P = \Re\{\overline{S_{app}}\}$ represents the average eddy-current losses in PMs.

3. Approach of PM Eddy-Current Loss Calculation

3.1. Magnetic Flux Density vs PM Thickness

Once the 2-D nonlinear adaptive MEC was validated on different stator and rotor paths. The radial component of $\boldsymbol{B}$ in the middle of North and South PMs according to the different tangential paths was evaluated. It should be noted that the paths number is imposed by the radial discretization number applied in the PMs region. The waveform of $\boldsymbol{B}$ in the PMs is done by stepwise simulation to get the radial component evolution according to the time or the rotor position.

Under the assumptions that the continuous magnitude B_m of $B_{\sigma y}^{mp}$ does not contribute to the eddy-current losses and that only the waveform ΔB_r is subjected to the FFT. Hence, the most influential harmonics can be used to estimate the PMs eddy-current losses. Knowing that:

$$B_{\sigma y}^{mp} = B_m + \Delta B_r \tag{11a}$$

$$\frac{\partial B_{\sigma y}^{mp}}{\partial t} = \frac{\partial B_m}{\partial t} + \frac{\partial(\Delta B_r)}{\partial t} \approx \frac{\partial(\Delta B_r)}{\partial t} \tag{11b}$$

Figure 17a provides the comparison between the developed model and 2-D FEA, for the radial component evolution in the middle of the North PM according to the time. The radial component variation for different path parallel to the tangential direction is illustrated on Figure 18a. Figure 17b presents the ripples in the middle of PMs. The validation is done by comparing the results to those of 2-D FEA. Figure 18b gives the ripples according to the symmetric axis of the PM, which are subjected to the FFT.

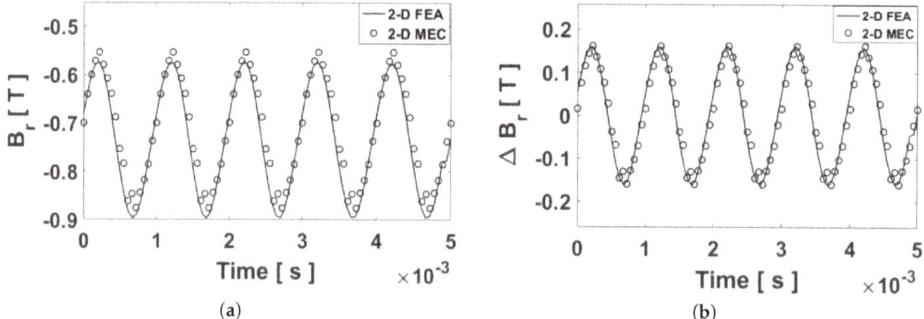

Figure 17. Validation of the radial component of $\boldsymbol{B}$ in the middle of the North PM: (**a**) with and (**b**) without average value.

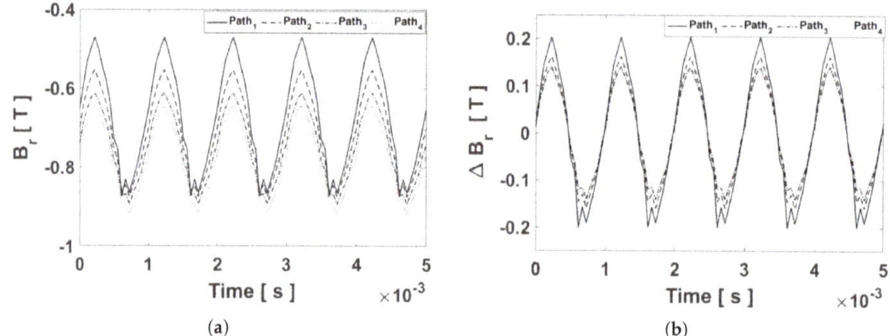

Figure 18. Evolution of the radial component of $\boldsymbol{B}$ in the middle of the North PM according to the time for different tangential path: (**a**) with and (**b**) without the average value.

3.2. Eddy-Current Loss Evolution in Different Paths

Figures 19 and 20 show the frequency spectrum of the radial component of $\boldsymbol{B}$ for different speeds. The spectral components are due to: (i) the supply (i.e., the current waveform), (ii) the slotting effect, and (iii) the spatial distribution of winding. Using the results given by the FFT for different paths in the PM, the PMs eddy-current losses can be estimated for each path as shown in Figure 21a,b.

By applying the sum, the total eddy-current losses due to the different paths are obtained as shown in Figure 21c. It should be noted that the eddy-current loss repartition in the PMs is greater at the top of PMs (whatever the PM magnetization) and decrease with to the PMs height to achieve the lower losses in the bottom of PMs. This distribution can be explained by the variation of the magnetic flux density amplitude that is more important in the region which is close to the air-gap, contrary to the bottom of PMs where the amplitude is relatively less important.

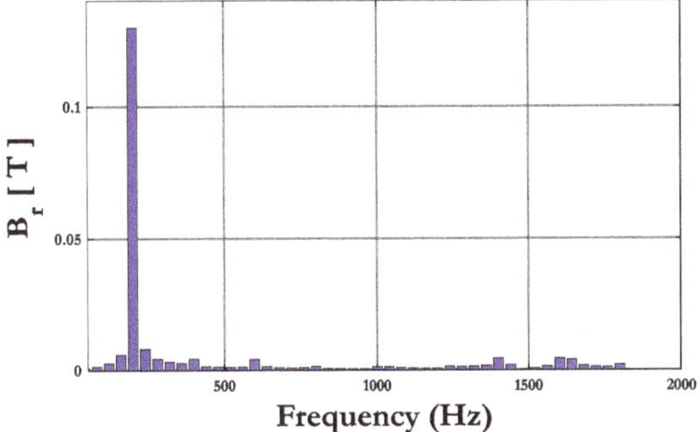

Figure 19. The frequency spectrum of the radial component of B in the middle of the North PM at 600 rpm.

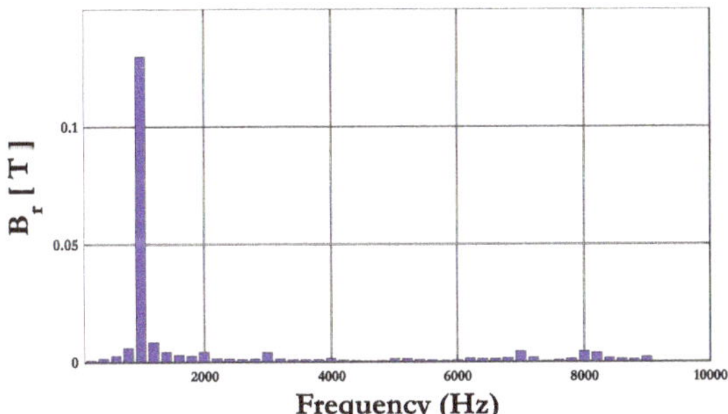

Figure 20. The frequency spectrum of the radial component of B in the middle of the North PM at 3000 rpm.

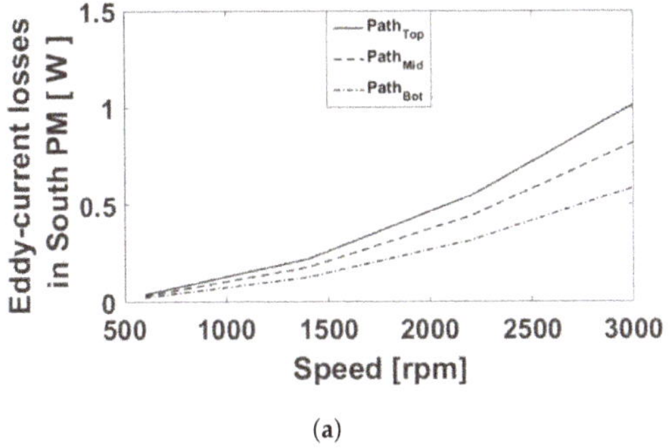

(a)

Figure 21. *Cont.*

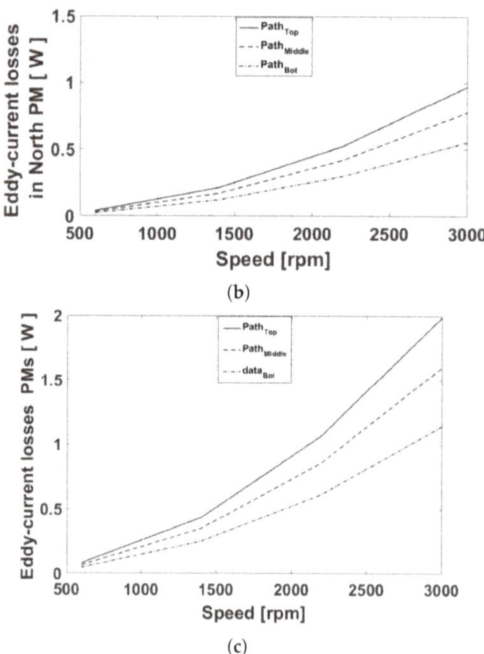

Figure 21. Non-uniform volume repartition of the eddy current losses vs. speed for different paths: (**a**) South PMs, (**b**) North PMs, and (**c**) Total.

4. 3-D nonlinear FEA Validation With(out) PM Segmentation

In this section, the results of the hybrid multi-layer model are compared with those obtained by 3-D nonlinear FEA [21]. The PMs eddy-current losses were computed at different speed ranges for different segmentations as follow: (i) one segment (no segmentation), (ii) five axial segments, and (iii) two circumferential segments.

To prove the validation of this approach, the results are compared with those of 3-D nonlinear FEA in transient state [21]. Due to the boundary conditions (i.e., periodicity and Dirichlet conditions), the electrical machine can be reduced into 5-slots/4-poles. The mesh generation for 3-D nonlinear FEA is equal to 2,263,818 second order elements. It is given the eddy-current loss evolution in all PMs by using different segmentations with two different approaches (see Figure 22):

- 1st approach: based on all paths;
- 2nd approach: based on the middle path of PMs.

Figure 23 represents the relative error between each approach and 3-D nonlinear FEA. These errors are due to medium mesh in the 3-D nonlinear FEA as well as the magnetic field evaluation on all paths of the 2-D nonlinear adaptive MEC which is a data input the analytical model based on the Maxwell–Fourier method permitting the PM eddy-current loss calculation. In both cases, the results provide a good agreement with a maximal error of 16% that is reached for the 2nd approach based on the middle path with two circumferential segments. While the maximal relative error for the 1st approach is estimated at 11%, which is achieved for one segment (no-segmentation) at 3000 rpm.

The developed model returns both results based on to the two different approaches. It is better to use the 1st approach to estimate the eddy-current losses in the circumferential segmentation case.

Concerning the computation time, the 3-D nonlinear FEA takes 16 days for one simulation while the hybrid multi-layer model takes only 12 min. For an operating point, the computation time is then divided by 1920 with respect to 3-D nonlinear FEA.

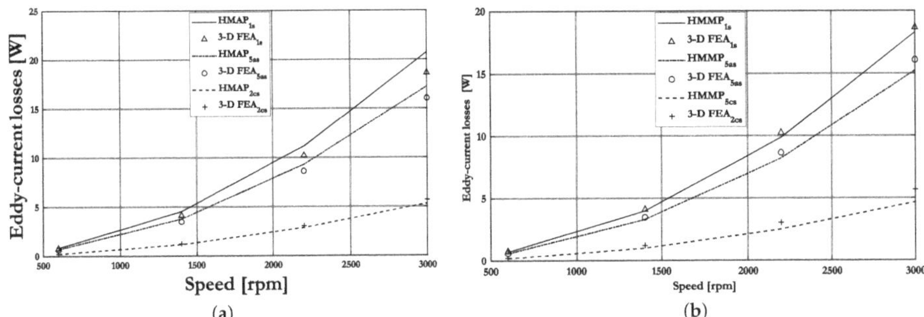

Figure 22. Eddy-current losses comparison with 3-D FEA: (**a**) all paths, and (**b**) middle path of PMs.

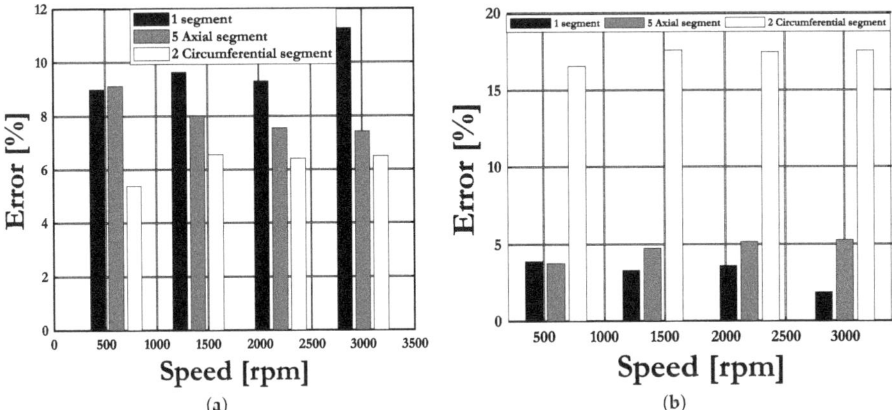

Figure 23. Relative error between the hybrid model and 3-D FEA: (**a**) all paths, and (**b**) middle path of PMs.

5. Conclusions

In this work, a hybrid multi-layer model, combining a 2-D generic MEC with a 2-D analytical model based on the Maxwell–Fourier method was performed. This developed model was aimed to compute the volumic PM eddy-current losses in multi-phase PMSM. It is clearly seen that the PM eddy-current losses distribution present no-uniformity according to the PMs height. In this study case, these losses are greater at the top of PMs. This can be justified by the amplitude of the radial component of $\boldsymbol{B}$, which is the highest in the top of PMs. Then, they decrease to achieve the minimal in the PM bottom. As a conclusion, the temperature repartition can be affected by the non-uniformity of the PM eddy-current loss distribution over the PM volume.

Author Contributions: Y.B. and D.O. prepared the structure and wrote the paper, F.D. and M.H. corrected and improved its content, F.D. and D.O. corrected the final version of the paper. All authors have read and agreed to the published version of the manuscript.

Acknowledgments: This work was supported by RENAULT-SAS, Guyancourt France. This scientific study is related to the project "Conception optimale des chaines de Traction Electrique" (COCTEL) financed by the "Agence De l'Environnement et de la Maîtrise de l'Énergie" (ADEME).

Conflicts of Interest: The authors declare no conflict of interest.

References

1. Cassat, A.; Espanet, C.; Coleman, R.; Burdet, L.; Leleu, E.; Torregrossa, D.; M'Boua, J.; Miraoui, A. A practical solution to mitigate vibrations in industrial PM motors having concentric windings. *IEEE Trans. Ind. Appl.* **2012**, *48*, 1526–1538. doi:10.1109/TIA.2012.2210172. [CrossRef]
2. Levi, E. Multiphase Electric Machines for Variable-Speed Applications. *IEEE Trans. Ind. Electron.* **2008**, *55*, 1893–1909. doi:10.1109/TIE.2008.918488. [CrossRef]
3. Scuiller, F.; Charpentier, J.F.; Semail, E.; Clenet, S. Comparison of two 5-phase permanent magnet machine winding configurations. Application on naval propulsion specifications. In Proceedings of the 2007 IEEE International Electric Machines & Drives Conference, Antalya, Turkey, 3–5 May 2007; Volume 1, pp. 34–39. doi:10.1109/IEMDC.2007.383548. [CrossRef]
4. de la Barriere, O.; Hlioui, S.; Ben Ahmed, H.; Gabsi, M. An Analytical Model for the Computation of No-Load Eddy-Current Losses in the Rotor of a Permanent Magnet Synchronous Machine. *IEEE Trans. Magn.* **2016**, *52*, 1–13. doi:10.1109/TMAG.2013.2257825. [CrossRef]
5. Dubas, F.; Espanet, C. Semi-Analytical Solution of 2-D Rotor Eddy-Current Losses due to the Slotting Effect in SMPMM. In Proceedings of the 17th Conference on the Computation of Electromagnetic Fields COMPUMAG 2009, Florianopolis, Brasil, 22–26 November 2009; pp. 20–25.
6. Markovic, M.; Perriard, Y. A simplified determination of the permanent magnet (PM) eddy current losses due to slotting in a PM rotating motor. In Proceedings of the 2008 International Conference on Electrical Machines and Systems (ICEMS), Wuhan, China, 17–20 October 2008; pp. 309–313.
7. Dubas, F.; Espanet, C.; Miraoui, A. Field Diffusion Equation in High-Speed Surface-Mounted Permanent Magnet Motors, Parasitic Eddy-Current Losses. In Proceedings of the 6th International Symposium on Advanced Electromechanical Motion Systems, Lausanne, Switzerland, 27–29 September 2005; pp. 1–6.
8. Deng, F.; Nehl, T. Analytical modeling of eddy-current losses caused by pulse-width-modulation switching in permanent-magnet brushless direct-current motors. *IEEE Trans. Magn.* **1998**, *34*, 3728–3736. doi:10.1109/20.718535. [CrossRef]
9. Nuscheler, R. Two-dimensional analytical model for eddy-current rotor loss calculation of PMS machines with concentrated stator windings and a conductive shield for the magnets. In Proceedings of the XIX International Conference on Electrical Machines (ICEM 2010), Rome, Italy, 6–8 September 2010; pp. 1–6. doi:10.1109/ICELMACH.2010.5608050. [CrossRef]
10. Mirzaei, M.; Binder, A.; Funieru, B.; Susic, M. Analytical Calculations of Induced Eddy Currents Losses in the Magnets of Surface Mounted PM Machines With Consideration of Circumferential and Axial Segmentation Effects. *IEEE Trans. Magn.* **2012**, *48*, 4831–4841. doi:10.1109/TMAG.2012.2203607. [CrossRef]
11. Yoshida, K.; Hita, Y.; Kesamaru, K. Eddy-current loss analysis in PM of surface-mounted-PM SM for electric vehicles. *IEEE Trans. Magn.* **2000**, *36*, 1941–1944. doi:10.1109/20.877827. [CrossRef]
12. El-Hasan, T. Rotor eddy current determination using finite element analysis for High-Speed Permanent Magnet Machines. In Proceedings of the 2014 IEEE 23rd International Symposium on Industrial Electronics (ISIE), Istanbul, Turkey, 1–4 June 2014; pp. 885–889. doi:10.1109/ISIE.2014.6864728. [CrossRef]
13. Saban, D.M.; Lipo, T.A. Hybrid Approach for Determining Eddy-Current Losses in High-Speed PM Rotors. In Proceedings of the 2007 IEEE International Electric Machines & Drives Conference, Antalya, Turkey, 3–5 May 2007; Volume 1, pp. 658–661. doi:10.1109/IEMDC.2007.382745. [CrossRef]
14. Gerlach, T.; Rabenstein, L.; Dietz, A.; Kremser, A.; Gerling, D. Determination of eddy current losses in permanent magnets of SPMSM with concentrated windings: A hybrid loss calculation method and experimental verification. In Proceedings of the 2018 Thirteenth International Conference on Ecological Vehicles and Renewable Energies (EVER), Monte-Carlo, Monaco, 10–12 April 2018; pp. 1–8. doi:10.1109/EVER.2018.8362395. [CrossRef]
15. Dubas, F.; Rahideh, A. Two-Dimensional Analytical Permanent-Magnet Eddy-Current Loss Calculations in Slotless PMSM Equipped With Surface-Inset Magnets. *IEEE Trans. Magn.* **2014**, *50*, 54–73. doi:10.1109/TMAG.2013.2285525. [CrossRef]
16. Ouamara, D.; Dubas, F. Permanent-Magnet Eddy-Current Losses: A Global Revision of Calculation and Analysis. *Math. Comput. Appl.* **2019**, *24*, 67. doi:10.3390/mca24030067. [CrossRef]

17. Pfister, P.D.; Yin, X.; Fang, Y. Slotted Permanent-Magnet Machines: General Analytical Model of Magnetic Fields, Torque, Eddy Currents, and Permanent-Magnet Power Losses Including the Diffusion Effect. *IEEE Trans. Magn.* **2016**, *52*, 1–13. doi:10.1109/TMAG.2015.2512528. [CrossRef]
18. Benlamine, R.; Dubas, F.; Randi, S.A.; Lhotellier, D.; Espanet, C. 3-D Numerical Hybrid Method for PM Eddy-Current Losses Calculation: Application to Axial-Flux PMSMs. *IEEE Trans. Magn.* **2015**, *51*, 1–10. doi:10.1109/TMAG.2015.2405053. [CrossRef]
19. Masmoudi, A.; Masmoudi, A. 3-D Analytical Model With the End Effect Dedicated to the Prediction of PM Eddy-Current Loss in FSPMMs. *IEEE Trans. Magn.* **2015**, *51*, 1–11. doi:10.1109/TMAG.2014.2356647. [CrossRef]
20. Ouamara, D.; Dubas, F.; Randi, S.A.; Benallal, M.N.; Espanet, C. Optimal Design of Multi-phases Permanent-Magnet Synchronous Machines Using Genetic Algorithms. *Mediterr. J. Model. Simul.* **2018**, *10*, 29–44.
21. Altair Flux. Electromagnetic, Electric, and Thermal Analysis. Available online: https://www.altair.com/flux/ (accessed on 4 March 2020).
22. Dubas, F..; Benlamine, R.; Randi, S.A.; Lhotellier, D.; Espanet, C. 2-D or quasi 3-D nonlinear adaptative magnetic equivalent circuit, Part I: Generalized modeling with air-gap sliding-line technic. *Appl. Energy*, under review.
23. Benlamine, R.; Dubas, F.; Randi, S.A.; Lhotellier, D.; Espanet, C. 2-D or quasi 3-D nonlinear adaptive magnetic equivalent circuit, Part II: Application to axial-flux interior permanent-magnet synchronous machines. *Appl. Energy*, under review.
24. Benlamine, R.; Dubas, F.; Randi, S.A.; Lhotellier, D.; Espanet, C. Modeling of an axial-flux interior PMs machine for an automotive application using magnetic equivalent circuit. In Proceedings of the 2015 18th International Conference on Electrical Machines and Systems (ICEMS), Pattaya, Thailand, 25–28 October 2015; pp. 1266–1271. doi:10.1109/ICEMS.2015.7385234. [CrossRef]
25. Benlamine, R.; Hamiti, T.; Vangraefschepe, F.; Dubas, F.; Lhotellier, D. Modeling of a coaxial magnetic gear equipped with surface mounted PMs using nonlinear adaptive magnetic equivalent circuits. In Proceedings of the 2016 XXII International Conference on Electrical Machines (ICEM), Lausanne, Switzerland, 4–7 September 2016, pp. 1888–1894. doi:10.1109/ICELMACH.2016.7732781. [CrossRef]
26. Benlamine, R.; Benmessaoud, Y.; Dubas, F.; Espanet, C. Nonlinear Adaptive Magnetic Equivalent Circuit of a Radial-Flux Interior Permanent-Magnet Machine Using Air-Gap Sliding-Line Technic. In Proceedings of the 2017 IEEE Vehicle Power and Propulsion Conference (VPPC), Belfort, France, 11–14 December 2017; Volume 24, pp. 1–6. doi:10.1109/VPPC.2017.8330894. [CrossRef]
27. Utegenova, S.; Dubas, F.; Jamot, M.; Glises, R.; Truffart, B.; Mariotto, D.; Lagonotte, P.; Desevaux, P. An Investigation into the Coupling of Magnetic and Thermal Analysis for a Wound-Rotor Synchronous Machine. *IEEE Trans. Ind. Electron.* **2018**, *65*, 3406–3416. doi:10.1109/TIE.2017.2756597. [CrossRef]
28. Stoll, R.L. *The Analysis of Eddy Currents*; Clarendon Press: Oxford, UK, 1974.

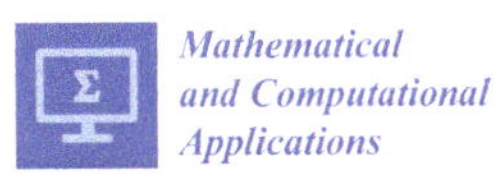

Review

Permanent-Magnet Eddy-Current Losses: A Global Revision of Calculation and Analysis

Daoud Ouamara *,† and Frédéric Dubas †

Département ENERGIE, FEMTO-ST, CNRS, Univ. Bourgogne Franche-Comté, F90000 Belfort, France
* Correspondence: daoud.ouamara@gmail.com
† These authors contributed equally to this work.

Received: 21 May 2019; Accepted: 8 July 2019; Published: 9 July 2019

Abstract: Eddy-current analysis is an important research field. This phenomenon occurs in multiple areas and has several applications: electromagnetic braking, repulsive effects, levitation, etc. Thereby, this paper is limited to eddy-current study in rotating electrical machines. In the design process, if the permanent-magnet (PM) loss calculation is very important, the overheating due to eddy-currents must be taken into account. The content of this paper includes sources, calculation methods, reduction techniques, and thermal analysis of PM eddy-current losses. This review aims to act as a guide for the reader to learn about the different aspects and points to consider in studying the eddy-current.

Keywords: analytical method; eddy-current; finite-element analysis; loss reduction; permanent-magnet losses; thermal analysis

---

1. Introduction

1.1. Context of this Paper

Eddy-currents are induced currents that originate, for example, in a moving conductor in a constant magnetic field or in a stationary conducting material subjected to a time-dependent magnetic field. According to [1], the term "eddy" originates from the fact that these induced currents create magnetic field vortices inside the conductors. Eddy-currents are used and exploited in many applications such as:

- Induction furnace: The rapid variation in the magnetic field generates very large eddy-currents, and the heat produced is sufficient to melt a metal;
- Electric brakes: The brakes expose the wheels (metal) to a magnetic field, which generates eddy-currents. The magnetic interaction between the applied field and the eddy-currents acts to slow the wheels down.
- Detection of flaws and cracks in materials.

In spite of eddy-currents' advantages, they can be undesirable. This is due to the fact that they generate thermal heat, which causes losses. This is very problematic in electrical machines.

A permanent-magnet synchronous machine (PMSM) is one of the most studied electrical machines. They are very attractive for different applications, thanks to their high efficiency, good torque density, lower maintenance costs, and easy manufacturing [2,3]. These interests have led designers to increase the frequency of electromechanical conversion, thus leading to a more detailed consideration of the various difficulties, PM loss being one of them [4,5]. Indeed, temperature issues can arise in high-speed applications due to eddy-currents [6]. In the ideal case, where the air-gap field contains only synchronized space harmonics (i.e., rotating in synchrony with the rotor), the magnetic eddy-current losses can be neglected (no losses). In real situations, the magnetic flux density in the air-gap has spatio-temporal variation: therefore, unsynchronized space harmonics are generated. For this reason, significant losses by eddy-currents are induced in the rotor [7]. In the PMSM design process, eddy-current losses' calculation and reduction have the primary objective to improve their performance. Several researchers have developed analytical and numerical models in order to better estimate these losses. In addition, techniques to reduce losses have been realized such as PM segmentation.

1.2. Objective of This Paper

The purpose of this work is to give a global revision of the calculation and analysis of PM eddy-current losses. Section 2 identifies the PM losses sources, then it explains their global causes in electrical machines. In Section 3, the methods of PM loss calculation are summarized, including 2D/3D analytical models, finite element analysis (FEA) methods, and hybrid models. PM loss reduction is analyzed in Section 4 in several categories, viz., PM segmentation, rotor shape, material type effect, and spatial filter. In Section 5, the methods of PM loss measurement are explained. Finally, Section 6 treats the thermal behavior of electrical machines generated by eddy-currents.

2. Sources of PM Losses

Before listing the methods of PM eddy-current losses' calculation, we will identify the different sources of these losses. Four major sources may be listed:

2.1. Slotting Effect

The harmonic content in the flux of PMSM due to the slotting of the stator was established in [8]. Eddy-currents in a solid rotor of a PMSM due to stator teeth were calculated in [9], by considering the magnetic saturation. It was shown that teeth harmonics decreased with saturation, and so, the losses were lower.

In [10], the no-load tooth-ripple due to the distortion of the fundamental flux density wave by the stator slotting was described. The permeance modulation that resulted from the teeth, in a tubular PM motor, was a major cause of the loss [11]. The calculation in the conducting regions of a rotor of PMSMs was presented in [12]. The determination of the PM eddy-current losses due to the slotting effect was investigated in [13–16] and in fractional-slot surface-mounted PMSMs in [17–19].

2.2. Winding Distribution

Space harmonics of the magnetomotive force (MMF) induce losses in the PM of electrical machines, due to the flux variation. In [10], the "on-load: term was used to denote the total harmonics losses occurring in the load condition. The non-uniform rotation of the armature MMFs induce eddy-current losses [20], referred to as commutation losses in [21,22]. The fundamental and lower order MMF harmonics can give rise to significant rotor eddy-currents [23,24]. According to [25], losses are almost double in concentric alternate teeth winding (single-layer winding) compared to concentric all teeth winding (two-layer winding). Eddy-currents are related to the asynchronous components of the MMF spatial harmonics [26], and the low order of unsynchronized spatial harmonics induces a large amount of eddy-current losses [27].

Based on the contents of space harmonics in the air-gap MMF distribution, a combination of slots and poles of a PMSM was done by [28–31] to get a topology with minimal PM losses. Combinations with a large number of poles and a small number of slots are characterized by large rotor losses [32]. For the same purpose, the impact of the number of phases is quantified to design lower eddy-current PM machines [33], and several windings were compared in [34]. Windings with different turns per coil side, in fractional-slot PMSMs, lead to reducing and/or canceling some space harmonics, resulting in lower rotor losses by the armature reaction field [35]. In [36,37], the interaction between the wavelengths of the space harmonics and PM pole dimension was studied. The analysis of MMF the harmonics of machines with a specific pole/slot ratio (ratio of 2/3) shows that second- and fourth-order space harmonics are dominant, which induces significant PM eddy-current losses [38].

2.3. Supply and Control

The stator currents cause asynchronous components in the air-gap field and induce rotor losses in electrical machines [39–41]. Control of PMSMS by pulse width modulation (PWM) can lead to PM losses due to the high frequency of stator magnetic field variations [42–48]. The eddy-current losses are mainly produced by the carrier harmonics of the PWM inverter [49]. A single-phase PM brushless DC motor supplied by a 180° square current waveform presents more eddy-current losses than when it is supplied with only the fundamental component of current [50]. In [51], by considering the carrier harmonics of PMW inverters, it was shown that the PM eddy-current losses in the concentrated winding motor were larger than the distributed winding motor. By comparing eddy-current losses per unit induced by PWM with sine wave supply, the authors of [52–54] deduced that PM losses were higher in the case of PWM supply. This was due to more harmonics content compared to the sine waveform. To calculate power loss in PMSMs in [55], two winding and rectifier topologies were considered (three-phase bridge rectifier and two three-phase bridges rectifiers connected in series). In [56], the harmonics caused by the PWM were incorporated into a series of steps to calculate the stator MMF. The eddy-current losses caused by the fundamental time harmonic of the winding have been calculated [57].

2.4. PM Hysteresis Losses

The behavior of PM hysteresis losses was experimentally investigated in [58]. The results showed that the hysteresis losses were larger than eddy-current losses when the AC field due to a slot ripple was of the order of several hundred hertz. The authors of [59] claimed that hysteresis losses in PM materials had no significant influence on rotating electrical machines' design. This conclusion was based on measurement results because the PM materials operated in the second quadrant of the $B(H)$ curve and mostly without crossing the J-axis.

A comparison study based on the hysteresis loss of different PMs (i.e., ferrite, samarium-cobalt, and neodymium PMs) has been established [60]. Rare-earth PMs exhibit less hysteresis losses than ferrite PMs, and the comparison, between samarium-cobalt and neodymium PMs, shows that samarium-cobalt PMs, at small field strength variation, have larger hysteresis losses. The investigation of the hysteresis behavior of the ferrite PMs in the second and first quadrants of the $B(H)$ curve confirms that hysteresis losses have a minor role in electrical machines' design [61].

3. Calculation of PM Losses

3.1. Two-Dimensional Analytical Models

Early in the 1970s, analytical models for calculating eddy-currents were developed. The variational methods were applied for thing purpose in thin conducting plates [62], and the surface impedance has been used to predict eddy-current fields in conductors [63]. To estimate PM eddy-current losses and retaining ring losses of a PMSM, the authors of [8] proposed a linear model, by taking as infinity the magnetic permeability of the rotor core and stator iron. The calculation was possible by representing PM by resistances in the equivalent circuits [64], where end effects were neglected, where the field was considered as one-dimensional, and by using the magnetic equivalent circuit (MEC) [65].

To predict eddy-current losses, a mathematical model based on a 2D electromagnetic field analysis in polar coordinates was developed in [41,43,44], while the stator and rotor cores were assumed to be infinitely permeable in [23,66], and an extended model considering time harmonics in the stator MMF distribution was proposed in [25]. To take into account the reaction field, an improved analytical model was proposed in [24,67]. A completely analytical solution of the losses generated by eddy-currents, in a cylindrical PM rotating inside a hollow conducting cylinder, was proposed by [68]. Based on the magneto-static flux density distribution in the air-gap, it is possible to calculate the eddy-current losses [69]. The excitation field may be replaced by a current sheet for a slotless structure [70]. To consider the slotting, the PM field is calculated in the slotless structure and then adapted to the structure with slots. The conformal mapping method allows the modulation function [14]. To take into account the 3D flow of eddy-current, a correction factor for the 2D analytical model was proposed in [71]:

$$F_{cn}(w_n) = 1 + \frac{2/(aL)}{\coth(\lambda L/2) + (a/\gamma)\coth(\lambda L/2) - 2a/Ly^2} \tag{1}$$

where L is the axial length of the PMs. a, λ, and γ for the nth harmonic are given by:

$$a = np_s/R_s; \lambda = \sqrt{j\omega_n\mu_0\mu_r\sigma}; \gamma = \sqrt{a^2 + \frac{\lambda}{g\mu_i}}$$

where μ_r and σ are respectively the relative permeability of the coil and magnets. ω_n is the angular frequency of the nth harmonic.

A nonlinear MEC applied to an interior PMSM was exposed in [72]; the method takes into account magnetic saturation and PM eddy-currents. In [73], an expanded analytical loss model was presented where the PM may be replaced by a retaining sleeve to study the behavior of the new rotor configuration. An analytical approach for the simultaneous calculation of stator and no-load field was performed. The finite PM dimensions was considered by introducing a correction factor for endless dimensions [15]. At low frequency, the skin effect did not influence the results of PM losses. Nevertheless, at higher frequencies, the flux density in the PM was non-homogenous, and other formulas must be applied [74]. A method for calculating eddy-current loss using MEC (or reluctance network analysis) was presented in [75], associated with an electric network model for a surface-mounted PMSM [76], and applied to an axial-flux PMSM [54,77].

The interaction between eddy-current harmonics having the same frequency, but different spatial order may not be neglected, otherwise the PM losses can be under- or over-estimated [78]. The authors of [79] proposed a simple analytic model based on Carter's and surface impedance theories to calculate the PM losses. By considering the finite permeability of the stator and rotor cores and accurate permeability of the PM, eddy-current losses were analytically calculated for a slotless PMSMs where the eddy-current reaction field was neglected and the induced eddy-currents were resistance limited [80]. The eddy-current reaction field in the slot was considered by solving Helmholtz's equation [81]. To take the stator teeth geometry into account regardless of the magnetic field of the slotless structure, the relative permeance function may be used [82]. A 2D subdomain method in polar coordinates was proposed in [83]. The method was applied for a slotless PMSM with surface-inset magnets by considering the eddy-current reaction field. In [84], an approach in Cartesian coordinates based on a harmonic method was used to calculate PM losses. The MMF was decomposed into Fourier series where the period was the PM width. The Cartesian coordinates were chosen in [85] for the PM loss calculation by using a simplified rectangular geometry for the analytical model.

An exact subdomain model was presented in [86], and it was applied to a slotted PMSM, considering that the diffusion effect and eddy-currents were not assumed resistance limited. In [87], an analytical method taking into account the effect of the reaction field was presented. To consider the diffusion phenomenon and the finite length of a magnet along the x-direction, the work in [16] solved a Fredholm integral equation for the computation of the no-load PM losses.

3.2. Three-Dimensional Analytical Models

PM loss calculation based on a quasi-3D analytical model was presented in [88]. The method was performed in Cartesian coordinates, which considered the reaction eddy-current. A 3D analytical model that took into account the end-effect was proposed in [89]. The reaction field and the end-effect of an interior PMSMs was well considered by [90]. The model was based on the Fourier transform of the armature reaction, the armature, and the PM slotting effect. The authors of [91] proposed an analytical model based on the generalized image theory. The model was established in 3D rectangular coordinates; the slotting effect was neglected; and the cores were assumed infinitely permeable.

The method was applied for the PMs with rectangular shapes. The magnetic field distribution may be calculated from analytical or FEA, and then, a 3D Fourier series was performed without including the eddy-current reaction field, so the accuracy of the results was only visible at low frequency. A modified generalized image theory was proposed in [92] to predict 3D high frequency eddy-current losses for surface-mounted PMSMs and applied to an eight-pole/18-slot interior PMSM [93,94]. A 3D subdomain model was used to englobe the slotting effect [95,96], where the method of variable separation was used to get the 3D eddy-currents in PMs. The work in [18] used an analytical method in polar coordinates for any pole-slot combinations of a surface-mounted PMSM under any conditions of load.

3.3. Finite-Elements Analysis

FEA was used to study the eddy-currents due to slotting effect in PMSMs [9] and loss calculation in both magnetic and non-magnetic sleeves [97]. Eddy-current losses were investigated on a tubular PM motor by using FEA [11]. Often, magnets are divided to decrease the eddy-current losses. However, the 3D FEA calculation of eddy-current in PM with slits is more difficult, and the computer resources (computation time) increase. The $A - \phi$ method with double nodes at slits was used to overcome this problem and increase the accuracy of calculation [98,99]. The side-insulation between adjacent PMs causes a discontinuity of the eddy-current distribution. This case was modeled by 2D FEA for surface-mounted PMSM [100]. PM eddy-current loss calculation by considering the end-effect and by using 3D FEA takes vast amounts of time. Yamazaki et al. [101] proposed a method where firstly a 2D nonlinear time-domain analysis was applied with the PWM voltage waveform. Next, the 3D frequency domain analysis was used for each remarkable harmonic. The total PM eddy-current losses were calculated by summing the results of the two steps, as shown in Figure 1.

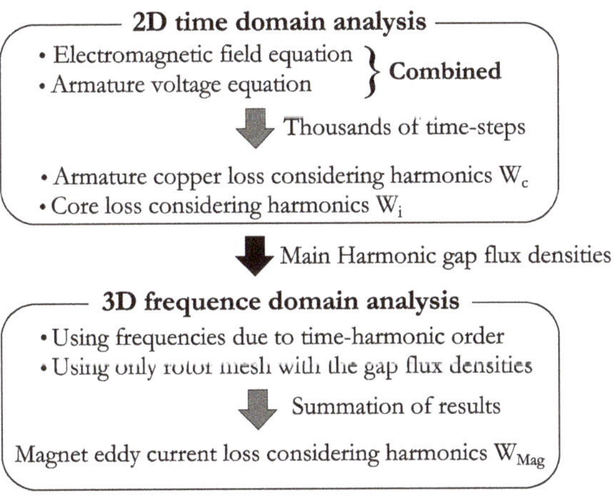

Figure 1. Outline of the proposed method [101].

Flux-switching PMSMs are characterized by a significant flux leakage. The latter induce significant eddy-currents that were investigated by FEA [102]. To take into account a motor's commutations ad time and space harmonics, FEA for surface-mounted PMSMs was used in [103]. Three different methods were explained in [104] to calculate eddy-current losses in PMSMs with concentrated windings:

1. FE magneto-static method with analytical post-processing:
   Loss was calculated analytically from the numerically-obtained flux density values. The inducing effect was neglected, as well as the eddy-current reaction field.

2. FE magneto-static method:
FEA software was used to calculate directly the PMs losses without the need for additional post-processing. The reaction field was not considered.
3. FE magneto-dynamic method:
The time-stepping method solved Maxwell's equation for a moving rotor and for any current waveform. It took into account the eddy-current reaction field, but it was very time consuming.

In [105], a combined 2D/3D method considering the harmonics and the magnetic saturation of the rotor was proposed to calculate eddy-current losses. The 2D time-domain analysis allowed the calculation of the flux density distribution. From this calculation, the differential permeability was determinate, which allowed the 3D linear frequency-domain analyses at each remarkable harmonic. The total rotor loss of the interior PMSM was obtained by the summation of the harmonic losses. According to the authors of this method, the calculation time of this method was less than 1/100 of the conventional 3D time-domain FEA.

End-effects are ignored in 2D FEA, causing large error in the results of eddy-current loss compared to 3D FEA. An adjustment was proposed in [106], where losses calculated trough 2D FEA were corrected by the factor F:

$$F = \frac{3}{4}\frac{L^2}{w^2 + L^2} \tag{2}$$

Fir the same purpose, to compensate the end-effect neglected in 2D FEA, another correction factor was proposed in [107]:

$$C_3 = \left(\frac{l_c}{l_c + \tau_m}\right)^2 \tag{3}$$

where l_c and τ_m are respectively the length and pitch of the magnet.

Eddy-current losses were analyzed for axial-flux PMSMs with concentrated windings using 3D FEA [108,109] and for two cases, when PMs were insulated or non-insulated [110]. To calculate eddy-current losses in interior PMSMs by FEA with less time calculation, a coupled 2D and 3D method was presented in [111]. The 3D eddy-current analysis only of the PM was calculated from the flux distribution obtained by 2D FEA. An extended coupled method applied to surface-mounted PMSM was presented in [112], where the uniform air-gap was replaced by a variable equivalent air-gap to consider the space variation of the air-gap width. The authors of [113,114] proposed a multilayer-2D-2D coupled model to estimate eddy-current losses in PMs of axial-flux PMSMs. Firstly, static 2D FEA for different rotor positions was used to calculate the 1D flux density data. The build up of this latter provided a 2D air-gap flux density distribution. These data were transformed into the frequency-domain and finally were used to calculate the eddy-current in PM by 2D time harmonic FEA.

In [115], eddy-current losses were calculated by a 2D magnetic vector potential solver ($2D - A - \phi - Solver$), which was coupled to a modified axial $2D - T - \Omega - Solver$ to include the influence of the PM axial length. The influence of the pole coverage (i.e., ratio of magnet width and pole pitch) on PM power losses was investigated for both cylindrical and linear arrangement [116]. It was shown that the influence of the pole coverage on power losses was visible only at lower harmonic waves. 3D FEA was used to investigate the influence of multi-phase and multi-layer windings of surface-mounted PMSMs on the PMs eddy-current losses [117]. Different windings layouts and layers numbers were compared.

3.4. Hybrid Models

A hybrid method to calculate eddy-current losses in PMs was presented in [118]. It consisted of a derived current sheet by 2D analytical method combined with a 3D FEA. The magnetic field at the stator inner diameter was calculated, and then, the 2D current sheet was determined. This latter was extended axially, and 3D FEA was performed. A nodal method was developed in [119], and it was based on network-field coupled time-stepping FEA (NF-TS-FEA). In [120], 2D FEA for a non-segmented magnet machine was done to measure the magnetic field. The data array containing the flux density on a path was obtained. These data were used to obtain the induced eddy-current density by solving analytically an equation derived from Maxwell's second equation. The work in [121] performed a frequency analysis of the data using FFT analysis. The eddy-current loss summation for each frequency gave the total eddy-current loss with consideration of the skin effect and the harmonics of the air-gap magnetic flux density. The computationally-efficient FEA (CE-FEA) developed by [122] calculated the eddy-current loss from the numerically-obtained flux density through a path. By applying Faraday's law and integrating over a path, the total losses of the PM were obtained. This method incorporated the 3D end-effects. The algorithm for mapping eddy-current loss within surface-mounted PMSM over a wide range of operating conditions was presented in [123,124]. The method required four FEA simulations for open-circuit at rated speed (n_R), at particular reference speed (n_w) in the field weakening region, rated current in the quadrature axis (I_q), rated current in the direct axis (I_d), and reduced current in the d-axis. Coefficients a, b, c, and d were calculated from those simulations and introduced in the following equation to calculate PM losses:

$$P_{PM} = \left(a \left(\frac{n_w}{n_R} \right)^2 I_q^2 + b I_d^2 + C I_d + d \left(\frac{n_w}{n_R} \right)^2 \right) \left(\frac{n}{n_w} \right)^2 \tag{4}$$

In [125], a 3D numerical hybrid method (NHM) of the PM eddy-current loss in axial-flux PMSM was described. The NHM is based on 3D FEA, where the PM magnetic flux density was determined in resistance-limited conditions. Then, the 3D PM eddy-current loss was calculated by the 3D finite-difference method. The semi-analytical model combined with 2D FEA was proposed in [126,127] to estimate the PM eddy-current losses. The flux density variation seen by the PMs was obtained from 2D FEA and then processed by an analytical model.

4. PM Loss Reduction

4.1. PM Segmentation

In [128–130], the effect of the number of PM segments per pole was investigated for conventional and modular PM brushless machines. It was shown that the circumferential segmentation allowed the reduction of the eddy-current losses in both machines. The use of a solid rotor core reduced the PMs eddy-current losses compared to a laminated one [131], and PMs' segmentation allowed the eddy-current loss reduction, but the losses in the rotor yoke increased. The segmentation was only useful for laminated rotors [45]. Losses caused by the slotting effect decreased with the number of PM segmentation for all motors. However, this was not the same case for losses caused by the inverter carrier. The carrier losses in the surface-mounted PM topology were more important compared to the interior and inset PMSM [132]. According to [133], in order to reach lowest eddy-current losses, the shortest PM side should be segmented. The concept of partial segmentation was introduced in [134], viz., single-sided partial PM segmentation (SS-PMS), double-sided partial PM segmentation (DS-PMS), and partial rotor yoke segmentation (PRYS). In the studied case, it was shown that the application of DS-PMS with four segments per PM and combined with PRYS with 128 segments gave a satisfactory loss reduction. In [135], it was proven that the eddy-current losses due to spatial the harmonics decreased by increasing the segmentation number, while those due temporal harmonics did not increase by decreasing the segmentation number. The non-uniform PM segmentation showed more reduction of PM eddy-current losses compared to the classical one; especially electrical machines with an integer number of slots per phase per pole [136].

Concentrated windings are characterized by higher orders of slot-harmonics and multi-loop eddy-currents distribution. In this case, the loss reduction effect by segmentation in surface-mounted PMSMs is smaller compared to interior PMSMs. The axial PM segmentation in the surface mounted PMSM is more effective than the circumferential one, while both segmentations have an identical effect on the interior PM topology [137].

4.2. Rotor Shape

Loss reduction due to the grooving was studied in [138]. It was shown that the grooved rotor surface of PMSM had less ripple loss compared to the ungrooved one. This technique of grooving was applied in [139], where it was associated with a pulse width modulation (PWM) technique. To decrease the PM eddy-current losses for interior PMSMs with concentrated windings, the authors of [140,141] optimized the shapes of the stator teeth and rotor bridges. Reduction of slot opening had the effect of decreasing of PM eddy-current losses [142]. An optimal adjustment of the number of PM segmentation in the x- and z-direction was proposed in [143,144] to reach the optimized reduction of parasitic eddy-current losses. The authors of [145] realized cuts in the rotor yoke to increase the reluctance without modifying the PM flux path. Three types of cut have been realized to reach a maximum limitation of MMF subharmonic flux. Results showed that this modification of the rotor yoke geometry allowed a rotor loss reduction.

The authors of [146] proposed a special rotor shape to reduce PMs' eddy-current losses. They used flux barriers and slits on the rotor surface of multi-layer interior PMSMs. According to [147], the introduction of flux barriers into the rotor yoke along the d-axis led to lower eddy-current losses.

A tooth-coil open-slot axial-flux machine was designed in [148], to reduce PMs eddy-current losses. It was demonstrated via the prototype that steel laminations on top of the PMs allowed the PM flux linkage maximization and eddy-current loss minimization.

4.3. Material Type Effect

The materials of the retaining sleeves influence the rotor losses of PMSMs. The use of a carbon-fiber/epoxy sleeve gave 5.9-times lower rotor losses than the Inconel718 sleeve according to [149]. The PM losses decreased when a copper layer was put between a carbon fiber sleeve and PM ring [150].

4.4. Spatial Filter

A comparison based on MMF harmonics of SPM and IPM machines, with a pole/slot ratio of 2/3 and concentrated windings, was made in [38]. It was shown that the spatial filter effect of the rotor yoke in IPM machines allowed a significant reduction of PM losses. The increase of the PM depth is a viable solution to reduce PM losses.

In [151], optimized auxiliary slots were added to the structure. This technique allowed canceling partially the asynchronous harmonics produced by the armature field, which resulted in the PM loss reduction.

5. PM Loss Measurement

The major difficulty of measuring PM losses in electrical machines is to separate them from the total losses. Indeed, the measurement of the open- or short-circuit core loss generates the total losses in the machine [152]. By using the thermometric method, it is possible to measure PM eddy-current losses. Figure 2 shows a manipulation realized in [153]. The sintered PM was inserted into a solenoid coil, and its temperature was measured by thermocouples. The system was supplied by an alternating magnetic field, and the eddy-current losses were calculated following:

$$Q = CVD\frac{dT}{dt} \tag{5}$$

The same apparatus principle was used in [154] to study the PM segmentation effect experimentally.

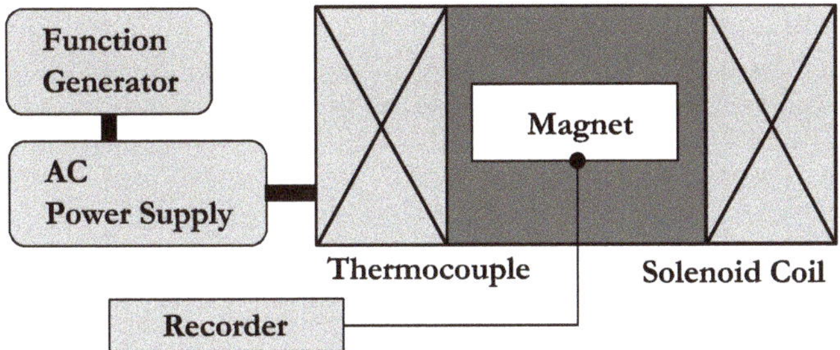

Figure 2. Schematic apparatus of eddy-current measurement [153].

To measure rotor losses of an axial-flux PMSMs due to slot-opening, the machine with magnetized PMs was driven without current. The impact of current/MMF harmonics frequency and the resultant losses was obtained when the machine was driven by AC/DC current with demagnetized PMs [145].

In [155], the rotor losses of an axial-flux PMSM were investigated by the measurement of different operating conditions. Static and rotary tests were done in [156] to measure rotor losses. However, the measurements were done for the total rotor losses, not for PM losses.

6. Thermal Analysis

The eddy-current losses in electrical machines are dissipated by radiation and natural convection. The temperature of the machine and of PM materials must be considered in electrical machines' design [157]. An analytical lumped-circuit method was used in [158], to analyze the thermal field of PM motor and generator. It was shown that the PWM duty ratio and the DC supply influenced directly the difference in temperature rises.

In [159], the influence of the axial cooling air in the air-gap on rotor air-friction losses was studied. A subminiature noncontact infrared thermometer was used to measure the rotor temperature.

7. Conclusions

This paper is a synthesis of the analysis and methods of eddy-current loss prediction in PMSMs. The different sources of these losses were outlined and divided into several categories. After the identification of the losses, their calculation was approached. The precision of the results and the time consumption are the compromises of the PMs eddy-current loss computation. Indeed, the 2D/3D analytical models allow a fast calculation, but a lower precision, considering the associated simplifying assumptions. On other hand, FEA allows more accuracy results, but with a greater time consumption. Hybrid models, as well as the introduction of correction coefficients are often proposed as a solution.

One of the consequences of eddy-currents in PMs is their demagnetization by a high temperature. Thereby, loss reduction by limiting these eddy-currents is necessary. This can be done with several methods, viz., PMs' segmentation, modifying the rotor shape, selecting the material type, and using a spatial filter.

The PM loss measurement is very difficult because the collective losses' separation is not easy. The most common methods to perform the measurement of losses are collective losses' separation and thermometric methods. This has been little discussed in the literature. Finally, the thermal analysis of PMs was discussed, and it must be taken into account in the design of PM electrical machines.

Author Contributions: D.O. and F.D. prepared the structure of the paper; D.O. wrote the paper, F.D. corrected and improved its content.

Acknowledgments: This work was supported by RENAULT-SAS, Guyancourt, France. This scientific study is related to the project "Conception optimale des chaines de Traction Electrique" (COCTEL) financed by the "Agence De l'Environnement et de la Maîtrise de l'Énergie" (ADEME) in the program "Véhicule du future des Investissements de l'Avenir".

Conflicts of Interest: The authors declare no conflict of interest.

References

1. Stoll, R.L. *The Analysis of Eddy Currents*; Oxford University Press: Oxford, UK, 1974.
2. Cassat, A.; Espanet, C.; Coleman, R.; Burdet, L.; Leleu, E.; Torregrossa, D.; M'Boua, J.; Miraoui, A. A practical solution to mitigate vibrations in industrial PM motors having concentric windings. *IEEE Trans. Ind. Appl.* **2012**, *48*, 1526–1538. [CrossRef]
3. Ouamara, D.; Dubas, F.; Randi, S.A.; Benallal, M.N.; Espanet, C. Electromagnetic Comparison of 3-, 5- and 7-Phases Permanent-Magnet Synchronous Machines: Mild Hybrid Traction Application. *Mediterr. J. Model. Simul.* **2016**, *6*, 012–022.
4. Huang, W.Y.; Bettayeb, A.; Kaczmarek, R.; Vannier, J.C. Optimization of magnet segmentation for reduction of eddy-current losses in permanent magnet synchronous machine. *IEEE Trans. Energy Convers.* **2010**, *25*, 381–387. [CrossRef]
5. Belahcen, A.; Arkkio, A. Permanent magnets models and losses in 2D FEM simulation of electrical machines. In Proceedings of the 19th International Conference on Electrical Machines, Rome, Italy, 6–8 September 2010. [CrossRef]
6. Alexandrova, J.; Jussila, H.; Nerg, J.; Pyrhönen, J. Comparison between models for eddy-current loss calculations in rotor surface-mounted permanent magnets. In Proceedings of the 19th International Conference on Electrical Machines, Rome, Italy, 6–8 September 2010. [CrossRef]
7. Tessarolo, A. A survey of state-of-the-art methods to compute rotor eddy-current losses in synchronous permanent magnet machines. In Proceedings of the 2017 IEEE Workshop on Electrical Machines Design, Control and Diagnosis, Nottingham, UK, 20–21 April 2017; pp. 12–19. [CrossRef]
8. Boules, N. Impact of slot harmonics on losses of high-speed permanent magnet machines with a magnet retaining ring. *Electr. Mach. Power Syst.* **1981**, *6*, 527–539. [CrossRef]
9. Bouillault, F.; Razek, A. Eddy-currents due to stator teeth in synchronous machine rotors. *IEEE Trans. Magn.* **1984**, *20*, 1939–1941. [CrossRef]
10. Abu Sharkh, S.M.; Harris, M.R.; Irenji, N.T. Calculation of rotor eddy-current loss in high-speed pm alternators. In Proceedings of the Eighth International Conference on Electrical Machines and Drives, Cambridge, UK, 1–3 September 1997; pp. 170–174. [CrossRef]
11. Hor, P.; Zhu, Z.; Howe, D. Eddy-current loss in a moving-coil tubular permanent magnet motor. *IEEE Trans. Magn.* **1999**, *35*, 3601–3603. [CrossRef]
12. Wills, D.A.; Kamper, M.J. Analytical prediction of rotor eddy-current loss due to stator slotting in PM machines. In Proceedings of the 2010 IEEE Energy Conversion Congress and Exposition, Atlanta, GA, USA, 12–16 September 2010; pp. 992–995. [CrossRef]
13. Dubas, F.; Espanet, C. Semi-Analytical Solution of 2D Rotor Eddy-Current Losses due to the Slotting Effect in SMPMM. In Proceedings of the 17th Conference on the Computation of Electromagnetic Fields, Florianópolis, Brazil, 22–26 November 2009; pp. 20–25.
14. Markovic, M.; Perriard, Y. A simplified determination of the permanent magnet (PM) eddy-current losses due to slotting in a PM rotating motor. In Proceedings of the International Conference on Electrical Machines and Systems (ICEMS), Wuhan, China, 17–20 October 2008; pp. 309–313.
15. Bode, C.; Canders, W.R. Advanced calculation of eddy-current losses in PMSM with tooth windings. In Proceedings of the XIX International Conference on Electrical Machines, Rome, Italy, 6–8 September 2010; Volume 1, pp. 1–6. [CrossRef]
16. de la Barriere, O.; Hlioui, S.; Ben Ahmed, H.; Gabsi, M. An Analytical Model for the Computation of No-Load Eddy-Current Losses in the Rotor of a Permanent Magnet Synchronous Machine. *IEEE Trans. Magn.* **2016**, *52*, 1–13. [CrossRef]
17. Fang, Z.X.; Zhu, Z.Q.; Wu, L.J.; Xia, Z.P. Simple and accurate analytical estimation of slotting effect on magnet loss in fractional-slot surface-mounted PM machines. In Proceedings of the 2012 XXth International Conference on Electrical Machines, Marseille, France, 2–5 September 2012; Volume 37, pp. 464–470. [CrossRef]
18. Nair, S.S.; Wang, J.; Chin, R.; Chen, L.; Sun, T. Analytical Prediction of 3D Magnet Eddy Current Losses in Surface Mounted PM Machines Accounting Slotting Effect. *IEEE Trans. Energy Convers.* **2017**, *32*, 414–423. [CrossRef]

19. Wu, L.J.; Zhu, Z.Q.; Staton, D.; Popescu, M.; Hawkins, D. Analytical model for predicting magnet loss of surface-mounted permanent magnet machines accounting for slotting effect and load. *IEEE Trans. Magn.* **2012**, *48*, 107–117. [CrossRef]
20. Nagarkatti, A.K.; Mohammed, O.A.; Demerdash, N.A. Special Losses in Rotors of Electronically Commutated Brushless DC Motors Induced by Non-Uniformly Rotating Armature MMFS. *IEEE Power Eng. Rev.* **1982**, *PER-2*, 33. [CrossRef]
21. Deng, F. Improved analytical modeling of commutation losses including space harmonic effects in permanent magnet brushless DC motors. In Proceedings of the 1997 IEEE International Electric Machines and Drives Conference Record, Milwaukee, WI, USA, 18–21 May 1997; pp. WB2/4.1–WB2/4.3. [CrossRef]
22. Deng, F. Commutation-caused eddy-current losses in permanent-magnet brushless DC motors. *IEEE Trans. Magn.* **1997**, *33*, 4310–4318. [CrossRef]
23. Atallah, K.; Howe, D.; Mellor, P.; Stone, D. Rotor loss in permanent-magnet brushless AC machines. *IEEE Trans. Ind. Appl.* **2000**, *36*, 1612–1618. [CrossRef]
24. Zhu, Z.; Ng, K.; Schofield, N.; Howe, D. Improved analytical modelling of rotor eddy-current loss in brushless machines equipped with surface-mounted permanent magnets. *IEE Proc. Electr. Power Appl.* **2004**, *151*, 641. [CrossRef]
25. Ishak, D.; Zhu, Z.; Howe, D. Eddy-current loss in the rotor magnets of permanent-magnet brushless machines having a fractional number of slots per pole. *IEEE Trans. Magn.* **2005**, *41*, 2462–2469. [CrossRef]
26. Dubas, F.; Espanet, C.; Miraoui, A. Field Diffusion Equation in High-Speed Surface-Mounted Permanent Magnet Motors, Parasitic Eddy-Current Losses. In Proceedings of the 6th International Symposium on Advanced Electromechanical Motion Systems, Lausanne, Switzerland, 27–29 September 2005; pp. 1–6.
27. Nakano, M.; Kometani, H.; Kawamura, M. A study on Eddy-current losses in rotors of surface permanent-magnet synchronous machines. *IEEE Trans. Ind. Appl.* **2006**, *42*, 429–435. [CrossRef]
28. Bianchi, N.; Bolognani, S.; Fomasiero, E. A General Approach to Determine the Rotor Losses in Three-Phase Fractional-Slot PM Machines. In Proceedings of the 2007 IEEE International Electric Machines & Drives Conference, Antalya, Turkey, 3–5 May 2007; Volume 1, pp. 634–641. [CrossRef]
29. Bianchi, N.; Fornasiero, E. Impact of MMF Space Harmonic on Rotor Losses in Fractional-Slot Permanent-Magnet Machines. *IEEE Trans. Energy Convers.* **2009**, *24*, 323–328. [CrossRef]
30. Bianchi, N.; Fornasiero, E. Index of rotor losses in three-phase fractional-slot permanent magnet machines. *IET Electr. Power Appl.* **2009**, *3*, 381. [CrossRef]
31. Bianchi, N.; Bolognani, S.; Fornasiero, E. An Overview of Rotor Losses Determination in Three-Phase Fractional-Slot PM Machines. *IEEE Trans. Ind. Appl.* **2010**, *46*, 2338–2345. [CrossRef]
32. Li, J.; Choi, D.W.; Son, D.H.; Cho, Y.H. Effects of MMF Harmonics on Rotor Eddy-Current Losses for Inner-Rotor Fractional Slot Axial Flux Permanent Magnet Synchronous Machines. *IEEE Trans. Magn.* **2012**, *48*, 839–842. [CrossRef]
33. El-Refaie, A.; Shah, M.; Qu, R.; Kern, J. Effect of Number of Phases on Losses in Conducting Sleeves of Surface PM Machine Rotors Equipped With Fractional-Slot Concentrated Windings. *IEEE Trans. Ind. Appl.* **2008**, *44*, 1522–1532. [CrossRef]
34. Fornasiero, E.; Bianchi, N.; Bolognani, S. Slot Harmonic Impact on Rotor Losses in Fractional-Slot Permanent-Magnet Machines. *IEEE Trans. Ind. Electron.* **2012**, *59*, 2557–2564. [CrossRef]
35. Dajaku, G.; Gerling, D. Eddy-current loss minimization in rotor magnets of PM machines using high-efficiency 12-teeth/10-slots winding topology. In Proceedings of the 2011 International Conference on Electrical Machines and Systems, Beijing, China, 20–23 August 2011. [CrossRef]
36. Aslan, B.; Semail, E.; Legranger, J. Analytical model of magnet eddy-current volume losses in multi-phase PM machines with concentrated winding. In Proceedings of the 2012 IEEE Energy Conversion Congress and Exposition (ECCE), Raleigh, NC, USA, 15–20 September 2012; Volume L, pp. 3371–3378. [CrossRef]
37. Aslan, B.; Semail, E.; Legranger, J. General Analytical Model of Magnet Average Eddy-Current Volume Losses for Comparison of Multiphase PM Machines With Concentrated Winding. *IEEE Trans. Energy Convers.* **2014**, *29*, 72–83. [CrossRef]
38. Li, J.; Xu, Y.; Zou, J.; Wang, Q.; Liang, W. Analysis and Reduction of Magnet Loss by Deepening Magnets in Interior Permanent-Magnet Machines With a Pole/Slot Ratio of 2/3. *IEEE Trans. Magn.* **2015**, *51*, 1–4. [CrossRef]

39. van der Veen, J.; Offringa, L.; Vandenput, A. Minimising rotor losses in high-speed high-power permanent magnet synchronous generators with rectifier load. *IEE Proc. Electr. Power Appl.* **1997**, *144*, 331–337. [CrossRef]
40. Polinder, H.; Hoeijmakers, M. Eddy-current losses in the segmented surface-mounted magnets of a PM machine. *IEE Proc. Electr. Power Appl.* **1999**, *146*, 261–266. [CrossRef]
41. Zhu, Z.; Ng, K.; Schofield, N.; Howe, D. Analytical prediction of rotor eddy-current loss in brushless machines equipped with surface-mounted permanent magnets. I. Magnetostatic field model. In Proceedings of the Fifth International Conference on Electrical Machines and Systems, Shenyang, China, 18–20 August 2001; Volume 2, pp. 806–809. [CrossRef]
42. Schofield, N.; Ng, K.; Zhu, Z.Q.; Howe, D. Parasitic rotor losses in a brushless permanent magnet traction machine. In Proceedings of the Eighth International Conference on Electrical Machines and Drives, Cambridge, UK, 1–3 September 1997; pp. 200–204.
43. Deng, F.; Nehl, T. Analytical modeling of eddy-current losses caused by pulse-width-modulation switching in permanent-magnet brushless direct-current motors. *IEEE Trans. Magn.* **1998**, *34*, 3728–3736. [CrossRef]
44. Bellara, A.; Bali, H.; Belfkira, R.; Amara, Y.; Barakat, G. Analytical Prediction of Open-Circuit Eddy-Current Loss in Series Double Excitation Synchronous Machines. *IEEE Trans. Magn.* **2011**, *47*, 2261–2268. [CrossRef]
45. Sergeant, P.; Van den Bossche, A. Segmentation of Magnets to Reduce Losses in Permanent-Magnet Synchronous Machines. *IEEE Trans. Magn.* **2008**, *44*, 4409–4412. [CrossRef]
46. Marashi, A.N.; Abbaszadeh, K.; Alam, F.R. Analysis and reduction of magnet eddy-current losses in surface mounted permanent magnet machines. In Proceedings of the 2014 22nd Iranian Conference on Electrical Engineering (ICEE), Tehran, Iran, 20–22 May 2014; pp. 782–786. [CrossRef]
47. Jumayev, S.; Merdzan, M.; Boynov, K.O.; Paulides, J.J.H.; Pyrhonen, J.; Lomonova, E.A. The Effect of PWM on Rotor Eddy-Current Losses in High-Speed Permanent Magnet Machines. *IEEE Trans. Magn.* **2015**, *51*, 1–4. [CrossRef]
48. Cheng, M.; Zhu, S. Calculation of PM Eddy Current Loss in IPM Machine Under PWM VSI Supply With Combined 2D FE and Analytical Method. *IEEE Trans. Magn.* **2017**, *53*, 1–12. [CrossRef]
49. Yamazaki, K.; Abe, A. Loss Investigation of Interior Permanent-Magnet Motors Considering Carrier Harmonics and Magnet Eddy Currents. *IEEE Trans. Ind. Appl.* **2009**, *45*, 659–665. [CrossRef]
50. Zhu, Z.Q.; Chen, Y.; Howe, D.; Gliemann, J.H. Rotor Eddy Current Loss in Single-Phase Permanent Magnet Brushless DC Motor. In Proceedings of the 2007 IEEE Industry Applications Annual Meeting, New Orleans, LA, USA, 23–27 September 2007; pp. 537–543. [CrossRef]
51. Yamazaki, K.; Fukushima, Y.; Sato, M. Loss Analysis of Permanent-Magnet Motors with Concentrated Windings—Variation of Magnet Eddy-Current Loss Due to Stator and Rotor Shapes. *IEEE Trans. Ind. Appl.* **2009**, *45*, 1334–1342. [CrossRef]
52. Ding, X.; Mi, C. Modeling of eddy-current loss in the magnets of permanent magnet machines for hybrid and electric vehicle traction applications. In Proceedings of the 2009 IEEE Vehicle Power and Propulsion Conference, Dearborn, MI, USA, 7–10 September 2009; Volume 20, pp. 419–424. [CrossRef]
53. Jumayev, S.; Borisavljevic, A.; Boynov, K.; Lomonova, E.A.; Pyrhonen, J. Analysis of rotor eddy-current losses in slotless high-speed permanent magnet machines. In Proceedings of the 2014 16th European Conference on Power Electronics and Applications, Lappeenranta, Finland, 26–28 August 2014. [CrossRef]
54. Hemeida, A.; Sergeant, P.; Vansompel, H. Comparison of Methods for Permanent Magnet Eddy-Current Loss Computations with and without Reaction Field Considerations in Axial Flux PMSM. *IEEE Trans. Magn.* **2015**, *51*, 1–11. [CrossRef]
55. Qazalbash, A.A.; Sharkh, S.M.; Irenji, N.T.; Wills, R.G.; Abusara, M.A. Rotor Eddy Current Power Loss in Permanent Magnet Synchronous Generators Feeding Uncontrolled Rectifier Loads. *IEEE Trans. Magn.* **2014**, *50*, 1–9. [CrossRef]
56. Balamurali, A.; Lai, C.; Mollaeian, A.; Loukanov, V.; Kar, N.C. Analytical Investigation Into Magnet Eddy Current Losses in Interior Permanent Magnet Motor Using Modified Winding Function Theory Accounting for Pulsewidth Modulation Harmonics. *IEEE Trans. Magn.* **2016**, *52*, 1–5. [CrossRef]
57. Narjes, G.; Ponick, B. Novel Method for the Determination of Eddy Current Losses in the Permanent Magnets of a High-Speed Synchronous Machine. In Proceedings of the 2018 XIII International Conference on Electrical Machines (ICEM), Alexandroupoli, Greece, 3–6 September 2018; pp. 1285–1290. [CrossRef]

58. Fukuma, A.; Kanazawa, S.; Miyagi, D.; Takahashi, N. Investigation of AC loss of permanent magnet of SPM motor considering hysteresis and eddy-current losses. *IEEE Trans. Magn.* **2005**, *41*, 1964–1967. [CrossRef]
59. Pyrhonen, J.; Ruoho, S.; Nerg, J.; Paju, M.; Tuominen, S.; Kankaanpaa, H.; Stern, R.; Boglietti, A.; Uzhegov, N. Hysteresis Losses in Sintered NdFeB Permanent Magnets in Rotating Electrical Machines. *IEEE Trans. Ind. Electron.* **2015**, *62*, 857–865. [CrossRef]
60. Petrov, I.; Egorov, D.; Link, J.; Stern, R.; Ruoho, S.; Pyrhonen, J. Hysteresis Losses in Different Types of Permanent Magnets Used in PMSMs. *IEEE Trans. Ind. Electron.* **2017**, *64*, 2502–2510. [CrossRef]
61. Egorov, D.; Petrov, I.; Pyrhonen, J.; Link, J.; Stern, R. Hysteresis Loss in Ferrite Permanent Magnets in Rotating Electrical Machinery. *IEEE Trans. Ind. Electron.* **2018**, *65*, 9280–9290. [CrossRef]
62. Sikora, R.; Purczynski, J.; Lipinski, W.; Gramz, M. Use of variational methods to the eddy-currents calculation in thin conducting plates. *IEEE Trans. Magn.* **1978**, *14*, 383–385. [CrossRef]
63. Davey, K.; Turner, L. Prediction of transient eddy-current fields using surface impedance methods. *IEEE Trans. Magn.* **1989**, *25*, 4156–4158. [CrossRef]
64. Polinder, H.; Hoeijmakers, M. Eddy-current losses in the permanent magnets of a PM machine. In Proceedings of the 1997 Eighth International Conference on Electrical Machines and Drives, Cambridge, UK, 1–3 September 1997; pp. 138–142.
65. Spooner, E.; Williamson, A. Parasitic Losses in Modular Permanent-Magnet Generators. *IEE Proc. Electr. Power Appl.* **1998**, *145*, 485–496. [CrossRef]
66. Wang, J.; Atallah, K.; Chin, R.; Arshad, W.M.; Lendenmann, H. Rotor Eddy-Current Loss in Permanent-Magnet Brushless AC Machines. *IEEE Trans. Magn.* **2010**, *46*, 2701–2707. [CrossRef]
67. Zhu, Z.; Ng, K.; Schofield, N.; Howe, D. Analytical Prediction of Rotor Eddy Current Loss in Brushless Machines Equipped with Surface-Mounted Permanent Magnets. II: Accounting for Eddy Current Reaction Field. In Proceedings of the Fifth International Conference on Electrical Machines and Systems, Shenyang, China, 18–20 August 2001; Volume 2, pp. 810–813. [CrossRef]
68. Markovic, M.; Perriard, Y. An Analytical Determination of Eddy-Current Losses in a Configuration With a Rotating Permanent Magnet. *IEEE Trans. Magn.* **2007**, *43*, 3380–3386. [CrossRef]
69. Miljavec, D.; Zidarič, B. Eddy-current losses in permanent magnets of the BLDC machine. *COMPEL Int. J. Comput. Math. Electr. Electron. Eng.* **2007**, *26*, 1095–1104. [CrossRef]
70. Markovic, M.; Perriard, Y. Analytical Solution for Rotor Eddy-Current Losses in a Slotless Permanent-Magnet Motor: The Case of Current Sheet Excitation. *IEEE Trans. Magn.* **2008**, *44*, 386–393. [CrossRef]
71. Wang, J.; Papini, F.; Chin, R.; Arshad, W.M.; Lendenmann, H. Computationally efficient approaches for evaluation of rotor eddy current loss in permanent magnet brushless machines. In Proceedings of the 2009 International Conference on Electrical Machines and Systems, Tokyo, Japan, 15–18 November 2009. [CrossRef]
72. Tariq, A.R.; Nino-Baron, C.E.; Strangas, E.G. Iron and Magnet Losses and Torque Calculation of Interior Permanent Magnet Synchronous Machines Using Magnetic Equivalent Circuit. *IEEE Trans. Magn.* **2010**, *46*, 4073–4080. [CrossRef]
73. Nuscheler, R. Two-dimensional analytical model for eddy-current rotor loss calculation of PMS machines with concentrated stator windings and a conductive shield for the magnets. In Proceedings of the XIX International Conference on Electrical Machines, Rome, Italy, 6–8 September 2010. [CrossRef]
74. Bettayeb, A.; Jannot, X.; Vannier, J.C. Analytical calculation of rotor magnet eddy-current losses for high speed IPMSM. In Proceedings of the XIX International Conference on Electrical Machines, Rome, Italy, 6–8 September 2010. [CrossRef]
75. Yoshida, Y.; Nakamura, K.; Ichinokura, O. A Method for Calculating Eddy Current Loss Distribution Based on Reluctance Network Analysis. *IEEE Trans. Magn.* **2011**, *47*, 4155–4158. [CrossRef]
76. Yoshida, Y.; Nakamura, K.; Ichinokura, O. Basic examination of eddy-current loss estimation in SPM motor based on electric and magnetic networks. In Proceedings of the 2012 XXth International Conference on Electrical Machines, Marseille, France, 2–5 September 2012; Volume 48, pp. 1586–1591. [CrossRef]
77. Hemeida, A.; Sergeant, P. Analytical modeling of eddy-current losses in Axial Flux PMSM using resistance network. In Proceedings of the 2014 International Conference on Electrical Machines (ICEM), Berlin, Germany, 2–5 September 2014; pp. 2688–2694. [CrossRef]

78. Wu, L.J.; Zhu, Z.Q.; Staton, D.; Popescu, M.; Hawkins, D. Analytical Modeling and Analysis of Open-Circuit Magnet Loss in Surface-Mounted Permanent-Magnet Machines. *IEEE Trans. Magn.* **2012**, *48*, 1234–1247. [CrossRef]
79. Pyrhonen, J.; Jussila, H.; Alexandrova, Y.; Rafajdus, P.; Nerg, J. Harmonic Loss Calculation in Rotor Surface Permanent Magnets—New Analytic Approach. *IEEE Trans. Magn.* **2012**, *48*, 2358–2366. [CrossRef]
80. Rahideh, A.; Korakianitis, T. Analytical magnetic field distribution of slotless brushless permanent magnet motors – Part I. Armature reaction field, inductance and rotor eddy-current loss calculations. *IET Electr. Power Appl.* **2012**, *6*, 628. [CrossRef]
81. Arumugam, P.; Hamiti, T.; Gerada, C. Estimation of Eddy Current Loss in Semi-Closed Slot Vertical Conductor Permanent Magnet Synchronous Machines Considering Eddy Current Reaction Effect. *IEEE Trans. Magn.* **2013**, *49*, 5326–5335. [CrossRef]
82. Gotovac, G.; Lampic, G.; Miljavec, D. Analytical Model of Permeance Variation Losses in Permanent Magnets of the Multipole Synchronous Machine. *IEEE Trans. Magn.* **2013**, *49*, 921–928. [CrossRef]
83. Dubas, F.; Rahideh, A. Two-Dimensional Analytical Permanent-Magnet Eddy-Current Loss Calculations in Slotless PMSM Equipped With Surface-Inset Magnets. *IEEE Trans. Magn.* **2014**, *50*, 54–73. [CrossRef]
84. Martin, F.; Zaim, M.E.H.; Tounzi, A.; Bernard, N. Improved Analytical Determination of Eddy Current Losses in Surface Mounted Permanent Magnets of Synchronous Machine. *IEEE Trans. Magn.* **2014**, *50*, 1–9. [CrossRef]
85. Paradkar, M.; Bocker, J. 2D analytical model for estimation of eddy-current loss in the magnets of IPM machines considering the reaction field of the induced eddy currents. In Proceedings of the 2015 IEEE International Electric Machines & Drives Conference (IEMDC), Coeur d'Alene, ID, USA, 10–13 May 2015; pp. 1096–1102. [CrossRef]
86. Pfister, P.D.; Yin, X.; Fang, Y. Slotted Permanent-Magnet Machines: General Analytical Model of Magnetic Fields, Torque, Eddy Currents, and Permanent-Magnet Power Losses Including the Diffusion Effect. *IEEE Trans. Magn.* **2016**, *52*, 1–13. [CrossRef]
87. Qazalbash, A.A.; Sharkh, S.M.; Irenji, N.T.; Wills, R.G.; Abusara, M.A. Rotor eddy loss in high-speed permanent magnet synchronous generators. *IET Electr. Power Appl.* **2015**, *9*, 370–376. [CrossRef]
88. Mirzaei, M.; Binder, A.; Funieru, B.; Susic, M. Analytical Calculations of Induced Eddy Currents Losses in the Magnets of Surface Mounted PM Machines With Consideration of Circumferential and Axial Segmentation Effects. *IEEE Trans. Magn.* **2012**, *48*, 4831–4841. [CrossRef]
89. Masmoudi, A.; Masmoudi, A. 3D Analytical Model With the End Effect Dedicated to the Prediction of PM Eddy-Current Loss in FSPMMs. *IEEE Trans. Magn.* **2015**, *51*, 1–11. [CrossRef]
90. Paradkar, M.; Bocker, J. 3D analytical model for estimation of eddy-currentlosses in the magnets of IPM machine considering the reaction field of the induced eddy currents. In Proceedings of the 2015 IEEE Energy Conversion Congress and Exposition (ECCE), Montreal, QC, Canada, 20–24 September 2015; pp. 2862–2869. [CrossRef]
91. Chen, L.; Wang, J.; Nair, S.S. An Analytical Method for Predicting 3D Eddy Current Loss in Permanent Magnet Machines Based on Generalized Image Theory. *IEEE Trans. Magn.* **2016**, *52*, 1–11. [CrossRef]
92. Nair, S.S.; Wang, J.; Chen, L.; Chin, R.; Manolas, I.; Svechkarenko, D. Prediction of 3D High-Frequency Eddy Current Loss in Rotor Magnets of SPM Machines. *IEEE Trans. Magn.* **2016**, *52*, 1–10. [CrossRef]
93. Nair, S.S.; Wang, J.; Chen, L.; Chin, R.; Manolas, I.; Svechkarenko, D. Computationally Efficient 3D Eddy Current Loss Prediction in Magnets of Interior Permanent Magnet Machines. *IEEE Trans. Magn.* **2016**, *52*, 1–10. [CrossRef]
94. Nair, S.S.; Wang, J.; Chen, L.; Chin, R.; Manolas, I.; Svechkarenko, D. Computationally efficient 3D rotor eddy-current loss prediction in permanent magnet machines. In Proceedings of the 2016 XXII International Conference on Electrical Machines (ICEM), Lausanne, Switzerland, 4–7 September 2016; pp. 1426–1432. [CrossRef]
95. Chin, R.; Chen, L.; Manolas, I.; Wang, J.; Nair, S.; Svechkarenko, D. 3D Analytical Slotting-Effect Model for Magnet Loss Prediction in SPM Machines. In Proceedings of the 8th IET International Conference on Power Electronics, Machines and Drives (PEMD 2016), Glasgow, UK, 19–21 April 2016.
96. Nair, S.S.; Chen, L.; Wang, J.; Chin, R.; Manolas, I.; Svechkarenko, D. Computationally efficient 3D analytical magnet loss prediction in surface mounted permanent magnet machines. *IET Electr. Power Appl.* **2017**, *11*, 9–18. [CrossRef]

97. Mecrow, B.; Masterman, J.M. Determination of Rotor Eddy Current Losses in Permanent Magnet Machines. In Proceedings of the Sixth International Conference on Electrical Machines and Drives, Oxford, UK, 8–10 September 1993; pp. 299–304.
98. Kamiya, Y.; Onuki, T. 3D eddy-current analysis by the finite element method using double nodes technique. *IEEE Trans. Magn.* **1996**, *32*, 741–744. [CrossRef]
99. Kawase, Y.; Ota, T.; Fukunaga, H. 3D eddy-current analysis in permanent magnet of interior permanent magnet motors. *IEEE Trans. Magn.* **2000**, *36*, 1863–1866. [CrossRef]
100. Yoshida, K.; Hita, Y.; Kesamaru, K. Eddy-current loss analysis in PM of surface-mounted-PM SM for electric vehicles. *IEEE Trans. Magn.* **2000**, *36*, 1941–1944. [CrossRef]
101. Yamazaki, K.; Watari, S. Loss analysis of permanent-magnet motor considering carrier harmonics of PWM inverter using combination of 2D and 3D finite-element method. *IEEE Trans. Magn.* **2005**, *41*, 1980–1983. [CrossRef]
102. Pang, Y.; Zhu, Z.; Howe, D.; Iwasaki, S.; Deodhar, R.; Pride, A. Eddy Current Loss in the Frame of a Flux-Switching Permanent Magnet Machine. *IEEE Trans. Magn.* **2006**, *42*, 3413–3415. [CrossRef]
103. Al-Naemi, F.I.; Moses, A.J. FEM Modeling of Rotor Losses in PM Motors. *J. Magn. Magn. Mater.* **2006**, *304*, 794–797. [CrossRef]
104. Deak, C.; Petrovic, L.; Binder, A.; Mirzaei, M.; Irimie, D.; Funieru, B. Calculation of eddy-current losses in permanent magnets of synchronous machines. In Proceedings of the 2008 International Symposium on Power Electronics, Electrical Drives, Automation and Motion, Ischia, Italy, 11–13 June 2008; Volume 30, pp. 26–31. [CrossRef]
105. Yamazaki, K.; Kanou, Y. Rotor Loss Analysis of Interior Permanent Magnet Motors Using Combination of 2D and 3D Finite Element Method. *IEEE Trans. Magn.* **2009**, *45*, 1772–1775. [CrossRef]
106. Ruoho, S.; Santa-Nokki, T.; Kolehmainen, J.; Arkkio, A. Modeling Magnet Length In 2D Finite-Element Analysis of Electric Machines. *IEEE Trans. Magn.* **2009**, *45*, 3114–3120. [CrossRef]
107. Fadriansyah, T.; Strous, T.D.; Polinder, H. Axial segmentation and magnets losses of SMPM machines using 2D FE method. In Proceedings of the 2012 XXth International Conference on Electrical Machines, Marseille, France, 2–5 September 2012; pp. 577–581. [CrossRef]
108. Yang, X.; Patterson, D.; Hudgins, J.; Colton, J. FEA estimation and experimental validation of solid rotor and magnet eddy-current loss in single-sided axial flux permanent magnet machines. In Proceedings of the 2013 IEEE Energy Conversion Congress and Exposition, Denver, CO, USA, 15–19 September 2013; pp. 3202–3209. [CrossRef]
109. El-Hasan, T. Rotor eddy-current determination using finite element analysis for High-Speed Permanent Magnet Machines. In Proceedings of the 2014 IEEE 23rd International Symposium on Industrial Electronics (ISIE), Istanbul, Turkey, 1–4 June 2014; pp. 885–889. [CrossRef]
110. Li, J.; Choi, D.W.; Cho, C.H.; Koo, D.H.; Cho, Y.H. Eddy-Current Calculation of Solid Components in Fractional Slot Axial Flux Permanent Magnet Synchronous Machines. *IEEE Trans. Magn.* **2011**, *47*, 4254–4257. [CrossRef]
111. Okitsu, T.; Matsuhashi, D.; Muramatsu, K. Method for Evaluating the Eddy Current Loss of a Permanent Magnet in a PM Motor Driven by an Inverter Power Supply Using Coupled 2D and 3D Finite Element Analyses. *IEEE Trans. Magn.* **2009**, *45*, 4574–4577. [CrossRef]
112. Okitsu, T.; Matsuhashi, D.; Gao, Y.; Muramatsu, K. Coupled 2D and 3D Eddy Current Analyses for Evaluating Eddy Current Loss of a Permanent Magnet in Surface PM Motors. *IEEE Trans. Magn.* **2012**, *48*, 3100–3103. [CrossRef]
113. Vansompel, H.; Sergeant, P.; Dupré, L. Effect of segmentation on eddy-current loss in permanent-magnets of axial-flux PM machines using a multilayer-2D–2D coupled model. In Proceedings of the 2012 XXth International Conference on Electrical Machines, Marseille, France, 2–5 September 2012; pp. 228–232. [CrossRef]
114. Vansompel, H.; Sergeant, P.; Dupré, L. A Multilayer 2D–2D Coupled Model for Eddy Current Calculation in the Rotor of an Axial-Flux PM Machine. *IEEE Trans. Energy Convers.* **2012**, *27*, 784–791. [CrossRef]
115. Steentjes, S.; Boehmer, S.; Hameyer, K. Permanent Magnet Eddy-Current Losses in 2D FEM Simulations of Electrical Machines. *IEEE Trans. Magn.* **2015**, *51*, 1–4. [CrossRef]

116. Schmidt, E.; Kaltenbacher, M.; Wolfschluckner, A. Eddy-current losses in permanent magnets of surface mounted permanent magnet synchronous machines—Analytical calculation and high order finite element analyses. *e & i Elektrotechnik und Informationstechnik* **2017**, *134*, 148–155. [CrossRef]
117. Chen, Q.; Liang, D.; Jia, S.; Wan, X. Analysis of Multi-Phase and Multi-Layer Factional-Slot Concentrated-Winding on PM Eddy Current Loss Considering Axial Segmentation and Load Operation. *IEEE Trans. Magn.* **2018**, *54*, 1–6. [CrossRef]
118. Saban, D.M.; Lipo, T.A. Hybrid Approach for Determining Eddy-Current Losses in High-Speed PM Rotors. In Proceedings of the 2007 IEEE International Electric Machines & Drives Conference, Antalya, Turkey, 3–5 May 2007; Volume 1; pp. 658–661. [CrossRef]
119. Niu, S.; Chau, K.T.; Li, J.; Li, W. Eddy-Current Analysis of Double-Stator Inset-Type Permanent Magnet Brushless Machines. *IEEE Trans. Appl. Supercond.* **2010**, *20*, 1097–1101. [CrossRef]
120. Madina, P.; Poza, J.; Ugalde, G.; Almandoz, G. Magnet eddy-current loss calculation method for segmentation analysis on permanent magnet machines. In Proceedings of the 2011 14th European Conference on Power Electronics and Applications (EPE 2011), Birmingham, UK, 30 August–1 September 2011; pp. 1–9.
121. Lee, S.-Y.; Jung, H.-K. Eddy-current loss analysis in the rotor of permanent magnet traction motor with high power density. In Proceedings of the 2012 IEEE Vehicle Power and Propulsion Conference, Seoul, South Korea, 9–12 October 2012; pp. 210–214. [CrossRef]
122. Zhang, P.; Sizov, G.Y.; He, J.; Ionel, D.M.; Demerdash, N.A.O. Calculation of Magnet Losses in Concentrated-Winding Permanent-Magnet Synchronous Machines Using a Computationally Efficient Finite-Element Method. *IEEE Trans. Ind. Appl.* **2013**, *49*, 2524–2532. [CrossRef]
123. Wu, X.; Wrobel, R.; Mellor, P.H.; Zhang, C. A computationally efficient PM power loss derivation for surface-mounted brushless AC PM machines. In Proceedings of the 2014 International Conference on Electrical Machines (ICEM), Berlin, Germany, 2–5 September 2014; Volume 105, pp. 17–23. [CrossRef]
124. Wu, X.; Wrobel, R.; Mellor, P.H.; Zhang, C. A Computationally Efficient PM Power Loss Mapping for Brushless AC PM Machines With Surface-Mounted PM Rotor Construction. *IEEE Trans. Ind. Electron.* **2015**, *62*, 7391–7401. [CrossRef]
125. Benlamine, R.; Dubas, F.; Randi, S.A.; Lhotellier, D.; Espanet, C. 3D Numerical Hybrid Method for PM Eddy-Current Losses Calculation: Application to Axial-Flux PMSMs. *IEEE Trans. Magn.* **2015**, *51*, 1–10. [CrossRef]
126. Daanoune, A.; Lateb, R.; da Silva, J. Semi analytical model for eddy current losses calculation in high speed permanent magnet synchronous machines. In Proceedings of the International Conference on Electrical Machines and Systems (ICEMS), Chiba, Japan, 13–16 November 2016; pp. 1–6.
127. Gerlach, T.; Rabenstein, L.; Dietz, A.; Kremser, A.; Gerling, D. Determination of eddy-current losses in permanent magnets of SPMSM with concentrated windings: A hybrid loss calculation method and experimental verification. In Proceedings of the 2018 Thirteenth International Conference on Ecological Vehicles and Renewable Energies (EVER), Monte-Carlo, Monaco, 10–12 April 2018; pp. 1–8. [CrossRef]
128. Toda, H.; Xia, Z.; Wang, J.; Atallah, K.; Howe, D. Rotor Eddy-Current Loss in Permanent Magnet Brushless Machines. *IEEE Trans. Magn.* **2004**, *40*, 2104–2106. [CrossRef]
129. Ede, J.D.; Atallah, K.; Jewell, G.W.; Wang, J.B.; Howe, D. Effect of Axial Segmentation of Permanent Magnets on Rotor Loss in Modular Permanent-Magnet Brushless Machines. *IEEE Trans. Ind. Appl.* **2007**, *43*, 1207–1213. [CrossRef]
130. Mirzaei, M.; Binder, A.; Deak, C. 3D analysis of circumferential and axial segmentation effect on magnet eddy-current losses in permanent magnet synchronous machines with concentrated windings. In Proceedings of the XIX International Conference on Electrical Machines, Rome, Italy, 6–8 September 2010; pp. 1–6. [CrossRef]
131. Nuscheler, R. Two-dimensional analytical model for eddy-current loss calculation in the magnets and solid rotor yokes of permanent magnet synchronous machines. In Proceedings of the 2008 18th International Conference on Electrical Machines, Vilamoura, Portugal, 6–9 September 2008; pp. 1–6. [CrossRef]
132. Yamazaki, K.; Shina, M.; Kanou, Y.; Miwa, M.; Hagiwara, J. Effect of Eddy Current Loss Reduction by Segmentation of Magnets in Synchronous Motors: Difference Between Interior and Surface Types. *IEEE Trans. Magn.* **2009**, *45*, 4756–4759. [CrossRef]
133. Klotzl, J.; Pyc, M.; Gerling, D. Permanent magnet loss reduction in PM-machines using analytical and FEM calculation. In Proceedings of the SPEEDAM 2010, Pisa, Italy, 14–16 June 2010; pp. 98–100. [CrossRef]

134. Wills, D.A.; Kamper, M.J. Reducing PM eddy-current rotor losses by partial magnet and rotor yoke segmentation. In Proceedings of the XIX International Conference on Electrical Machines, Rome, Italy, 6–8 September 2010; pp. 1–6. [CrossRef]
135. Li, B.; Li, M. Calculation and Analysis of Permanent Magnet Eddy Current Loss Fault with Magnet Segmentation. *Math. Probl. Eng.* **2016**, *2016*, 1–6. [CrossRef]
136. Madina, P.; Poza, J.; Ugalde, G.; Almandoz, G. Analysis of non-uniform circumferential segmentation of magnets to reduce eddy-current losses in SPMSM machines. In Proceedings of the 2012 XXth International Conference on Electrical Machines, Marseille, France, 2–5 September 2012; Volume 14, pp. 79–84. [CrossRef]
137. Yamazaki, K.; Fukushima, Y. Effect of Eddy-Current Loss Reduction by Magnet Segmentation in Synchronous Motors With Concentrated Windings. *IEEE Trans. Ind. Appl.* **2011**, *47*, 779–788. [CrossRef]
138. Moriyasu, S.; Endo, K. The Ripple Loss at the Rotor Surface of Synchronous Machines. *IEEE Trans. Power Apparatus Syst.* **1980**, *PAS-99*, 2393–2399. [CrossRef]
139. Chaithongsuk, S.; Takorabet, N.; Kreuawan, S. Reduction of Eddy-Current Losses in Fractional-Slot Concentrated-Winding Synchronous PM Motors. *IEEE Trans. Magn.* **2015**, *51*, 1–4. [CrossRef]
140. Yamazaki, K.; Kanou, Y.; Fukushima, Y.; Ohki, S.; Nezu, A.; Ikemi, T.; Mizokami, R. Reduction of Magnet Eddy-Current Loss in Interior Permanent-Magnet Motors With Concentrated Windings. *IEEE Trans. Ind. Appl.* **2010**, *46*, 2434–2441. [CrossRef]
141. Yamazaki, K.; Kitayuguchi, K. Teeth shape optimization of surface and interior permanent-magnet motors with concentrated windings to reduce magnet eddy-current losses. In Proceedings of the International Conference on Electrical Machines and Systems, Incheon, South Korea, 10–13 October 2010; pp. 990–995.
142. Sharkh, S.M.; Qazalbash, A.A.; Irenji, N.T.; Wills, R.G. Effect of slot configuration and airgap and magnet thicknesses on rotor electromagnetic loss in surface PM synchronous machines. In Proceedings of the 2011 International Conference on Electrical Machines and Systems, Beijing, China, 20–23 August 2011. [CrossRef]
143. Bode, C.; May, H. Optimized Reduction of Parasitic Eddy Current Losses in High Speed Permanent Magnet Motors based on 2D and 3D Field Calculations. In Proceedings of the XV ISEF, International Symposium on Electromagnetic Fields in Mechatronics, Electrical and Electronic Engineering, Madeira, Portugal, 1–3 September 2011.
144. Belli, Z.; Mekideche, M.R. Optimization of magnets segmentation for eddy-current losses reduction in permanent magnets electrical machines. In Proceedings of the 2013 Eighth International Conference and Exhibition on Ecological Vehicles and Renewable Energies (EVER), Monte Carlo, Monaco, 27–30 March 2013; Volume 34, pp. 1–7. [CrossRef]
145. Alberti, L.; Fornasiero, E.; Bianchi, N. Impact of the Rotor Yoke Geometry on Rotor Losses in Permanent-Magnet Machines. *IEEE Trans. Ind. Appl.* **2012**, *48*, 98–105. [CrossRef]
146. Yamazaki, K.; Kato, Y.; Ikemi, T.; Ohki, S. Reduction of Rotor Losses in Multilayer Interior Permanent-Magnet Synchronous Motors by Introducing Novel Topology of Rotor Flux Barriers. *IEEE Trans. Ind. Appl.* **2014**, *50*, 3185–3193. [CrossRef]
147. Choi, G.; Jahns, T.M. Reduction of Eddy-Current Losses in Fractional-Slot Concentrated-Winding Synchronous PM Machines. *IEEE Trans. Magn.* **2016**, *52*, 1. [CrossRef]
148. Jara, W.; Lindh, P.; Tapia, J.A.; Petrov, I.; Repo, A.K.; Pyrhonen, J. Rotor Eddy Current Losses Reduction in an Axial Flux Permanent Magnet Machine. *IEEE Trans. Ind. Electron.* **2016**, *63*, 1. [CrossRef]
149. Cho, H.W.; Jang, S.M.; Choi, S.K. A Design Approach to Reduce Rotor Losses in High-Speed Permanent Magnet Machine for Turbo-Compressor. *IEEE Trans. Magn.* **2006**, *42*, 3521–3523. [CrossRef]
150. Etemadrezaei, M.; Wolmarans, J.J.; Polinder, H.; Ferreira, J.A. Precise calculation and optimization of rotor eddy-current losses in high speed permanent magnet machine. In Proceedings of the 2012 XXth International Conference on Electrical Machines, Marseille, France, 2–5 September 2012; pp. 1399–1404. [CrossRef]
151. Ma, J.; Zhu, Z.Q. Magnet eddy-current loss reduction in a 3-slot 2-pole permanent magnet machine. In Proceedings of the 2017 IEEE International Electric Machines and Drives Conference (IEMDC), Miami, FL, USA, 21–24 May 2017. [CrossRef]
152. Gilbert, A. A method of measuring loss distribution in electrical machines. *Proc. IEE A Power Eng.* **1961**, *108*, 239. [CrossRef]
153. Aoyama, Y.; Miyata, K.; Ohashi, K. Simulations and experiments on eddy-current in Nd-Fe-B magnet. *IEEE Trans. Magn.* **2005**, *41*, 3790–3792. [CrossRef]

154. Yamazaki, K.; Shina, M.; Miwa, M.; Hagiwara, J. Investigation of Eddy Current Loss in Divided Nd–Fe–B Sintered Magnets for Synchronous Motors Due to Insulation Resistance and Frequency. *IEEE Trans. Magn.* **2008**, *44*, 4269–4272. [CrossRef]
155. Alberti, L.; Fornasiero, E.; Bianchi, N.; Bolognani, S. Rotor Losses Measurements in an Axial Flux Permanent Magnet Machine. *IEEE Trans. Energy Convers.* **2011**, *26*, 639–645. [CrossRef]
156. Liu, D.; Jassal, A.; Polinder, H.; Ferreira, J. Validation of eddy-current loss models for permanent magnet machines with fractional-slot concentrated windings. In Proceedings of the 2013 International Electric Machines & Drives Conference, Chicago, IL, USA, 12–15 May 2013; Volume 16, pp. 678–685. [CrossRef]
157. Liu, Z.; Binns, K.; Low, T. Analysis of eddy-current and thermal problems in permanent magnet machines with radial-field topologies. *IEEE Trans. Magn.* **1995**, *31*, 1912–1915. [CrossRef]
158. Zhao, N.; Zhu, Z.Q.; Liu, W. Rotor Eddy Current Loss Calculation and Thermal Analysis of Permanent Magnet Motor and Generator. *IEEE Trans. Magn.* **2011**, *47*, 4199–4202. [CrossRef]
159. Huang, Z.; Fang, J.; Liu, X.; Han, B. Loss Calculation and Thermal Analysis of Rotors supported by Active Magnetic Bearings for High-speed Permanent Magnet Electrical Machines. *IEEE Trans. Ind. Electron.* **2015**, *63*, 2027–2035. [CrossRef]

MDPI
St. Alban-Anlage 66
4052 Basel
Switzerland
Tel. +41 61 683 77 34
Fax +41 61 302 89 18
www.mdpi.com

Mathematical and Computational Applications Editorial Office
E-mail: mca@mdpi.com
www.mdpi.com/journal/mca

www.ingramcontent.com/pod-product-compliance
Lightning Source LLC
LaVergne TN
LVHW070116170726
843515LV00007B/1700